COMPUTING FOR NUMERICAL METHODS USING VISUAL C++

THE WILEY BICENTENNIAL—KNOWLEDGE FOR GENERATIONS

Each generation has its unique needs and aspirations. When Charles Wiley first opened his small printing shop in lower Manhattan in 1807, it was a generation of boundless potential searching for an identity. And we were there, helping to define a new American literary tradition. Over half a century later, in the midst of the Second Industrial Revolution, it was a generation focused on building the future. Once again, we were there, supplying the critical scientific, technical, and engineering knowledge that helped frame the world. Throughout the 20th Century, and into the new millennium, nations began to reach out beyond their own borders and a new international community was born. Wiley was there, expanding its operations around the world to enable a global exchange of ideas, opinions, and know-how.

For 200 years, Wiley has been an integral part of each generation's journey, enabling the flow of information and understanding necessary to meet their needs and fulfill their aspirations. Today, bold new technologies are changing the way we live and learn. Wiley will be there, providing you the must-have knowledge you need to imagine new worlds, new possibilities, and new opportunities.

Generations come and go, but you can always count on Wiley to provide you the knowledge you need, when and where you need it!

WILLIAM J. PESCE
PRESIDENT AND CHIEF EXECUTIVE OFFICER

PETER BOOTH WILEY
CHAIRMAN OF THE BOARD

COMPUTING FOR NUMERICAL METHODS USING VISUAL C++

Shaharuddin Salleh
Universiti Teknologi Malaysia
Skudai, Johor, Malaysia

Albert Y. Zomaya
University of Sydney
Sydney, New South Wales, Australia

Sakhinah Abu Bakar
National University of Malaysia
Bangi, Selangor, Malaysia

WILEY-INTERSCIENCE

A JOHN WILEY & SONS, INC., PUBLICATION

Wiley Bicentennial Logo: Richard J. Pacifico

Library of Congress Cataloging-in-Publication Data:

Salleh Shaharuddin, 1956-
   Computing for numerical methods using Visual c++ / by Shaharuddin Salleh, Albert Y.
Zomaya, Sakhinah Abu Bakar.
       p.   cm. – (Wiley series on parallel and distributed computing)
    Includes index.
    ISBN 978-0-470-12795-7 (cloth)
    1. Microsoft Visual C# .NET.   2. Numerical analysis–Data processing.
I. Zomaya, Albert Y.   II. Bakar, Sakhinah Abu, 1982–   III. Title.

   QA76.73.C154S28 2007
   005.2′768–dc22

                                                            2007023224

To our families,
for their help, courage, support, and patience.

■ TRADEMARKS

Microsoft is the trademark of Microsoft Corporation Redmond, WA.
Visual Studio and Visual Studio.Net are the trademarks of Microsoft Corporation.
Visual C++ and Visual C++.Net are the trademarks of Microsoft Corporation.
Microsoft Foundation Classes is the trademark of Microsoft Corporation.
Matlab is the trademark of The Mathworks, Inc., Natick, MA.
Maple is the trademark of Waterloo Maple, Inc., Waterloo, Ontario, Canada.
Mathematica is the trademark of Wolfram Research, Inc., Champaign, IL.

CONTENTS

Computing for Numerical Methods Using Visual C++ has been written to promote the use of Visual C++ in scientific computing. C++ is a beautiful language that has contributed to shaping the modern world today. The language has contributed to many device drivers in electronic equipment, as a tool in the development of many computer software programs, and as a tool for both research and teaching. Therefore, its involvement in providing the solution for numerical methods is very much expected.

Today, research has no boundary. A problem for study in a topic in research may involve people from several disciplines. A typical problem in engineering for studying the effect of chemical spills in a lake may involve engineers, chemists, biologists, medical doctors, mathematicians, economists, urban planners, and politicians. A comprehensive solution that satisfies all parties can only be produced if people from these disciplines cooperate, rather than having them acting as rivals.

Numerical computing is an important area of research in science and engineering. The topic is widely implemented in the modeling of a problem and its simulation. Additional work involves visualization, which makes the problem and its solution acceptable to the general audience. In the early days of computing in the 1960s and 1970s, the solutions to problems were mostly presented as text and numbers. A programming language like FORTRAN was the dominant tool, and there were no friendly interfaces to present the solutions. Things have improved much since then, as new advancements in hardware and software produce friendly tools based on Microsoft Windows. Numerical computing benefits much from Windows as the results from computation can now be visualized as graphs, numbers, moving images, as well as text.

We select numerical methods as the main title in the book as the concepts in this topic serve as the fundamentals in science and engineering. The importance of numerical methods has been proven as nearly all problems involving mathematical modeling and simulations in science and engineering have their roots in numerical methods. All numeric-intensive applications involving arrays and vectors have their concepts defined in numerical methods. Numerical methods discusses vital and critical techniques in implementing algorithms for providing fast, reliable, and stable solutions to these problems.

Numerical analysis is a branch of mathematics that studies the numerical solutions to problems involving nonlinear equations, systems of linear equations, interpolation and approximation for curve fittings, differentiation, integrals, ordinary differential equations, and partial differential equations. Numerical solutions are needed for these problems as an alternative to the exact solutions, which may be difficult to obtain. The exact solutions for these problems are studied using analytical methods based on mathematical techniques and theorems. The numerical solutions produced may not be exact, but they are good enough for acceptance as they only differ by a small margin.

The scope for numerical analysis is very broad. The study involves the analytical derivation of the methods or techniques using mathematical principles and rules. The study also involves a detailed analysis of the errors between the approximated solutions and the exact solutions, so as to provide faster convergence as well as more accurate solutions.

Numerical methods is different from numerical analysis. Numerical methods is a branch of numerical analysis that specially deals with the implementation of the methods for solving the problems. The details about the derivation of algorithms and techniques for solving the problems and the analysis of errors are not in the main agenda of numerical methods. The main objective in numerical methods is applying the given methods for solving the problems. It is the implementation of the numerical methods that attracts interest from the practitioners who comprise the biggest consumer market. Engineers, scientists, and technologists belong to this group of people who view numerical methods as an important tool for solving their problems. On the other hand, numerical analysis is mostly confined to die-hard mathematicians who love further challenges in developing new numerical techniques for solving the problems.

There are several objectives in developing *Computing for Numerical Methods Using Visual C++*. First, no books on the market today discuss the visual solutions to problems in numerical methods using C++. There are similar books using software packages such as Matlab, Maple, and Mathematica. These software packages are not really primitive programming languages. They have been developed to hide the programming details and to implement the solution as a black box. In other words, software packages do not really teach the mathematical concepts and principles in solving a problem. For example, the inverse of a matrix can be computed using a single line of command in these packages. The user only needs to know the format and syntax of the command in order to produce the desired solution. It is not important for the user to know the underlying concepts in solving the problem. *Computing for Numerical Methods Using Visual C++* is one effort to integrate C++ with the visual solution to problems using numerical methods.

A student cannot be too reliant on software packages. There are cases where software packages fail to provide a solution because of the lack of special routines. For example, a software package may only support a maximum of five levels of the rectangular grids in a boundary-value problem involving partial differential equations. To produce 10 levels, the user will have to use C++ as the language because it is more flexible. Flexibility and versatility are two features in C++ that cannot be matched by any software package.

Our second objective is to promote C++ as a language for numerical computing. C++ has all the necessary ingredients for numerical computing because of its flexible language format, its object-oriented methodology, and its support for high numerical precisions. However, in the past, the popularity of C++ has suffered from the emergence of several new languages. Among them are Java, Python, and C#. These new languages have been developed with the main objective to handle Web and network programming requirements. Other than that, C++ is still dominant and practical for implementation. Because of this reason, C++ is still popular in schools and universities, mostly for teaching and research purposes. C++ is also used widely in the manufacturing sectors such as in the design of device drivers for electronic components.

Our third objective is to make numerical problems friendly and approachable. This goal is important as the general public perception about mathematics is that it is tough, unfriendly, boring, and not applicable in daily life. A mathematician should not be placed in the basement floor of a building under the feeling that he is not important for people to meet. A mathematician can become a role model if he can exert his usefulness in a friendly and acceptable way, which can be done by making mathematics interesting and approachable through a series of friendly interfaces. A weak or average student can become motivated with mathematics if the right tools for understanding mathematics are provided.

A visual approach based on Windows in *Computing for Numerical Methods Using Visual C++* is our step in achieving this objective. The book teaches the reader on the friendly interfaces in tackling problems in numerical methods. The interfaces include buttons, dialog boxes, menus, and mouse clicks. The book also provides a very useful tool called *MyParser*, which can be used to develop various friendly numerical applications on Windows. MyParser is an equation parser that reads an equation input by the user in the form of a string, processes the string, and produces its solution. In promoting its use, we hide all technical details in the development of the parser so that the reader can concentrate on producing the solution to the problem. But the best part is MyParser is free for distribution for those who are interested.

Our last objective is to maintain links with the Microsoft family of products through the .Net platform. Microsoft is unarguably the driver in providing visual solutions based on Windows, and the .Net platform provides a common multilanguage program development for applications on Windows. As Visual C++ is one product supported in the .Net platform, there is a guarantee of continued support from Microsoft for its users. A Visual C++ follower can also enjoy the benefit of integrating her product with other products within the .Net platform with very minimum effort. This flexibility is important as migrating from one system to a different system by bringing along data and programs can be a very expensive, time-consuming, and resource-dependent affair.

In providing the solutions, this book does not provide detailed coverage of each topic in numerical methods. There are already many books on the market that do cover these topics, and we do not wish to compete against them. Instead, we focus on the development stages of each topic from the practical point of view, using Visual C++ as the tool. Knowing how to write the visual interfaces for the numerical problems will definitely contribute to guiding the reader toward the more ambitious numerical modeling and simulation projects. This objective is the main benefit that can be expected from the book. The reader can take advantage of the supplied codes to create several new projects for high-performance computing.

The book is accompanied by source codes that can be downloaded from the given Wiley website. This website will be maintained and updated by the authors from time to time. The codes have been designed to be as compact as possible to make them easy to understand. They are based on the Microsoft Foundation Classes library for providing the required user-friendliness tools. In designing the codes, we opted for the unguided (or non-wizard) approach in order to show the detailed steps for producing the output. This is necessary as the guided (or wizard) approach does not teach some key steps, and it causes the program to be extremely long in size. However, the reader should be able to convert each unguided program code to the guided code if the need arises.

In preparing the manuscript for *Computing for Numerical Methods Using Visual C++*, the authors would like to thank several people who have been directly or indirectly involved. The authors would like to thank Tan Sri Prof. Dr. Ir. Mohd. Zulkifli Tan Sri Mohd. Ghazali, Vice Chancellor of Universiti Teknologi Malaysia, for his forward vision in leading the university towards becoming a world-class university by 2010. Special thanks also to Professor Dr. Alias Mohd. Yusof and Professor Dr. Md. Nor Musa from Universiti Teknologi Malaysia, and Professor Stephan Olariu from Old Dominion University, for their support and encouragement.

Shaharuddin Salleh
Albert Y. Zomaya
Sakhinah Abu Bakar
April 2007

CODES FOR DOWNLOAD

All codes discussed in this book can be downloaded from the following URL:

ftp://ftp.wiley.com/public/sci_tech_med/computing_numerical

The files in the URL are organized into a directory called SALLCode. The codes for the program are located in the folders bearing the chapter numbers; for example, Code4 is the project for Chapter 4. The files included are the executable (.exe), header (.h), C++ (.cpp) and the parser object file, MyParser.obj.

The ftp site will constantly be maintained and updated. Any questions, comments, and suggestions should be addressed to the first author at ss@utm.my.

The system requirements for the codes are as follows:

Intel Pentium-based Personal computer with 256 MB RM and above.
Microsoft Windows 1998 and above.
Microsoft Visual C++ version 6, and above.

All code files have been tested using Microsoft Visual C++.Net version 2003. The same files should be compatible with Visual C++ version 6 and below and with Microsoft Visual C++.Net version 2005 and above.

CHAPTER 1

Modeling and Simulation

1.1 NUMERICAL APPROXIMATIONS

Numerical methods is an area of study in mathematics that discusses the solutions to various mathematical problems involving differential equations, curve fittings, integrals, eigenvalues, and root findings through approximations rather than exact solutions. This discussion is necessary because the exact solutions to these problems are difficult to obtain through the analytical approach. For example, it may be wise to evaluate

$$\int_{-1}^{5} x \sin x \, dx,$$

as the exact solution can be obtained through a well-known technique in calculus called *integration by parts*. However, it is not possible to apply the same method or any other analytical method to solve

$$\int_{0}^{2} \frac{e^{-x}}{3 \sin x + x^2} \, dx.$$

The given equation in the above integral is difficult to solve as it is not subject to the exact methods discussed in ordinary calculus. Therefore, a numerical method is needed to produce a reasonably good approximated solution. A good approximated solution, whose value may differ from the exact solution by some fractions, is definitely better than nothing.

Numerical methods are also needed in cases where the mathematical function for a given problem is not given. In many practical situations, the governing equations for a given problem cannot be determined. Instead, an engineer may have a set of n data (x_i, y_i) collected from the site to analyze in order to produce a working model. In this case, a numerical method is applied to fit a curve that corresponds to this set of data for modeling the scenario.

Solutions to numerical problems can be obtained on the computer in two ways: using a ready-made software program or programming using a primitive language. A ready-made software program is a commercial package that has been designed to solve specific problems without the hassle of going through programming. The software provides quick solutions to the problems with just a few commands. The solution to a problem, such as finding the inverse of a matrix, is obtained by typing just one or two lines of command. Matlab, Maple, and Mathematica are some of the most common examples of ready-made software programs that are tailored to solve numerical problems.

The easy approach using ready-made software programs has its drawback. The software behaves like a black box where the user does not need to know the details of the method for solving the problem. The underlying concepts in solving the problem are hidden in the software, and this approach does not really test the mathematical skill of the user. Very often, the user does not understand how the method works as all he or she gets is the generated solution. In addition, the user may face difficulty in trying to figure out why a particular solution fails because of problems such as singularity in the domain.

A ready-made software program also does not provide the flexibility of customizing the solution according to the user's requirement. The user may need special features to visualize the solution, but these features may not be supported in the software. In addition, a ready-made software program generates files that are relatively large in size. The large size is the result of the large number of program modules from its library that are stored in order to run the program.

The real challenge in solving a numerical problem is through the native language programming. It is through programming that a person will understand the whole method comprehensively. The developer will need to start from scratch and will need to understand all the fundamental concepts for solving a given problem before the solution to the problem can be developed. The whole process in providing the solution may take a long time, but a successful solution indicates the programmer fully understands the whole process.

It is also important for us to accept both approaches. Ready-made software programs are needed in cases where the program needs to be delivered fast. A ready-made software program can produce the desired solution within a short period of time if all required routines are available in the software, which is done by following the right commands and procedures in handling the software. In cases where some required actions are not supported in the software, it may be necessary to integrate the solution with a programming language such as C++. Conversely, if one starts from a programming language, it may be necessary call a ready-made software program to handle some difficult tasks. For example, C++ may be used to draw up the numerical solution to the heat distribution problem. To see the solution in the form of surface graph, it may be wise to call a few routines from Matlab as the same feature in C++ will require a long time to develop.

1.2 C++ FOR NUMERICAL MODELING

C++ is a language that has its origin from C, developed in early 1980s. C++ retains all the procedural structure of C but adds the object-oriented features in order to meet the new requirements in computing. Both C and C++ are heavily structured high-level languages that produce small executable files. The two languages are also suitable for producing low-level routines that run the device drivers of many electronic components.

C++ is popular because of its general-purpose features to support a wide variety of applications, such as data processing, numerical, scientific, and engineering. C++ is available in all computing platforms, including Windows, UNIX, Macintosh, and operating systems for mainframe and minicomputers. C++ is a revolutionary language that has a very strong following from students, practitioners, researchers, and software developers all over the world. The language is taught in most universities and colleges in the world as a one- or two-semester subject to support numerical and general-purpose applications.

C++ is a language that strongly supports object-oriented programming. *Object-oriented programming* is a programming approach based on objects. An *object* is an instance of a class. A *class* is a set of entities that share the same parent. As it stood, C++ is one of the most popular object-oriented programming languages in the world. The main reason for its popularity is because it is a high-level language, but at the same time, it runs as powerful as the assembly language. But the real strength of C++ lies in its takeover from C to move to the era of object-oriented programming in the late 1980s. This conquest provides C++ with the powerful features of the procedural C and an added flavor for object-oriented programming.

The original product from Microsoft consists of the C compiler that runs under the Microsoft DOS (disk operating system), and it has been designed to compete against Turbo C, which was produced by the Borland Corp. In 1988, C++ was added to C and the compiler was renamed Microsoft C++. In early 1989, Microsoft launched the Microsoft Windows operating system, which includes the Windows API (application programming interface). This interface is based on 16 bits and it supports the procedural mode of programming using C.

Improvements were made over the following years that include the Windows Software Development Kit (SDK). This development takes advantage of the API for the graphical user interface (GUI) applications with the release of the Microsoft C compiler. As this language is procedural, the demands in the applications require an upgrade to the object-oriented language design approach, and this contributes to the release of the Microsoft C++ compiler. With the appearance of the 32-bit Windows API (or Win32 API) in the early 1990s, C++ was reshaped to tackle the extensive demands on Windows programming and this brings about the release of the Microsoft Foundation Classes (MFC) library. The library is based on C++, and it has been tailored with the object-oriented methodology for supporting the application architecture and implementation.

The main reason why Visual C++.Net is needed in numerical methods is its powerful simulation and visualization tools. The Net platform refers to a huge collection of library functions and objects for creating full-featured applications on both the desktop and the enterprise Web. The classes and objects provide support for friendly user interface functions like multiple windows, menus, dialog boxes, message boxes, buttons, scroll bars, and labels. Besides, the platform also includes several tedious task-handling jobs like file management, error handling, and multiple threading. This platform also supports advanced frameworks and environments such as the Passport, Windows XP, and the Tablet PC. The strength of the Net platform is obvious in providing the Internet and Web enterprise solutions. Web services include information sharing, e-commerce, HTTP, XML, and SOAP. XML, or Extensible Markup Language, is a platform-independent approach for creating markup languages needed in a Web application.

A new approach in Visual C++.Net is the Managed Extension, which performs automatic garbage collection for optimizing the code. Garbage collection involves the removal of memory and resources unused any more in the application, which is often neglected by the programmer. The managed extension is a more structured way in programming, and it is now the default in Visual C++.Net. Central to the .Net platform is the Visual Studio integrated development environment (IDE). It is in this platform that applications are built from a choice of several powerful programming languages that include Visual Basic, Visual C++, Visual C#, and Visual J++.

In addition, IDE also provides the integration of these languages in tackling a particular problem under the .Net banner. Visual C++.Net is one of the high-performance compilers that makes up the .NET platform. This highly popular language has its root in C and was improved to include the object-oriented elements; now with the .Net extension, it is capable of creating solutions for the Web enterprise requirements. A relatively new language called Visual C# in the .Net family was developed by taking the best features from Visual Basic visual tools with the programming power of Visual C++.

In addition to its single-machine prowess, Visual C++.Net presents a powerful approach for building applications that interact with databases through ADO.NET. This product evolves from the earlier ActiveX Data Objects (ADO) technology, and it encompasses XML and other tools for accessing and manipulating databases for several large-scale applications. This feature makes possible Visual C++.Net as an ideal tool for several Web-based database applications.

1.3 MATHEMATICAL MODELING

Many problems arise in science and engineering that have their roots in mathematics. Problems of this nature are best described through mathematical models that provide the fundamental concepts needed in solving the problem. A successful mathematical modeling always leads to a successful implementation of the given project.

A *mathematical model* is an abstract model that uses mathematical language to describe the behavior of a system. It is an attempt to find the analytical solutions for enabling the prediction of the behavior of the system from a set of parameters, and their initial and boundary conditions in a given problem. A mathematical model is composed of variables and operators to represent a given problem. *Variables* are the abstractions of the quantities of interest in the described systems, whereas *operators* are the mechanisms that act on these variables. An operator can be expressed in the form of an algebraic operator, a function, a differential operator, and so on.

One good example of mathematical modeling is in the heat distribution problem in a two-dimensional plane, which is modeled as a Laplace function given by

$$\nabla^2 u = \frac{\partial^2 u}{\partial x^2} + \frac{\partial^2 u}{\partial y^2}.$$

In the above equation, $\nabla^2 = \frac{\partial^2}{\partial x^2} + \frac{\partial^2}{\partial y^2}$ is an operator that acts on the heat quantity u, whose independent variables are x and y. The model describes the analytical solution of heat distribution in a given domain subject to certain initial and boundary conditions. We will discuss the numerical solution to this problem later in Chapter 11.

A mathematical model is often represented as variables in terms of objective functions and constraints. If the objective functions and constraints in the problem are represented entirely by linear equations, then the model is regarded as a *linear model*. If one or more of the objective functions or constraints are represented with a nonlinear equation, then the model is known as a *nonlinear model*. Most problems in science and engineering today are modeled as nonlinear.

A *deterministic* model is one where every set of variable states is uniquely determined by parameters in the model and by sets of previous states of these variables. For example, the path defined by a delivery truck for distributing petrol in a city is defined as a deterministic model in the form of a graph. In this case, the petrol stations can be modeled as the nodes of the graph, whereas the path is defined as the edges between the nodes. Conversely, in a *probabilistic* or *stochastic* model, randomness is present, and the variable states are not described by unique values. Very often, the model is represented in the form of probability distribution functions.

A mathematical model can be classified as static or dynamic. A *static* model does not account for the element of time, whereas a dynamic model does. Dynamic models typically are represented with difference equations or differential equations. A model is said to be *homogeneous* if it is in a consistent state throughout the entire system. If the state varies according to certain controlling mechanism, then the model is *heterogeneous*. If the model is homogeneous, then the parameters are *lumped*, or confined to a central depository. A heterogeneous, model has its parameters distributed. Distributed parameters are typically represented with ordinary or partial differential equations.

Mathematical modeling problems are also classified into black-box or white-box models, according to how much *a priori* information is available in the system. A *black-box* model is a system of which no *a priori* information is available. A *white-box* model, also called glass box or clear box, is a system where all necessary information is available.

1.4 SIMULATION AND ITS VISUALIZATION

Mathematical modeling is often followed by a series of numerical simulations to support and verify its validity and correctness. *Simulation* is an imitation of some real thing, state of affairs, or process representing certain key characteristics or behaviors of a selected physical or abstract system. It is also the objective of a simulation to optimize the results by controlling the variables that make up the problem. By changing the variables during simulation, predictions may be made about the behavior of the system. Simulation is necessary to save time, cost, human capital, and other resources. Good results from a simulation contribute in some critical decision-making process.

Simulation can be implemented using three approaches: microscopic, macroscopic, and mesoscopic. In the *microscopic* simulation, the detail physical and performance characteristics, such as the properties of the elements that make up the problem, are considered. The simulation involves some tiny properties of the individuals or elements that make up the pieces. The results from a microscopic simulation are always reliable and accurate. However, this approach could be very costly and time consuming as data from the individuals or elements are not easy to obtain.

An easier approach is the *macroscopic simulation* that considers the deterministic factors of the whole population, rather than all the individuals or elements. In this approach, factors such as the governing mathematical equations and their macro data are considered. The steps in this approach may skip the detail components, and therefore, the results may not be as accurate as the one produced in the microscopic approach. However, this approach saves time and is not as costly as the microscopic approach.

A more realistic and practical approach is the *mesoscopic simulation* that combines the good parts of the microscopic and macroscopic approaches to produce a more versatile model. In this approach, some deterministic properties of the elements in the system are integrated with the detail information to produce a workable model.

A good simulation has several visualization features that accurately describe the elements in a system. Visualization is a form of graphical or textual presentation that easily describes the solution to a particular problem. An effective visualization includes components such as text, graphics, diagrams, images, animation, and sound in order to describe the system.

In most cases, numerical simulations are carried out effectively in a computer. A computer can accurately describe the functionality and behavior of the elements that make up the system. Today's computers are fast and have all the required resources to perform most college-level numerical simulations. Supercomputers also exist that are multiprocessor systems capable of processing numeric-intensive applications with a whopping gigaflop speed. Several parallel and distributed computer systems are also available in processing these numeric-intensive applications. Computers are also grouped into clusters to work cooperatively in grid computing networks that span across many countries in the globe.

1.5 NUMERICAL METHODS

Numerical methods is a branch of mathematics that consists of seven core areas, as follows:

- Nonlinear equations
- System of linear equations
- Interpolation and approximation
- Differentiation and integration
- Eigenvalues and eigenvectors
- Ordinary differential equations
- Partial differential equations

The basic problem in a nonlinear equation is in finding the zeros of a given function. The problem translates into finding the roots of the equation, or the points along the x-axis where the function crosses. The roots may exist as real numbers. In cases where the real roots do not exist, their corresponding imaginary roots may become a topic of study.

Linear equations are often encountered in various science and engineering problems. The problem arises frequently as the original problem reduces to linear equations at some stage in the solution. For example, mesh modeling using the finite-element method in a fluid dynamics results in a system of hundreds of linear equations. As the size of the matrix in this problem is large, the solution needs to be tackled using a fast numerical method.

Interpolation and approximation are curve fitting problems that contribute in things like designing the surface of an aircraft. The techniques are also applied in other problems, such as forecasting, pattern matching, and routing.

Differentiation and integration are fundamental topics that arise in many problems. Good approximations are needed to these two topics as their exact values may not be easy to obtain.

Problems involving ordinary differential equations arise in modeling and simulation. Exact solutions are difficult to obtain as the models are subject to variation because of the presence of constraints and nonlinear factors. Therefore, numerical methods are needed in their successful implementation.

Modeling and simulation involving partial differential equations commonly use numerical techniques as their fundamental elements. Numerical methods contribute to provide the desired solutions in most cases as the exact solutions are not practical for implementation on the computer.

1.6 NUMERICAL APPLICATIONS

Most problems in science and engineering are inherently nonlinear in nature. This nonlinearity is because the problems are dependent on variables and parameters that are nonlinear and are subject to many constraints. Many problems are also dynamic and nondeterministic with no *a priori* information. Because of their nature, the solution to these problems will not be a straightforward task.

In most cases, the normal approach for solving nonlinear problems in science and engineering is to start from the fundamental concepts that are based on mathematics. A mathematical model that describes the system needs to be developed to represent the problem. Data from the problem are collected as an input. Numerical simulation based on the theoretical model is then performed on the data. During the simulation, the results obtained are periodically compared and matched with some real values in order to verify the correctness of the simulation.

We discuss some common modeling and simulation work that makes use of numerical methods.

Bacteria Population Growth

Population growth of bacteria in a geographical region over a period of time has been successfully modeled using a differential equation, given as

$$\frac{dx}{dt} = kx,$$

where $x(t)$ represents the size of the population at time t and k is a constant. In a broader scope, $x(t)$ in the above equation may represent the number of bacteria in a sample container.

The numerical solution to this model consists of solving an initial value problem involving the first-order ordinary differential equation. A suitable solution is provided in the form of the Runge–Kutta method of order 4, as will be discussed in Chapter 10. However, the solution is not solely provided by this method alone. Constraints such as temperature and the acidity of the fluid in the container may affect the validity of the mathematical model. The bacteria may grow faster when the temperature is high and when the acidity of the fluid is low. Therefore, a numerical simulation is needed to integrate the mathematical model with other variables and parameters. Several parameters can be included in the simulation in order to produce the correct model for this problem.

Computational Fluid Dynamics

Computational fluid dynamics (CFD) is an area of research that deals with the dynamic behavior and movement of fluid under certain physical conditions. Numerical methods and algorithms are used extensively to solve and analyze problems involving fluid flow. For example, the interaction between particles in fluids and gases are studied through simulation on the computer. Millions of calculations are performed in this simulation, as the original problem reduces to numerical problems such as matrix multiplications, matrix inverse, system of linear equations, and the computation of eigenvalues.

One area of study in computational fluid dynamics is the blood flow modeling in stenosed artery of the human body. The study contributes in predicting the occurrence of cardiovascular diseases such as heart attack and stroke. CFD simulations have also been carried out in the aerospace and automotive industries for evaluating the air flow around moving aircrafts and cars.

The fundamental tool in CFD simulation is the Navier–Stokes equation, which describes a single-phase fluid flow. Also, a set of ordinary or partial differential equations with their initial and boundary conditions is given. The solution to these problems makes use of the finite-difference method or finite-element methods, which reduces the problem to several systems of linear equations.

Finite-Element Modeling

The finite-element method is a numerical technique for evaluating things like stresses and displacements in mechanical objects and systems. The finite-element method has been successfully applied for modeling problems involving heat transfer, fluid dynamics, electromagnetism, and solid state diffusion. At the National Aeronautics and Space Administration (NASA), the finite-element method has been applied in modeling the turbulence that occurs during the aircraft flight.

The finite-element method requires the domain in the problem to be divided into several elements in the form of line segments, triangles, rectangular meshes, volume, and so on. The method provides flexibility where the elements need not be of equal in terms of dimension and size. Solutions are obtained from these elements, and they are grouped to produce the overall solution.

The fundamentals of finite-element methods rest heavily on numerical methods. They include curve and surface fittings using interpolation or approximation techniques, systems of linear equation, and ordinary and partial differential equations.

Printed-Circuit Board Design

Massive simulations are carried out to produce optimal designs for printed-circuit boards (PCBs). A PCB forms the main circuitry of all electronic devices. A typical PCB can accommodate thousands or millions of microelectronic components such as pins, vias, transistors, processors, and memory chips. In addition, the PCB has massive wirings that connect these components.

As the space on a PCB is limited, the designer must optimize the placement of components and its routing (wiring) so that the board is capable of accommodating as many components as possible within the limited space area. We can imagine a PCB function like a city where the buildings and streets need to be designed properly so that the city will not be too congested with problems such as improper housing and traffic jams.

One technique commonly applied for routing in the PCB design is single-row routing. The problem is about designing non-crossing tracks between pairs of pins that are arranged in a single row in such a way that the tracks do not cross. Single-row routing has been known to be NP-complete with many interacting degrees of freedom. Figure 1.1 shows an optimal output from single-row routing involving 42 pins. The proven methods for solving this problem involve computer simulations using graph theory, simulated annealing, and genetic algorithm. A technique from the authors called ESSR in 2003 has been successful in producing optimal results for the problem.

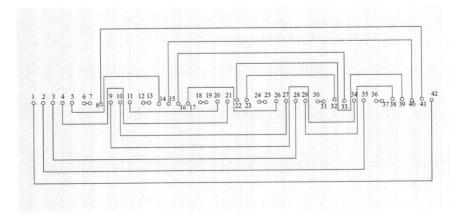

FIGURE 1.1. Optimal single-row routing involving 42 pins.

Wireless Sensor Networks

A sensor network is a deployment of massive numbers of small, inexpensive, self-powered devices called motes that can sense, compute, and communicate with other devices for the purpose of gathering local information to make global decisions about a physical environment. The object of the network is to detect certain items over a geographical region, such as the presence of harmful chemical, bacteria, electromagnetic field, and temperature distribution. In one scenario, thousands of tiny sensor motes are distributed from an aircraft over a region. Sensor network research originates from a DARPA project called Smartdust in late 1990s [2]. The research attracts interest from many disciplines because of several developments in the micro-electro mechanical system (MEMS) technology that produces many cheap and small sensor motes.

Locating sensor motes at their geographical location has become one major issue in forming the network, because many constraints need to be considered in the problem. Each node in a sensor network has a short lifetime based on a battery. To save energy, the node sleeps most of the time and becomes awake only occasionally. Therefore, the process of training the nodes in order to locate their correct location is a highly nonlinear problem with many constraints.

A coarse-grain model developed in Ref. 3 proposed a method called asynchronous training for locating the nodes in a sensor network. Figure 1.2 shows a model consisting of three networks for training the sensor nodes. Each network is indicated by concentric circles that originate from a center called *sink*. The sink is represented by a device called the *aggregating and forwarding node* (AFN), which has a powerful transmitter and receiver for reaching all nodes in its network. In this model, AFN is responsible for training the nodes, storing and retrieving data from the sensor nodes, and performing all the necessary computations on the data received to produce the desired results. The model makes use of the dynamic coordinate system based on coronas and wedges to locate the nodes through a series of transmission from the sink.

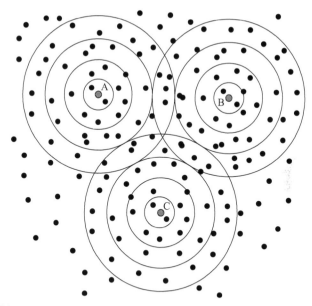

FIGURE 1.2. The asynchronous model for training the sensor nodes.

This theoretical model is not implemented yet in the real world, but it has a strong potential for applications. Modeling and simulation on the computer definitely helps in convincing the lawmakers to realize the benefit from this research.

Flood Control Modeling

The city of Kuala Lumpur, Malaysia, recently completed the construction of a tunnel that has dual purposes, as a road during the normal time and as a waterway tunnel whenever flood hits the city. The project is called SMART or the *Stormwater Management and Road Tunnel*. This project is the first of its kind in the world, and it contributes in controlling flash flood, which frequently disrupts communication in the city.

SMART is an underground tunnel about 10 km long that was constructed using tunnel-boring machines. The tunnel connects two rivers, one in the middle of the city and the other away from the city. During the normal days, the tunnel serves as two trunk roads going into and out of the city, and this contributes in reducing traffic jam in some parts of the city. During heavy rain, parts of the city get flooded as water overflows from the main river in the city. The tunnel is immediately closed to traffic, and it is converted into a waterway to divert the water from the flood area into the second river. This has the immediate effect of reducing the occurrence of flood in these areas. When the flood ends, the tunnel is cleaned, and it is open for traffic again.

A lot of work involving modeling and simulation has been carried out before the project gets started. As the project is costly, modeling and simulation using computer help in providing all the necessary input before a decision on the viability of the project is made. The heavy costs involved are well justified here.

REFERENCES

1. S. Salleh, B. Sanugi, H. Jamaluddin, S. Olariu, and A. Y. Zomaya, Enhanced simulated annealing technique for the single-row routing problem, *Journal of Supercomputing*, 21(3), 2002, 285–302.
2. S. Olariu and Q. Xu, A simple self-organization protocol for massively deployed sensor networks, *Computer Communications*, 28, 2005, 1505–1516.
3. Q. Xu, R. Ishak, S. Olariu, and S. Salleh, On asynchronous training in sensor networks, *Journal of Mobile Multimedia*, 3(1), 2007.

CHAPTER 2

Fundamental Tools for Mathematical Computing

2.1   C++ FOR HIGH-PERFORMANCE COMPUTING

Engineering problems involving large arrays with high-resolution graphics for visualization are some of the typical applications in high-performance computing. In most cases, high-performance computing is about intensive numerical computing over a large problem that requires good accuracy and high-precision results. High-performance computing is a challenging area requiring advanced hardware as well as a nicely crafted software that makes full use of the resources in the hardware.

Numerical applications often involve a huge amount of computation that requires a computer with fast processing power and large memory. Fast processing capability is necessary for calculating and updating several large size arrays, which represent the computational elements of the numerical problem. Large memory is highly desirable to hold the data, which also contributes toward speeding up the calculations.

C++ was originally developed in the early 1980s by Bjarne Stroustrup. The features in C++ have been laid out according to the specifications laid by ANSI (American National Standards Institute) and ISO (International Standards Organization). The language is basically an extension of C that was popular in the 1970s, replacing Pascal and Fortran. C is a highly modular and structured high-level language, which also supports low-level programming. The language is procedural in nature, and it has what it takes to perform numerical computing besides being used for general-purpose applications, such as in systems-level programming, string processing, and database construction.

Primarily, C++ extends what is lacking in C, namely, the object-oriented approach to programming, for supporting today's complex requirements. Object-oriented programming requires a problem to be broken down into objects, which can then be handled in a more practical and convenient manner. That is the standard put forward for meeting today's challenging requirements. The programming language must provide support for exploring the computer's resources in providing an easy and friendly-looking graphical user interface for the given applications. The resources are provided by the Windows operating system, which includes the mouse's left and right button clicks, text and graphics, keys from the keyboard, dialog boxes, and multimedia support. In addition, C++ is a scalable language that makes possible the extension of the language to support several new requirements, for meeting today's requirements, including the Internet, parallel computing, wireless computing, and communication with other electronic devices.

In performing scientific computing, several properties for good programming help in achieving efficient coding. They include

- Object-oriented methodology
- Modular and structured program design
- Dynamic memory allocation for arrays
- Maximizing the use of local variables
- Encouraging data passing between functions

In this chapter, we discuss several fundamental mathematical tools and their programming strategies that are commonly deployed in high-performance computing. They include strategies for allocating memory dynamically on arrays, passing array data between functions, performing algebraic computation on complex numbers, producing random numbers, and performing algebraic operations on matrices. As the names suggest, these tools provide a window of opportunity for high performance computing that involves vast areas of numeric-intensive applications.

2.2 DYNAMIC MEMORY ALLOCATION

A variable represents a single element that has a single value at a given time. An array, in contrast, is a set of variables sharing the same name, where each of them is capable of storing one value at a single time. An array in C++ can be in the form of one or more dimensions, depending on the program requirement. An array is suitable for storing quantities such as the color pixel values of an image, the temperature of the points in a rectangular grid, the pressure of particles in the air and the values returned by sensors in an electromagnetic field.

An array is considered large if its size is large. One typical example of large arrays is one that represents an image. A fine and crisp rectangular image of size 1 MB produced from a digital camera may have pixels formed from hundreds of rows and columns. Each pixel in the image stores an integer value representing its color in the red, green, and blue scheme.

A computer program having many large arrays often produces a tremendous amount of overhead to the computer. The speed of execution of this program is very much affected by the data it carries as each array occupies a substantial amount of memory in the computer. However, a good strategy for managing the computer's memory definitely helps in overcoming many outstanding issues regarding large arrays. One such strategy is *dynamic memory allocation* where memory is allocated to arrays only when they are active and is deallocated when they are inactive.

In scientific computing, a set of variables that have the same dimension and length is often represented as a single vector. *Vector* is an entity that has magnitude or length and direction. Opposite to that is a *scalar*, which has magnitude but does not have direction. For example, the mass of an object is a scalar, whereas its weight subject to the gravitational pull toward the center of the earth is a vector.

Allocation for One-Dimensional Arrays

In programming, a vector is represented as a one-dimensional array. A vector v of size N has N elements defined as follows:

$$v = [v_1 \quad v_2 \quad \cdots \quad v_N]^T = \begin{bmatrix} v_1 \\ v_2 \\ \cdots \\ v_N \end{bmatrix}. \tag{2.1}$$

The above vector is represented as a one-dimensional array, as follows:

$$v[1], v[2], \ldots, v[N].$$

The *magnitude* or *length* of this vector is

$$|v| = \sqrt{v_1^2 + v_2^2 + \cdots + v_N^2} = \sqrt{\sum_{i=1}^{N} v_i^2}. \tag{2.2}$$

In numerical applications, this vector can be declared as real or integer depending on the data it carries. The length of the vector needs to be declared as a floating-point as the value returned involves decimal points. It is important to declare the variable correctly according to the variable data type as wrong declaration could result in data loss.

As the size of an array can reach several hundreds, thousands, or even millions, an optimum mechanism for storing its data is highly desirable. For a small size array, static memory allocation is a normal way of storing the data. An array v with N+1 elements is declared statically as

```
int v[N+1];
```

This array consists of the elements v[0], v[1], ..., v[N], and memory from the computer's RAM has been set aside for this array no matter whether all elements in the array are active. Having only v[1], v[2], and v[3] active out of N = 100 from the above declaration is clearly a waste as memory has been allocated to all 100 elements in the array. This mode of allocating memory is called *static memory allocation.*

A more practical way of allocating memory is the dynamic memory allocation where memory from the computer is allocated to the active elements only in the array.

In dynamic memory allocation, the same array whose maximum size is N+1 is declared as follows:

```
int *v;
v=new int v[N+1];
```

Memory for a one-dimensional array is allocated dynamically as a pointer using the new directive. The allocated size in the declaration serves as the upper limit; memory is allocated according to the actual values assigned to the elements in the array. For example, the array v above is allocated with a maximum of N+1 elements, but memory is assigned to three variables that are active only, as follows:

```
int *v;                        // declaring a pointer to the
                               // integer array
v = new int [N+1];             // allocating memory of size N+1
v[1]=7; v[2]=-3; v[3]=1;       // assigning values to the array
...

...
delete v;                      // destroying the array
```

Once the array has completed its duty and is no longer needed in the program, it can be destroyed using the delete directive. The destruction means the array can no longer be used, and the memory it carries is freed. In this way, the memory can be used by other modules in the program, thus, making the system healthier.

Dynamic memory allocation for one-dimensional arrays is illustrated using an example in multiplying two vectors. The multiplication or *dot product* of two vectors of the same size $n, u = [u_1 \ u_2 \ \cdots \ u_N]^T$ and $v = [v_1 \ v_2 \ \cdots \ v_N]^T$ produces a scalar, as follows:

$$u \cdot v = u_1 v_1 + u_2 v_2 + \cdots + u_N v_N. \tag{2.3}$$

This task is implemented in C++ with the result stored in w in the following code:

```
w=0;
for (i=1;i<=N;i++)
        w += u[i]*v[i];
```

The above code has a complexity of $O(N)$. Code2A.cpp shows a full C++ program that allocates memory dynamically to the one-dimensional arrays u and v, and it computes the row product between the two arrays:

Code2A.cpp: dot-product of two vectors

```cpp
#include <iostream.h>
#define N 3
void main()
{
    int i;
    int *u, *v, w;
    u=new int [N+1];
    u[1]=4; u[2]=-5; u[3]=3;
    cout << endl << "Vector u:" << endl;
    for (i=1;i<=N;i++)
        cout << u[i] << " ";
    cout << endl;

    v=new int [N+1];
    v[1]=-2; v[2]=3; v[3]=7;
    cout << "Vector v:" << endl;
    for (i=1;i<=N;i++)
        cout << v[i] << " ";
    cout << endl;

    w=0;
    for (i=1;i<=N;i++)
        w += u[i]*v[i];
    cout << "the product w=u.v is " << w << endl;
    delete u,v;
}
```

Allocation for Higher Dimensional Arrays

A two-dimensional array represents a matrix that consists of a set of vectors placed as its columnar elements. Basically, a two-dimensional array q with M+1 rows and N+1 columns is declared as follows:

```cpp
int q[M+1][N+1];
```

The above declaration is the static way of allocating memory to the array. The dynamic memory allocation method for the same array is shown, as follows:

```cpp
int **q;                    // declaring a pointer to the
                            // integer array
q = new int *[M+1];         // allocating memory of size
                            // M+1 to the rows
for (int i=0;i<=M;i++)
```

```
q[i]=new int [N+1];      // allocating memory of size
                            N+1 to the columns
```

. . .

. . .


```
for (i=0;i<=N;i++)
     delete q[i];         // deallocating the columnar
                             array memory
delete q;                 // destroying the array
```

In the above method, a double pointer to the variable is needed because the array is two-dimensional. Memory is first allocated to the M+1 row vectors in q. Each row vector has N+1 columns, and therefore, another chunk of memory is allocated to each of these vectors. Hence, a maximum total of [M+1] [N+1] amount of memory is allocated to this array.

Case Study: Matrix Multiplication Problem

We illustrate an important tool in linear algebra involving the multiplication of two matrices whose memories are allocated dynamically. Suppose $A = [a_{ik}]$ and $B = [b_{kj}]$ are two matrices with $i = 1, 2, \ldots, M$ representing the rows of A, and $j = 1, 2, \ldots, N$ are the columns in B. The basic rule allowing two matrices to be multiplied is the number of columns in the first matrix must be equal to the number of rows of the second matrix, which is $k = 1, 2, \ldots, P$. The product of A and B is then matrix $C = [c_{ij}]$, or

$$AB = C.$$

For example, if A and B are matrices of size 3×4 and 4×2, respectively, then their product is matrix C with the size of 3×2, as follows:

$$C = \begin{bmatrix} c_{11} & c_{12} \\ c_{21} & c_{22} \\ c_{31} & c_{32} \end{bmatrix}$$

$$= \begin{bmatrix} a_{11} & a_{12} & a_{13} & a_{14} \\ a_{21} & a_{22} & a_{23} & a_{24} \\ a_{31} & a_{32} & a_{33} & a_{34} \end{bmatrix} \begin{bmatrix} b_{11} & b_{12} \\ b_{21} & b_{22} \\ b_{31} & b_{32} \\ b_{41} & b_{42} \end{bmatrix}$$

$$= \begin{bmatrix} a_{11}b_{11} + a_{12}b_{21} + a_{13}b_{31} + a_{14}b_{41} & a_{11}b_{12} + a_{12}b_{22} + a_{13}b_{32} + a_{14}b_{42} \\ a_{21}b_{11} + a_{22}b_{21} + a_{23}b_{31} + a_{24}b_{41} & a_{21}b_{12} + a_{22}b_{22} + a_{23}b_{32} + a_{24}b_{42} \\ a_{31}b_{11} + a_{32}b_{21} + a_{33}b_{31} + a_{34}b_{41} & a_{31}b_{12} + a_{32}b_{22} + a_{33}b_{32} + a_{34}b_{42} \end{bmatrix}.$$

Obviously, each element in the resultant matrix is the sum of four terms that can be written in a more compact form through their sequence. For example, c_{11} has the following entry:

$$c_{11} = a_{11}b_{11} + a_{12}b_{21} + a_{13}b_{31} + a_{14}b_{41},$$

which can be rewritten in a simplified form as

$$c_{11} = \sum_{k=1}^{4} a_{1k}b_{k1}.$$

Other elements can be expressed in the same general manner. The elements in matrix C become

$$C = \begin{bmatrix} c_{11} & c_{12} \\ c_{21} & c_{22} \\ c_{31} & c_{32} \end{bmatrix} = \begin{bmatrix} \sum_{k=1}^{4} a_{1k}b_{k1} & \sum_{k=1}^{4} a_{1k}b_{k2} \\ \sum_{k=1}^{4} a_{2k}b_{k1} & \sum_{k=1}^{4} a_{2k}b_{k2} \\ \sum_{k=1}^{4} a_{3k}b_{k1} & \sum_{k=1}^{4} a_{3k}b_{k2} \end{bmatrix}.$$

Looking at the sequential relationship between the elements, the following simplified expression represents the overall solution to the matrix multiplication problem:

$$C = [c_{ij}] = \left[\sum_{k=1}^{P} a_{ik} b_{kj} \right] \tag{2.4}$$

for $i = 1, 2, \ldots, M$, and $j = 1, 2, \ldots, N$. This compact form is also the algorithmic solution for the matrix multiplication problem as program codes can easily be designed from the solution.

We discuss the program design for the above problem. Three loops with the iterators i, j, and k are required. As k is the common variable representing both the row number of A and the column number of B, it is ideally placed in the innermost loop. The middle loop has j as the iterator because it represents the column number of C. The outermost loop then represents the row number of C, having i as the iterator.

The corresponding C++ code for the multiplication problem has the complexity of $O(MNP)$, and the code segments are

```
for (i=1; i<=M; i++)
      for (j=1; j<=N; j++)
      {
            c[i][j]=0;
            for (k=1; k<=P; k++)
                  c[i][j] += a[i][k]*b[k][j];
      }
```

Code2B.cpp is a full program for the multiplication problem between the matrices $A = [a_{ik}]$ and $B = [b_{kj}]$ with $i = 1, 2, \ldots, M$ and $j = 1, 2, \ldots, N$. The program demonstrates the full implementation of the dynamic memory allocation scheme for two-dimensional arrays.

Code2B.cpp: multiplying two matrices

```cpp
#include <iostream.h>
#define M 3
#define P 4
#define N 2

void main()
{
    int i,j,k;
    int **a,**b,**c;
    a = new int *[M+1];
    b = new int *[P+1];
    c = new int *[M+1];
    for (i=1;i<=M;i++)
        a[i]=new int [P+1];
    for (j=1;j<=P;j++)
        b[j]=new int [N+1];
    for (k=1;k<=M;k++)
        c[k]=new int [N+1];

    a[1][1]=2; a[1][2]=-3; a[1][3]=1; a[1][4]=5;
    a[2][1]=-1; a[2][2]=4; a[2][3]=-4; a[2][4]=-2;
    a[3][1]=0; a[3][2]=-3; a[3][3]=4; a[3][4]=2;
    cout << "Matrix A:" << endl;
    for (i=1; i<=M; i++)
    {
        for (j=1;j<=P;j++)
            cout << a[i][j] << " ";
        cout << endl;
    }

    b[1][1]=4; b[1][2]=-1;
    b[2][1]=3; b[2][2]=2;
    b[3][1]=1; b[3][2]=-1;
    b[4][1]=-2; b[4][2]=4;
    cout << endl << "Matrix B:" << endl;
    for (i=1; i<=P; i++)
    {
        for (j=1; j<=N; j++)

            cout << b[i][j] << " ";
```

```
            cout << endl;
    }

    cout << endl << "Matrix C (A multiplied by B):" << endl;
    for (i=1; i<=M; i++)
    {
            for (j=1;j<=N;j++)
            {
                    c[i][j]=0;
                    for (k=1;k<=P;k++)
                            c[i][j] += a[i][k]*b[k][j];
                    cout << c[i][j] << " ";
            }
            cout << endl;
    }
    for (i=1;i<=M;i++)
            delete a[i];
    for (j=1;j<=P;j++)
            delete b[j];
    for (k=1;k<=M;k++)
            delete c[k];
    delete a,b,c;
}
```

The declaration and coding for arrays with higher dimensions in dynamic memory allocation follow a similar format as those in the one- and two-dimensional arrays. For example, a three-dimensional array is declared statically as follows:

```
int r[M+1][N+1][P+1];
```

The following code shows the dynamic memory allocation method for the same array:

```
int ***r;                       // declaring a pointer
                                // to the integer array
r = new int **[M+1];            // allocation to the
                                // first parameter
for (int i=0;i<=M;i++)
{
    r[i]=new int *[N+1];        // allocation to the
                                // second parameter
    for (int j=0;j<=N;j++)
            r[i][j]=new int [P+1];  // allocation to the
                                    // third parameter
}
...
```

. . .

```
for (i=0;i<=N;i++)
{
    for (j=0;j<=P;j++)
        delete r[i][j];
    delete r[i];
}
delete r;                          // destroying the array
```

2.3 MATRIX REDUCTION PROBLEMS

One of the most important operations involving a matrix is its reduction to the upper triangular form U. The technique is called *row operations*, and it has a wide range of useful applications for solving problems involving matrices. Row operation is an *elimination* technique for reducing a row in a matrix into its simpler form using one or more basic algebraic operations from addition, subtraction, multiplication, and division. A reduction on one element in a row must be followed by all other elements in that row in order to preserve their values.

Vector and Matrix Norms

The norm of a vector is a scalar value for describing the size or length of the vector with respect to the dimension n of the vector. Norm values of a vector are extensively referred especially in evaluating the errors that arise from an operation involving the vector.

Definition 2.1. The *norm-n* of a vector of size m, $v = [v_1 \ v_2 \ \cdots \ v_m]^T$, is defined as the nth root of the sum to the power of n of the elements in the vector, or

$$\|v\|_n = \sqrt[n]{\sum_{i=1}^{m} v_i^n} = \sqrt[n]{x_1^n + x_2^n + \cdots + x_m^n}. \tag{2.5}$$

It follows from Definition 2.1 that the first norm of $v = [v_1 \ v_2 \ \cdots \ v_n]^T$, called the *sum of the vector magnitude norm*, is the sum of the absolute value of the elements, as follows:

$$\|v\|_1 = \sum_{i=1}^{m} |v_i| = |v_1| + |v_2| + \cdots + |v_m|. \tag{2.6}$$

The second norm of $v = [v_1 \ v_2 \ \cdots \ v_m]^T$ is called the Euclidean vector norm, which is expressed as follows:

$$\|v\| = \|v\|_2 = \sqrt{\sum_{i=1}^{m} v_i^2} = \sqrt{v_1^2 + v_2^2 + \cdots + v_m^2}. \tag{2.7}$$

Another important norm is the *maximum vector magnitude norm*, or $\|v\|_\infty$, of $v = [v_1 \ v_2 \ \cdots \ v_m]^T$, which is the maximum of the absolute value of the elements. This norm is defined as follows:

$$\|v\|_\infty = \max_{1 \le i \le m} |v_i| = \max\left(|v_1|, |v_2|, \ldots, |v_m|\right). \tag{2.8}$$

The rule for determining the *maximum matrix norm* follows from that of a vector because a matrix of size $m \times n$ is made up of m rows of n column vectors each. This is expressed as follows:

$$\|A\|_\infty = \max_{1 \le i \le m} \left(|a_{i1}| + |a_{i2}| + \cdots + |a_{in}|\right)$$

$$= \max(|a_{11}| + |a_{12}| + \cdots + |a_{1n}|, |a_{21}| + |a_{22}| + \cdots + |a_{2n}|, \ldots, |a_{m1}|$$

$$+ |a_{m2}| + \cdots + |a_{mn}|).$$

Example 2.1. Given $v = [-3, 5, 2]^T$ and $A = \begin{bmatrix} -4 & 1 & 2 \\ 3 & -1 & 5 \\ -2 & -1 & 1 \end{bmatrix}$, find $\|v\|_1$, $\|v\|_2$, $\|v\|_\infty$ and $\|A\|_\infty$.

Solution. The calculation is very straightforward, as follows:

$$\|v\|_1 = \sum_{i=1}^{3} |v_i| = |-3| + |5| + |2| = 3 + 5 + 2 = 10.$$

$$\|v\| = \|v\|_2 = \sqrt{\sum_{i=1}^{3} v_i^2} = \sqrt{(-3)^2 + 5^2 + 2^2} = \sqrt{38}.$$

$$\|v\|_\infty = \max_{1 \le i \le 3} |v_i| = \max\left(|-3|, |5|, |2|\right) = \max\,(3, 5, 2) = 5.$$

$$\|A\|_\infty = \max\left(|-4| + |1| + |2|, |3| + |-1| + |5|, |-2| + |-1| + |1|\right)$$

$$= \max\,(7, 9, 4) = 9.$$

Row Operations

Row operations are the key ingredients for reducing a given matrix into a simpler form that describes the behavior and properties of the matrix.

Definition 2.2. Row operation on row i with respect to row j is an operation involving addition and subtraction defined as $R_i \leftarrow R_i + mR_j$, where m is a constant and R_i denotes row i. In this definition, R_i on the left-hand side is a new value obtained from the update with R_i on the right-hand side as its old value.

Theorem 2.1. Row operation in the form $R_i \leftarrow R_i + m R_j$ on row i with respect to row j does not change the matrix values. In this case, the matrix with the reduced row is said to be *equivalent* to the original matrix.

Consider matrix $A = \begin{bmatrix} p & q & r \\ s & t & u \end{bmatrix}$ having two rows and three columns. Row operation on row 2 with respect to row 1 is defined as follows:

$$R_2 \leftarrow R_2 + m R_1.$$

The addition operation on R_2 produces

$$B = \begin{bmatrix} p & q & r \\ s + mp & t + mq & u + mr \end{bmatrix}.$$

From Theorem 2.1, the new matrix obtained, B, is said to have equal values with the original matrix A. Another operation is subtraction:

$$C = \begin{bmatrix} p & q & r \\ s - mp & t - mq & u - mr \end{bmatrix},$$

which results in matrix C having the same values as A and B. We say all three matrices are equivalent, or

$$\begin{bmatrix} p & q & r \\ s & t & u \end{bmatrix} \sim \begin{bmatrix} p & q & r \\ s + mp & t + mq & u + mr \end{bmatrix} \sim \begin{bmatrix} p & q & r \\ s - mp & t - mq & u - mr \end{bmatrix}.$$

A useful reduction is the case in which the constant m is based on a *pivot element*. A pivot element is an element in a row that serves as the key to the operation on that row. The normal goal of pivoting an element is to reduce the value of an element to zero. In matrix C above, the element a_{21} in A can be reduced to 0 by letting $m = \frac{s}{p}$. In this case, $a_{11} = p$ is the pivot element as this element causes other elements in the row to change their values as a result from the row operation. Row operation $R_2 \leftarrow R_2 - \frac{s}{p} R_1$ produces

$$C = \begin{bmatrix} p & q & r \\ 0 & t - \frac{s}{p}q & u - \frac{s}{p}r \end{bmatrix}.$$

Matrix Reduction to Triangular Form

Reduction to zero elements is the basis of a technique for reducing a matrix to its *triangular* form. A triangular matrix is a square matrix where all elements above or below the diagonal elements have zero values. The name triangular is reflected in the shape of the matrix with two triangles occupying the elements, one with zero elements of the matrix and another with other numbers (including zeros).

An *upper triangular matrix* is a triangular matrix denoted as $U = [u_{ij}]$, where $u_{ij} = 0$ for $j < i$. Other elements with $j \geq i$ can have any value including zero. For example, a 4×4 upper triangular matrix has the following form:

$$U = \begin{bmatrix} * & * & * & * \\ 0 & * & * & * \\ 0 & 0 & * & * \\ 0 & 0 & 0 & * \end{bmatrix}, \tag{2.9}$$

where * can be any number including 0.

Opposite to the upper triangular matrix is the *lower triangular matrix* $L = [l_{ij}]$, where $l_{ij} = 0$ for $i < j$. The following matrix is a general notation for the 4×4 lower triangular matrix L:

$$L = \begin{bmatrix} * & 0 & 0 & 0 \\ * & * & 0 & 0 \\ * & * & * & 0 \\ * & * & * & * \end{bmatrix}. \tag{2.10}$$

A square matrix A is reduced to U or L through a series of row operations. Row operation is a technique for reducing a row in a matrix into a simpler form by performing an algebraic operation with another row in the matrix. Matrix A, when reduced to U or L, is said to be equivalent to that matrix.

In general, a square matrix with N rows requires $N - 1$ row operations with respect to rows $1, 2, \ldots, N - 1$. A matrix is usually reduced by eliminating all the rows of one of its columns to zero with respect to the pivot element in the row. The *pivot elements* of matrix $A = [a_{ij}]$ are a_{ii}, which are the diagonal elements of the matrix. The reduction of a square matrix into its upper triangular matrix is best illustrated through a simple example.

Example 2.2. Find the upper triangular matrix U of the following matrix:

$$A = \begin{bmatrix} 8 & 2 & -1 & 2 \\ -2 & 1 & -3 & -8 \\ 2 & -1 & 7 & -1 \\ 1 & -8 & 1 & -2 \end{bmatrix}.$$

Solution. There are four rows in the matrix. Reduction of the matrix to its upper triangular form requires three row operations with respect to rows 1, 2, and 3.

Operations with Respect to the First Row. $a_{11} = 8.000$ is the pivot element of the first row, and let $m = a_{i1}/a_{11}$ for the rows $i = 2, 3, 4$. All elements in the second, third, and fourth rows are reduced to their corresponding values using the relationships $a_{ij} \leftarrow a_{ij} - m*a_{1j}$ for columns $j = 1, 2, 3, 4$.

8.000	2.000	−1.000	2.000	
0.000	1.500	−3.250	−7.500	$m = a_{21}/a_{11}, a_{2j} \leftarrow a_{2j} - m*a_{1j}$
0.000	−1.500	7.250	−1.500	$m = a_{31}/a_{22}, a_{3j} \leftarrow a_{3j} - m*a_{1j}$
0.000	−8.250	1.125	−2.250	$m = a_{41}/a_{22}, a_{4j} \leftarrow a_{4j} - m*a_{1j}$

Operations with Respect to the Second Row. $a_{22} = 1.500$ is the pivot element of the second row, and therefore, $m = a_{i2}/a_{22}$ for the rows $i = 3, 4$. All elements in the third and fourth rows are reduced to their corresponding values using the relationships $a_{ij} \leftarrow a_{ij} - m*a_{2j}$ for the columns $j = 1, 2, 3, 4$.

8.000	2.000	−1.000	2.000	
0.000	1.500	−3.250	−7.500	
0.000	0.000	4.000	−9.000	$m = a_{32}/a_{22},$
0.000	0.000	−16.750	−43.500	$m = a_{42}/a_{22},$

Operations with Respect to the Third Row. $a_{33} = 4.000$ is the pivot element of the third row. Therefore, $m = a_{i3}/a_{33}$ for the row $i = 4$. All elements in the fourth row are reduced to their corresponding values using the relationship $a_{ij} \leftarrow a_{ij} - m*a_{3j}$ for the columns $j = 1, 2, 3, 4$.

8.000	2.000	−1.000	2.000	
0.000	1.500	−3.250	−7.500	
0.000	0.000	4.000	−9.000	
0.000	0.000	0.000	−81.188	$m = a_{43}/a_{33},$

Row operations on A produce U:

$$U = \begin{bmatrix} 8 & 2 & -1 & 2 \\ 0 & 1.500 & -3.250 & -7.500 \\ 0 & 0 & 4.000 & -9.000 \\ 0 & 0 & 0 & -81.188 \end{bmatrix}.$$

The C++ program to reduce A to U is very brief and compact. The code is constructed from the equation $a_{ij} \leftarrow a_{ij} - m*a_{kj}$, where $m = a_{ik}/a_{kk}$, and $i = k + 1, k + 2, \ldots, N, k = 1, 2, \ldots, N - 1$ and $j = 1, 2, \ldots, N$, as follows:

```
double m;
for (k=1;k<=N-1;k++)
        for (i=k+1;i<=N;i++)
```

```
        {
                m=a[i][k]/a[k][k];
                for (j=1;j<=N;j++)
                        a[i][j] -= m*a[k][j];
        }
```

Code2C.cpp shows the full program for reducing A to U with Code2C. in as the input tile for A

Code2C.cpp: reducing a square matrix to its upper triangular form

```
#include <fstream.h>
#include <iostream.h>
#define N 4

void main()
{
        int i,j,k;
        double **a;
        a=new double *[N+1];
        for (i=0;i<=N;i++)
                a[i]=new double [N+1];

        ifstream InFile("Code2C.in");
        for (i=1;i<=N;i++)
                for (j=1;j<=N;j++)
                        InFile >> a[i][j];
        InFile.close();

        // row operations
        double m;
        for (k=1;k<=N-1;k++)
                for (i=k+1;i<=N;i++)
                {
                        m=a[i][k]/a[k][k];
                        for (j=1;j<=N;j++)
                                a[i][j] -= m*a[k][j];
                }
        for (i=1;i<=N;i++)
        {
                for (j=1;j<=N;j++)
                        cout << a[i][j] << " ";
                cout << endl;
        }
}
```

Computing the Determinant of a Matrix

As we will see later, the triangular form of a matrix says a lot about the properties of its original matrix, which includes the determinant of the matrix. Finding the determinant of a matrix proves to be an indispensable tool in numerical computing as it contributes toward determining some properties of the matrix. This includes an important decision such as whether the inverse of the matrix exists.

Definition 2.3. The determinant of a matrix $A = [a_{ij}]$ can only be computed if it is a square matrix and A is reducible to U.

Definition 2.4. A square matrix $A = [a_{ij}]$ is said to be *singular* if its determinant is zero. Otherwise, the matrix is nonsingular. A singular matrix is not reducible to U.

Definition 2.5. The inverse of a singular matrix does not exist, which implies that if A is nonsingular, then its inverse, or A^{-1}, exists.

The determinant of matrix A, denoted by $|A|$ or $\det(A)$, is computed in several different ways. By definition, the determinant is obtained by computing the cofactor matrix from the given matrix. This method is easy to implement, but it requires many steps in its calculations.

A more practical approach to computing the determinant of a matrix is to reduce the given matrix to its upper or lower triangular matrix based on the same elimination method discussed above.

Theorem 2.2. If a square matrix is reducible to its triangular matrix, the determinant of this matrix is the product of the diagonal elements of its reduced upper or lower triangular matrix.

The above theorem states that if a square matrix $A = [a_{ij}]$ of size $N \times N$ is reducible to its upper triangular matrix $U = [u_{ij}]$, then

$$|A| = |U| = \prod_{i=1}^{N} u_{ii} = u_{11} \cdot u_{22}. \, . \, . \, . \, . u_{NN}. \tag{2.11}$$

Example 2.3. Find the determinant of the matrix defined in Example 2.2.

Solution. From Theorem 2.2, the determinant of matrix A in Example 2.2 is the product of the diagonal elements of matrix U, as follows:

$$|A| = |U| = u_{11}.u_{22}.u_{33}.u_{44}$$
$$= (8)(1.5)(4)(-81.188) = -3897.024.$$

The following code shows how a determinant is computed:

```
double Product, m;
for (k=1;k<=N-1;k++)
     for (i=k+1;i<=N;i++)
     {
            m=a[i][k]/a[k][k];
            for (j=1;j<=N;j++)
                  a[i][j] -= m*a[k][j];
     }
Product=1;
for (i=1;i<=N;i++)
     Product *= a[i][i];
```

The full program for computing the determinant of a square matrix is listed in Code2D.cpp, as follows:

Code2D.cpp: computing the determinant of a matrix

```
#include <fstream.h>
#include <iostream.h>
#define N 4

void main()
{
     int i,j,k;
     double A[N+1][N+1];

     cout.setf(ios::fixed);
     cout.precision(5);
     cout << "Input matrix A: " << endl;
     ifstream InFile("Code2D.in");
     for (i=1;i<=N;i++)
     {
            for (j=1;j<=N;j++)
            {
                  InFile >> A[i][j];
                  cout << A[i][j] << " ";
            }
            cout << endl;
     }
     InFile.close();

     // row operations
     double Product,m;
     for (k=1;k<=N-1;k++)
            for (i=k+1;i<=N;i++)
```

```
          {
                    m=A[i][k]/A[k][k];
                    for (j=1;j<=N;j++)
                              A[i][j]-=m*A[k][j];
          }

     cout << endl << "matrix U:" << endl;
     for (i=1;i<=N;i++)
     {
               for (j=1;j<=N;j++)
                         cout << A[i][j] << " ";
               cout << endl;
     }

     Product=1;
     for (i=1;i<=N;i++)
          Product *= A[i][i];

     // display results
     cout << endl << "det(A)=" << Product << endl;
}
```

Computing the Inverse of a Matrix

Finding the inverse of a matrix is a direct application of the row operations in the elimination method described earlier. Therefore, the problem is very much related to the row operations in reducing a given square matrix into its triangular form.

Theorem 2.3. The inverse of a square matrix A is said to exist if the matrix is not singular, that is, if its determinant is not zero, or $|A| \neq 0$.

The above theorem specifies the relationship between the inverse of a square matrix and its determinant. The product of a matrix A and its inverse A^{-1} is $AA^{-1} = I$, where I is the identity matrix. Hence, the problem of finding the inverse is equivalent to finding the values of matrix X in the following equation:

$$AX = I.$$

X from the above relationship can be found through row operations by defining an *augmented matrix* $A|I$. Row operations reduce A on the left-hand side to U and I on the right-hand side to a new matrix V. Continuing the process, U on the left-hand side of the augmented matrix is further reduced to I, whereas V on the right-hand side becomes $X = A^{-1}$, which is the solution.

Figure 2.1 depicts the row operations technique for finding the inverse of the matrix A. It follows that Example 2.3 illustrates an example in implementing this idea.

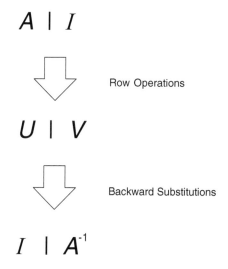

$$A \mid I$$

Row Operations

$$U \mid V$$

Backward Substitutions

$$I \mid A^{-1}$$

FIGURE 2.1. Schematic flow of the matrix inverse problem.

Example 2.4. Find the inverse of the matrix in Example 2.2.

Solution. The first step in solving this problem is to form the augmented matrix $A \mid I$, as follows:

8.000	2.000	−1.000	2.000	1.000	0.000	0.000	0.000
−2.000	1.000	−3.000	−8.000	0.000	1.000	0.000	0.000
2.000	−1.000	7.000	−1.000	0.000	0.000	1.000	0.000
1.000	−8.000	1.000	−2.000	0.000	0.000	0.000	1.000

Operations with Respect to Row 1. Start the operations with respect to row 1 with $m = a_{i1}/a_{11}$ by reducing a_{21}, a_{31} and a_{41} to 0 through the equations $a_{ij} \leftarrow a_{ij} - m*a_{1j}$ and $x_{ij} \leftarrow x_{ij} - m*x_{1j}$ for the columns $j = 1, 2, 3, 4$.

8.000	2.000	−1.000	2.000	1.000	0.000	0.000	0.000
0.000	1.500	−3.250	−7.500	0.250	1.000	0.000	0.000
0.000	−1.500	7.250	−1.500	−0.250	0.000	1.000	0.000
0.000	−8.250	1.125	−2.250	−0.125	0.000	0.000	1.000

Operations with Respect to Row 2. The next step is row operations with respect to row 2 with $m = a_{i2}/a_{22}$ for $i = 3, 4$ to reduce a_{32} and a_{42} to 0. This is achieved through $a_{ij} \leftarrow a_{ij} - m*a_{2j}$ and $x_{ij} \leftarrow x_{ij} - m*x_{2j}$ for the columns $j = 1, 2, 3, 4$, to produce the following values:

8.000	2.000	−1.000	2.000	1.000	0.000	0.000	0.000
0.000	1.500	−3.250	−7.500	0.250	1.000	0.000	0.000
0.000	0.000	4.000	−9.000	0.000	1.000	1.000	0.000
0.000	0.000	−16.750	−43.500	1.250	5.500	0.000	1.000

Operations with Respect to Row 3. Finally, U is produced from row operations with respect to row 3 with $m = a_{i3}/a_{33}$ for $i = 4$. Other elements in the row are updated according to $a_{ij} \leftarrow a_{ij} - m*a_{3j}$ and $x_{ij} \leftarrow x_{ij} - m*x_{3j}$ for $j = 1, 2, 3, 4$.

8.000	2.000	−1.000	2.000	1.000	0.000	0.000	0.000
0.000	1.500	−3.250	−7.500	0.250	1.000	0.000	0.000
0.000	0.000	4.000	−9.000	0.000	1.000	1.000	0.000
0.000	0.000	0.000	−81.188	1.250	9.688	4.188	1.000

Reducing U to I. The next step is to reduce U from the left portion of the augmented matrix to the identity matrix I. The strategy is perform row operations starting on row 4 to reduce a_{j4} to 0 for $j = 1, 2, 3, 4$.

8.000	2.000	−1.000	0.000	1.031	0.239	0.103	0.025
0.000	1.500	−3.250	0.000	0.135	0.105	−0.387	−0.092
0.000	0.000	4.000	0.000	−0.139	−0.074	0.536	−0.111
0.000	0.000	0.000	−81.188	1.250	9.688	4.188	1.000

This is followed by row operations with respect to row 3.

8.000	2.000	0.000	0.000	0.996	0.220	0.237	−0.003
0.000	1.500	0.000	0.000	0.022	0.045	0.048	−0.182
0.000	0.000	4.000	0.000	−0.139	−0.074	0.536	−0.111
0.000	0.000	0.000	−81.188	1.250	9.688	4.188	1.000

The next step is the row operations with respect to row 2.

8.000	0.000	0.000	0.000	0.967	0.160	0.172	0.240
0.000	1.500	0.000	0.000	0.022	0.045	0.048	−0.182
0.000	0.000	4.000	0.000	−0.139	−0.074	0.536	−0.111
0.000	0.000	0.000	−81.188	1.250	9.688	4.188	1.000

The last step is to reduce the diagonal matrix into an identity matrix by dividing each row with its corresponding diagonal element, as follows:

$$a_{ii} \leftarrow \frac{a_{ij}}{a_{ii}}, \text{ followed by } x_{ij} \leftarrow \frac{x_{ij}}{a_{ii}}, \text{ for } i = 1, 2, 3, 4 \text{ and } j = 1, 2, 3, 4.$$

1.000	0.000	0.000	0.000	0.121	0.020	0.022	0.030
0.000	1.000	0.000	0.000	0.015	0.030	0.032	−0.122
0.000	0.000	1.000	0.000	−0.035	−0.018	0.134	−0.028
0.000	0.000	0.000	1.000	−0.015	−0.119	−0.052	−0.012

We obtain the solution matrix

$$A^{-1} = X = \begin{bmatrix} 0.121 & 0.020 & 0.022 & 0.030 \\ 0.015 & 0.030 & 0.032 & -0.122 \\ -0.035 & -0.018 & 0.134 & -0.028 \\ -0.015 & -0.119 & -0.052 & -0.012 \end{bmatrix}.$$

The following program, Code2E.cpp, is the implementation of the elimination method for finding the inverse of a matrix.

Code2E.cpp: finding the inverse of a matrix

```cpp
#include <fstream.h>
#include <iostream.h>
#define N 4

void main()
{
        int i,j,k;
        double **A, **B, **X;
        A=new double *[N+1];
        B=new double *[N+1];
        X=new double *[N+1];
        for (i=0;i<=N;i++)
        {
                A[i]=new double [N+1];
                B[i]=new double [N+1];
                X[i]=new double [N+1];
        }
```

```
cout.setf(ios::fixed);
cout.precision(5);
ifstream InFile("Code2E.in");
for (i=1;i<=N;i++)
    for (j=1;j<=N;j++)
        {
            InFile >> A[i][j];
            if (i==j)
                B[i][j]=1;
            else
                B[i][j]=0;
        }
InFile.close();

//   row operations
double Sum,m;
for (k=1;k<=N-1;k++)
    for (i=k+1;i<=N;i++)
        {
            m=A[i][k]/A[k][k];
            for (j=1;j<=N;j++)
            {
                A[i][j] -= m*A[k][j];
                B[i][j] -= m*B[k][j];
            }
        }

//   backward substitutions
for (i=N;i>=1;i--)
    for (j=1;j<=N;j++)
    {
        Sum=0;
        for (k=i+1;k<=N;k++)
            Sum += A[i][k]*X[k][j];
        X[i][j]=(B[i][j]-Sum)/A[i][i];
    }
for (i=1;i<=N;i++)
{
    for (j=1;j<=N;j++)
        cout << X[i][j] << " ";
    cout << endl;
}
}
```

2.4 MATRIX ALGEBRA

Mathematical operations involving matrices can be very tedious and time consuming. This is especially true as the size of the matrices become large. As described, each matrix in a typical engineering application may occupy a large amount of computer memory. As a result, a mathematical operation involving several large matrices may slow down the computer and may cause the program to hang because of insufficient memory.

Efficient coding is required to handle algebraic operations involving many matrices. The preparation for a good coding includes the use of several functions to make the program modular and a distance separation of code from its data. The two requirements mean a good C++ program must implement an extensive data passing mechanism between functions as a way to simplify the execution of code.

Generally, algebraic operations on matrices involve all four basic tools: addition, subtraction, multiplication, and division. We discuss all four operations and our preparation for the large-scale algebraic operations. In achieving these objectives, we discuss methods for passing data between functions that serve as an unavoidable mechanism in making the program modular.

In making the program modular, several functions that represent the modules in the problem are created. Four main functions in matrix algebra are addition (with subtraction), multiplication, and matrix inverse (which indirectly represents matrix division). We discuss the modular technique for solving an algebraic operation involving matrices based on all of these primitive operations. To simplify the discussion, all matrices discussed are assumed to be square matrices of the same size and dimension.

Data Passing Between Functions

A good computer program considers data and program code as separate entities. In most cases, data should not be a part of the program. Data should be stored in an external storage medium, and it will only be called when required. This strategy is necessary as each unit of data occupies some precious space in the computer memory.

Data passing is important in making the program modular. Data passing encourages the variables to be declared locally instead of globally. A good C++ program should promote the use of local variables as much as possible in place of global variables as this contributes to the modularity of the program significantly. For example, if A is the input matrix for the addition function in a program, this same matrix should also be the input for the multiplication function. It is important to have a mechanism to allow data from the matrix to be passed from one function to another.

Matrix Addition and Subtraction

Two matrices of the same dimension, $A = [a_{ij}]$ and $B = [b_{ij}]$ are added to produce a new matrix $C = [c_{ij}]$, which also has the same dimension as the first two matrices. The matrix addition between two matrices is a straightforward addition between the elements in the two matrices; that is, $c_{ij} = a_{ij} + b_{ij}$. The task is represented as the function MatAdd() from a class called MatAlgebra, as follows:

```
void MatAlgebra::MatAdd(bool flag,double **c,double **a,
        double **b)
{
     int i,j,k;
     for (i=1;i<=N;i++)
           for (j=1;j<=N;j++)
               if (flag)
                       c[i][j]=a[i][j]+b[i][j];
               else
                       c[i][j]=a[i][j]-b[i][j];
}
```

In MatAdd(), the input matrices are a and b, which receive data from the calling function. The computed array is c, which returns the value to the calling function in the program. The Boolean variable, flag, is introduced in the code to allow the function to support subtraction as well as addition. Addition is performed when flag=1 (TRUE), whereas flag=0 (FALSE) means subtraction. The use of flag is necessary to avoid the creation of another function specifically for, subtraction, which proves to be redundant.

Matrix Multiplication

The matrix multiplication function is another indispensable module in matrix algebra. We call the function MatMultiply(), and its contents are very much similar to Code2B.cpp, as follows:

```
void MatAlgebra::MatMultiply(double **c,double **a, double **b)
{
     int i,j,k;
     for (i=1;i<=N;i++)
           for (j=1;j<=N;j++)
           {
                 c[i][j]=0;
                 for (k=1;k<=N;k++)
                       c[i][j] += a[i][k]*b[k][j];
           }
}
```

In the above code, a and b are two local matrices whose input values are obtained from the calling function. From these values, c is computed, and the results are passed to the matrix in the calling function.

Matrix Inverse

Matrix inverse is the reciprocal of a matrix that indirectly means division. Two matrices A and B are divided as follows:

$$C = \frac{A}{B} = AB^{-1}.$$

As demonstrated above, C can only be computed if B is not singular, that is, B^{-1} exists. Division means the first matrix is multiplied to the inverse of the second matrix. Therefore, a function for computing the matrix inverse is necessary as part of the overall matrix algebra operation.

Matrix inverse function, MatInverse() is reproduced from Code2E.cpp, as follows:

```
void MatAlgebra::MatInverse(double **x,double **a)
{
    int i,j,k;
    double Sum,m;
    double **b, **q;
    b=new double *[N+1];
    q=new double *[N+1];
    for (i=1; i<=N; i++)
    {
        b[i]=new double [N+1];
        q[i]=new double [N+1];
    }
    for (i=1;i<=N;i++)
        for (j=1;j<=N;j++)
        {
            b[i][j]=0;
            q[i][j]=a[i][j];
            if (i==j)
                b[i][j]=1;
        }
    //  Perform row operations
    for (k=1;k<=N-1;k++)
        for (i=k+1;i<=N;i++)
        {
            m=q[i][k]/q[k][k];
            for (j=1;j<=N;j++)
            {
                q[i][j]-=m*q[k][j];
                b[i][j]-=m*b[k][j];
            }
        }
```

```
//  Perform back substitution
for (i=N;i>=1;i--)
        for (j=1;j<=N;j++)
        {
                Sum=0;
                x[i][j]=0;
                for (k=i+1;k<=N;k++)
                        Sum += q[i][k]*x[k][j];
                x[i][j]=(b[i][j]-Sum)/q[i][i];
        }
        for (i=0;i<=N;i++)
                delete b[i], q[i];
        delete b, q;
}
```

MatInverse() requires two parameters, x and a, both of which are two-dimensional arrays. In this function, a is the input matrix that receives data from the calling function. The computed values in matrix x are passed to an array in the calling function.

Putting the Pieces Together

We are now ready to deploy a technique for performing an operation in matrix algebra. We discuss the case of computing $Z = A^2 B^{-1} + A^{-1} B - AB$, where the input matrices A and B have the same size and dimension. This expression consists of addition, subtraction, multiplication, and matrix inverse operations.

The expression $A^2 B^{-1} + A^{-1} B - AB$ is a tree that can be broken down into several smaller modules, as shown in Figure 2.2. The items in the expression are modularized into three algebraic components: addition, multiplication, and inverse. Several temporary arrays are created to hold the results from these operations, and they are P, Q, R, U, S, T, V, and Z. Groupings are made based on the priority level, as follows:

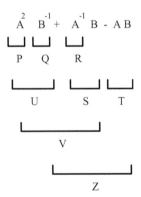

FIGURE 2.2. Tree diagram for $Z = A^2 B^{-1} + A^{-1} B - AB$.

First level grouping:

$$P = A^2, \ Q = B^{-1} \text{ and } R = A^{-1}.$$

This is followed by the second-level grouping:

$$U = PQ, \ S = RB \text{ and } T = AB.$$

Then, the third-level grouping:

$$V = U + S.$$

And the fourth-level grouping:

$$Z = V - T.$$

The diagram in Figure 2.2 is a tree with the root at Z. Each grouping described above is represented as a call to the corresponding function in main(). Supposing g is the object linked in main() to those functions, the C++ solution to the expression $Z = A^2B^{-1} + A^{-1}B - AB$ is given as follows:

```
g.ReadData(A,B);
g.MatMultiply(P,A,A);
g.MatInverse(Q,B);
g.MatInverse(R,A);
g.MatMultiply(U,P,Q);
g.MatMultiply(S,R,B);
g.MatMultiply(T,A,B);
g.MatAdd(1,V,U,S);
g.MatAdd(0,Z,V,T);
```

In the above solution, MatAdd(), MatMultiply(), and MatInverse() are the matrix addition, multiplication, and inverse, respectively. Prior to its implementation, data for the matrices are read and passed into main() through ReadData(). In this case, data for the matrices A and B are received from this function. The values for A become an input for MatMultiply(P,A,A), which performs the multiplication A*A and stores the result in a new matrix P. Similarly, MatInverse(Q,B) receives data for matrix B, computes its inverse of B, and stores the result into a new matrix Q. The function MatAdd(1,V,U,S) adds the matrices U and S and then stores the result into V. The Boolean parameter 1 here means addition, or V=U+S; with 0 the function performs subtraction, or V=U-S.

The above code is demonstrated in its complete form in Code2F.cpp with A and B matrices defined in a file called Code2F.in, as follows:

Code2F.cpp: computing $Z = A^2 B^{-1} + A^{-1} B - AB$.

```
#include <fstream.h>
#include <iostream.h>
#define N 3

class MatAlgebra
{
public:
        MatAlgebra() { }
        ~MatAlgebra() { }
        void ReadData(double **,double **);
        void MatAdd(bool,double **,double **,double **);
        void MatInverse(double **,double **);
        void MatMultiply(double **,double **,double **);
};

void MatAlgebra::ReadData(double **a,double **b)
{
        int i,j;
        ifstream InFile("Code2F.in");
        for (i=1;i<=N;i++)
                for (j=1;j<=N;j++)
                        InFile >> a[i][j];
          for (i=1;i<=N;i++)
                for (j=1;j<=N;j++)
                        InFile >> b[i][j];
          InFile.close();
}

void MatAlgebra::MatAdd(bool flag,double **c,double **a,
        double **b)
{
        int i,j,k;
        for (i=1;i<=N;i++)
                for (j=1;j<=N;j++)
                        c[i][j]=((flag)?a[i][j]+b[i][j]:
                                a[i][j]-b[i][j]);
}
void MatAlgebra::MatMultiply(double **c,double **a,double **b)
{
        int i,j,k;
        for (i=1;i<=N;i++)
                for (j=1;j<=N;j++)
```

```
            {
                 c[i][j]=0;
                 for (k=1;k<=N;k++)
                         c[i][j] += a[i][k]*b[k][j];
            }
}

void MatAlgebra::MatInverse(double **x,double **a)
{
       int i,j,k;
       double Sum,m;
       double **b, **q;
       b=new double *[N+1];
       q=new double *[N+1];
       for (i=0; i<=N; i++)
       {
            b[i]=new double [N+1];
            q[i]=new double [N+1];
       }
       for (i=1;i<=N;i++)
            for (j=1;j<=N;j++)
            {
                    b[i][j]=0;
                    q[i][j]=a[i][j];
                    if (i==j)
                            b[i][j]=1;
            }

       //  Perform row operations
       for (k=1;k<=N-1;k++)
            for (i=k+1;i<=N;i++)
            {
                    m=q[i][k]/q[k][k];
                    for (j=1;j<=N;j++)
                    {
                            q[i][j]-=m*q[k][j];
                            b[i][j]-=m*b[k][j];
                    }
            }

       //  Perform back substitution
       for (i=N;i>=1;i--)
            for (j=1;j<=N;j++)
            {
                    Sum=0;
                    x[i][j]=0;
```

```
                for (k=i+1;k<=N;k++)
                        Sum += q[i][k]*x[k][j];
                x[i][j]=(b[i][j]-Sum)/q[i][i];
        }
        for (i=0;i<=N;i++)
                delete b[i], q[i];
        delete b, q;
}

void main()
{
    int i,j;
    double **A,**B;
    double **P,**Q,**R,**S,**T,**U,**V,**Z;
    MatAlgebra g;

    A=new double *[N+1];
    B=new double *[N+1];
    P=new double *[N+1];
    Q=new double *[N+1];
    R=new double *[N+1];
    S=new double *[N+1];
    T=new double *[N+1];
    U=new double *[N+1];
    V=new double *[N+1];
    Z=new double *[N+1];
    for (i=0;i<=N;i++)
    {
            A[i]=new double [N+1];
            B[i]=new double [N+1];
            P[i]=new double [N+1];
            Q[i]=new double [N+1];
            R[i]=new double [N+1];
            S[i]=new double [N+1];
            T[i]=new double [N+1];
            U[i]=new double [N+1];
            V[i]=new double [N+1];
            Z[i]=new double [N+1];
    }
    cout.setf(ios::fixed);
    cout.precision(12);
    g.ReadData(A,B);
    g.MatMultiply(P,A,A);
    g.MatInverse(Q,B);
    g.MatInverse(R,A);
    g.MatMultiply(U,P,Q);
```

```
g.MatMultiply(S,R,B);
g.MatMultiply(T,A,B);
g.MatAdd(1,V,U,S);
g.MatAdd(0,Z,V,T);
for (i=1;i<=N;i++)
{
        for (j=1;j<=N;j++)
                cout << Z[i][j] << " ";
        cout << endl;
}
for (i=0;i<=N;i++)
        delete A[i],B[i],P[i],Q[i],R[i],S[i],T[i],U[i],V[i],
                Z[i];
        delete A,B,P,Q,R,S,T,U,V,Z;
}
```

2.5 ALGEBRA OF COMPLEX NUMBERS

Not all numbers in existence are real. There are also cases in which a number exists from an imagination based on some undefined entity. *Complex number* is a number representation based on the imaginary definition of

$$i = \sqrt{-1}.$$

This definition implies $i^2 = -1$, which is not possible in the real world. However, the existence of complex numbers cannot be denied as they provide a powerful foundation to several problems in mathematical modeling.

Complex numbers play an important role in numerical computations. For example, in one area of mathematics called *conformal mapping*, complex numbers are used to map one geometrical shape from a complex plane to another complex plane in such a way to preserve certain quantities. One such work is the mapping of a circular cylinder to a family of airfoil shapes by considering the pressure and velocity of particles around the cylinder. In another area of study called *chaotic theory*, complex numbers are used in modeling randomly displaced particles in a fluid that are subject to a motion called Brownian movement.

A complex number z has two parts, namely, *real* and *imaginary*, defined as follows:

$$z = a + bi. \tag{2.12}$$

In the above form, a is the real part, whereas b is the imaginary part. A complex number with $b = 0$ is a real number. Figure 2.3 shows a complex number $z = a + bi$ with respect to the real axis, R and the imaginary axis i, with the origin at $z = 0 + 0i = 0$.

Complex numbers are not represented directly in their standard form in the computer, because the computer supports integers only in its operations. Therefore, some specialized routines need to be created for manipulating the integers in the computer to enable complex number arithmetic.

In C++, a complex number can be defined using a structure whose contents consist of the real and imaginary parts of the number. For example, the following structure

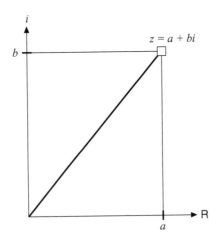

FIGURE 2.3. Complex number $z = a + bi$.

called Complex declares a complex number:

```
typedef struct
{
        double r;                    // real part
        double i;                    // imaginary part
}
    Complex;
```

Using the above structure, a complex number z can be created as an object through the declaration

```
Complex z;
```

It follows that z.r represents the real part of z (which is a) and z.i represents the imaginary part of z (which is b), or

```
z.r = a, and z.i = b.
```

Arithmetic on complex numbers obeys the same set of rules in mathematics as that of the real numbers. Basically, the operations involve addition, subtraction, multiplication, and division. We discuss the development of functions for each of the algebraic tools using a single class called ComAlgebra.

Addition and Subtraction

The addition of two complex numbers is carried out by adding their real part and imaginary part separately. The same rule also applies to subtraction. Suppose $z_1 =$

$a_1 + b_1 i$ and $z_2 = a_2 + b_2 i$ are two complex numbers. Then their sum is

$$z_1 + z_2 = (a_1 + a_2) + (b_1 + b_2)i.$$

Similarly, subtraction is carried out as follows:

$$z_1 - z_2 = (a_1 - a_2) + (b_1 - b_2)i.$$

In C++, addition between two complex numbers u and v is performed using the function Add() to produce w, as follows:

```
Complex ComAlgebra::Add(Complex u,Complex v)
{
        Complex w;
        w.r = u.r + v.r;
        w.i = u.i + v.i;
        return(w);
}
```

Subtraction is very much similar to addition, using the function Subtract(), as follows:

```
Complex ComAlgebra::Subtract(Complex u,Complex v)
{
        Complex w;
        w.r = u.r - v.r;
        w.i = u.i - v.i;
        return(w);
}
```

Multiplication

The multiplication of two complex numbers works in a similar manner as the dot-product between two vectors where the product of the real parts adds to the product of the imaginary parts. For two complex numbers $z_1 = a_1 + b_1 i$ and $z_2 = a_2 + b_2 i$, their product is

$$z_1 . z_2 = (a_1 + b_1 i)(a_2 + b_2 i)$$
$$= a_1 a_2 + b_1 b_2 i^2 + (a_1 b_2 + a_2 b_1)i.$$
$$= (a_1 a_2 - b_1 b_2) + (a_1 b_2 + a_2 b_1)i$$

Multiplication between two numbers, u and v, to produce w is demonstrated using the function `Multiply()`, as follows:

```
Complex ComAlgebra::Multiply (Complex u, Complex v)
{
        Complex w;
        w.r = u.r*v.r - u.i*v.i;
        w.i = u.r*v.i + v.r*u.i;
        return(w);
}
```

Conjugate

The *conjugate* of a complex number $z = a + bi$ is $\bar{z}$, which is a reflection of the number on the imaginary axis, denoted as $\bar{z} = a - bi$. The code segments for computing the conjugate w of the complex number u using the function `Conjugate()` are as follows:

```
Complex ComAlgebra::Conjugate(Complex u)
{
        Complex w;
        w.r = u.r;
        w.i = -u.i;
        return(w);
}
```

Division

When a complex number $z_1 = a_1 + b_1 i$ is divided by another complex number $z_2 = a_2 + b_2 i$, the result is also a complex number. This calculation is shown as

$$\frac{z_1}{z_2} = \frac{a_1 + b_1 i}{a_2 + b_2 i}$$

$$= \frac{a_1 + b_1 i}{a_2 + b_2 i} \cdot \frac{a_2 - b_2 i}{a_2 - b_2 i}$$

$$= \frac{(a_1 a_2 + b_1 b_2) + (a_2 b_1 - a_1 b_2)i}{a_2^2 + b_2^2}$$

$$= \frac{(a_1 a_2 + b_1 b_2)}{a_2^2 + b_2^2} + \frac{(a_2 b_1 - a_1 b_2)}{a_2^2 + b_2^2} i.$$

A function called `Divide()` shows the division of two complex numbers, u on v, to produce another complex number, w, as follows:

```
Complex ComAlgebra::Divide(Complex u, Complex v)
{
        Complex c, w, s;
        double denominator;
        c = Conjugate(v);
        s = Multiply (u, c);
        denominator = v.r*v.r + v.i*v.i + 1.2e-60; // to prevent
                                                      division
                                                      by zero

        w.r = s.r/denominator;
        w.i = s.i/denominator;
        return(w);
}
```

A variable called `denominator` in the above code segment has its value added with a very small number 1.2e-60, or 1.2×10^{-60}. The addition does not change the result significantly, and it is necessary in order to prevent the value from becoming zero, hence, avoiding the division by zero problem.

Inverse of a Complex Number

The inverse z^{-1} of a complex number $z = a + bi$ is computed as follows:

$$z^{-1} = \frac{1}{z} = \frac{1}{a + bi}$$

$$= \frac{1}{a + bi} \cdot \frac{a - bi}{a - bi}$$

$$= \frac{a - bi}{a^2 + b^2}$$

$$= \frac{a}{a^2 + b^2} - \frac{b}{a^2 + b^2} i.$$

The corresponding C++ code using the function `Inverse()` for computing the inverse of u is shown as follows:

```
Complex ComAlgebra::Inverse(Complex u)
{
        Complex w,v;
        v.r = 1.0;
        v.i = 0.0;
```

```
    w = Divide(v,u);
    return(w);
}
```

Performing Complex Number Arithmetic

An algebraic arithmetic involving complex numbers is easily performed using the basic operations discussed above: addition, subtraction, multiplication, division, conjugate, and inverse. The strategy is to have a function for each operation so that the whole process becomes structured and modular. As demonstrated earlier in the case of matrix operation, an algebraic operation on complex numbers should be based on a good programming principle. To achieve this objective, the operation must maximize the use of local variables, encourage data passing between functions, and encourage object-oriented methodology.

We illustrate complex number algebra using an example in evaluating the expression $z = \frac{2uv^{-1} - u^2 v}{3u + v}$, assuming u and v are complex numbers, and $3u + v \neq 0$. In solving this problem, we adopt the same strategy discussed in Section 2.4. The whole operation can be broken down into four levels of groupings based on their priorities with respect to the basic algebraic tools. The whole concept for this problem is depicted in the form of a priority tree in Figure 2.4.

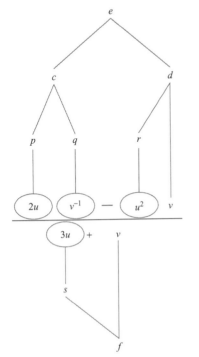

First-level grouping:
$p = 2u, q = v^{-1}, r = u^2$, and $s = 3u$.

Second-level grouping:
$c = pq$ and $d = rv$.

Third-level grouping:
$e = c - d$ and $f = s + v$.

Fourth-level grouping:
Compute $z = \frac{e}{f}$

FIGURE 2.4. Groupings according to the priority tree for $z = \frac{2uv^{-1} - u^2 v}{3u + v}$.

The priority tree in Figure 2.4 is a structured approach for solving the expression $z = \frac{2uv^{-1} - u^2 v}{3u + v}$. The solution is coded in `main()` based on calls to the respective functions described earlier, as follows:

```
Complex u,v,z,k;
Complex p,q,r,s,c,d,e,f;
ComAlgebra Compute;

k.r=2; k.i=0;
p=Compute.Multiply(k,u);
q=Compute.Inverse(v);
r=Compute.Multiply(u,u);
k.r=3; k.i=0;
s=Compute.Multiply(k,u);

c=Compute.Multiply(p,q);
d=Compute.Multiply(r,v);
e=Compute.Subtract(c,d);
f=Compute.Add(s,v);
z=Compute.Divide(e,f);
```

Code2G.cpp shows the full program for the algebraic operation in solving the expression $z = \frac{2uv^{-1} - u^2 v}{3u + v}$.

Code2G.cpp: complex number algebra for solving $z = \frac{2uv^{-1} - u^2 v}{3u + v}$.

```
#include <iostream.h>
#include <math.h>

typedef struct
{
     double r;          // real part
     double i;          // imaginary part
} Complex;

class ComAlgebra
{
public:
     ComAlgebra( );
     Complex Add(Complex,Complex);
     Complex Subtract(Complex,Complex);
     Complex Conjugate(Complex);
     Complex Multiply(Complex,Complex);
     Complex Divide(Complex,Complex);
     Complex Inverse(Complex);
     double Magnitude(Complex);
     double Angle(Complex);
};
```

```
ComAlgebra::ComAlgebra( )
{ }

Complex ComAlgebra::Add(Complex u,Complex v)
{
     Complex w;
     w.r = u.r + v.r;
     w.i = u.i + v.i;
     return(w);
}

Complex ComAlgebra::Subtract(Complex u,Complex v)
{
     Complex w;
     w.r = u.r - v.r;
     w.i = u.i - v.i;
     return(w);
}

Complex ComAlgebra::Conjugate(Complex u)
{
     Complex w;
     w.r = u.r;
     w.i = -u.i;
     return(w);
}

Complex ComAlgebra::Multiply (Complex u, Complex v)
{
     Complex w;
     w.r = u.r*v.r - u.i*v.i;
     w.i = u.r*v.i + v.r*u.i;
     return(w);
}

Complex ComAlgebra::Divide(Complex u, Complex v)
{
     Complex c, w, s;
     double denominator;

     c = Conjugate(v);
     s = Multiply (u, c);
     denominator = v.r*v.r + v.i*v.i + 1.2e-60;    //to prevent
                                                   division
                                                   by zero

     w.r = s.r/denominator;
```

```
        w.i = s.i/denominator;
        return(w);
}


Complex ComAlgebra::Inverse(Complex u)
{
        Complex w,v;
        v.r = 1.0;
        v.i = 0.0;
        w = Divide(v,u);
        return(w);
}


void main()
{
        Complex u,v,z,k;
        Complex p,q,r,s,c,d,e,f;
        ComAlgebra Compute;

        u.r=-4; u.i=7;
        v.r=6; v.i=3;

        k.r=2; k.i=0;
        p=Compute.Multiply(k,u);
        q=Compute.Inverse(v);
        r=Compute.Multiply(u,u);
        k.r=3; k.i=0;
        s=Compute.Multiply(k,u);

        c=Compute.Multiply(p,q);
        d=Compute.Multiply(r,v);
        e=Compute.Subtract(c,d);
        f=Compute.Add(s,v);
        z=Compute.Divide(e,f);

        cout << "z = " << z.r << " + " << z.i << "i" << endl;
}
```

2.6 NUMBER SORTING

One important feature in numbers is their order based on the values they carry. For example, 3 has a lower value than 7; therefore, 3 should be placed before 7. Arranging numbers in order according to their values helps in making the data more readable and organized. Besides, ordered data contribute toward an efficient searching mechanism and for use in their analysis. The main application involving number sorting can be found in the database management system.

Data can be sorted in the order from lowest to highest, or highest to lowest, based on their numeric or alphabetic values. An integer or double variable is ordered according its numeric value, whereas a string is sorted based on its alphabetical value.

Number sorting is a data structure problem that may require high computational time in its implementation. Number sorting helps in searching for a particular record based on its order.

The code below performs sorting with k steps with a complexity of $O(N^2)$, where at each step the numbers w[i] with w[i+1] are compared. If w[i] has a value greater than w[i+1], then the two values are swapped, using tmp as the temporary variable to hold the values before swapping.

```
for (int k=1;k<=N;k++)
     for (i=1;i<=N-1;i++)
     if (w[i]>w[i+1])          // swap for low to high
     {
             tmp=w[i];
             w[i]=w[i+1];
             w[i+1]=tmp;
     }
```

Example 2.5. Sort the numbers in the list given by 60, 74, 43, 57 and 45 in ascending order.

Solution. Applying the sorting algorithm above with $k = 1$:

	old	new
w[1]	60	60
w[2]	74	43
w[3]	43	57
w[4]	57	45
w[5]	45	74

i=1: 60<74, no change with w[1]=60, w[2]=74.
i=2: 74>43, swap with w[2]=43, w[3]=74.
i=3: 74>57, swap with w[3]=57, w[4]=74.
i=4: 74>45, swap with w[4]=45, w[5]=74.

Continue with $k = 2$:

	old	new
w[1]	60	43
w[2]	43	57
w[3]	57	45
w[4]	45	60
w[5]	74	74

i=1: 60>43, swap with w[1]=43, w[2]=60.
i=2: 60>57, swap with w[2]=57, w[3]=60.
i=3: 60>45, swap with w[3]=45, w[4]=60.
i=4: 60<74, no change with w[4]=60, w[5]=74.

Next iteration with $k = 3$:

	old	new
w[1]	43	43
w[2]	57	45
w[3]	45	57
w[4]	60	60
w[5]	74	74

i=1:43<57, no change with w[1]=43, w[2]=57.
i=2: 57>45, swap with w[2]=45, w[3]=57.
i=3:57<60, no change with w[3]=57, w[4]=60.
i=4: 60<74, no change with w[4]=60, w[5]=74.

The numbers have been fully sorted after $k = 3$. The final order from the given list is 43, 45, 57, 60, and 74.

Code2H.cpp shows the full C++ program for the sorting problem using random numbers as the input.

Code2H.cpp: sorting numbers

```cpp
#include <iostream.h>
#include <stdlib.h>
#include <time.h>
#define N 8

void main()
{
        int *v,*w,tmp;
        v = new int [N+1];
        w = new int [N+1];

        time_t seed=time(NULL);
        srand((unsigned)seed);

        for (int i=1;i<=N;i++)
        {
                v[i]=1+rand()%100;              // random numbers from 1
                                               // to 100

                w[i]=v[i];
        }
        for (int k=1;k<=N;k++)
                for (i=1;i<=N-1;i++)
                        if (w[i]>w[i+1])       // swap for low to high
                        {
                                tmp=w[i];
                                w[i]=w[i+1];
                                w[i+1]=tmp;
                        }
        cout << "the unsorted random numbers v[i] are:" << endl;
```

```
      for (i=1;i<=N;i++)
            cout << v[i] << " ";
      cout << endl << "sorted from lowest to highest w[i]:"
            << endl;
      for (i=1;i<=N;i++)
            cout << w[i] << " ";
      cout << endl;
}
```

2.7 SUMMARY

This chapter outlines the importance of good programming strategies for handling several mathematical-intensive operations involving massive data. The strategies include exploring the resources in the computer to the maximum in order to optimize the operations. This chapter highlights some important programming strategies such as breaking down the program into modules, allocating memory dynamically to arrays, maximizing the use of local variables, and encouraging data passing between functions. It is important to consider these issues seriously in designing the solution as a typical scientific problem often causes the computer to perform below its capability. In some cases, the computer performance is adversely affected through poor management of memory and improper techniques that originate from the spaghetti type of coding.

The main issue regarding scientific computing that affects the computer performance is the handling of arrays. A typical array in an application represents a matrix that often occupies a large amount of computer memory. A good strategy is to allocate memory dynamically to these arrays. Besides memory management, an array is also subject to its mathematical properties and behavior. Any operation involving arrays must be carefully executed by considering their mathematic properties in all steps of the operation. For example, if a given matrix is singular, then it is not possible to find its inverse as the computation will only lead to some bad results.

Data passing is an important issue that contributes toward a structured and modular program design. A good contribution can be seen from the feature in data passing, which encourages the arrays and variables to be declared local, and their values can be shared by another function through this mechanism. This is illustrated through a discussion on how a complex expression involving arrays and vectors can be solved easily through a modular technique that advocates data passing.

We discuss row operations, which is an important technique for reducing a matrix into its simpler form. Row operations involve some massive and tedious calculations that contribute to problems such as finding the determinant of a matrix, reducing a matrix to its triangular form, and finding the inverse of a matrix. Another core application involving row operations is in solving a system of linear equations, and this topic will be discussed thoroughly in the next chapter.

We also explore some mathematical problems involving complex numbers and number sorting. By default, complex numbers are not supported in the computer. But, it is possible to perform operations involving complex numbers by treating the numbers as members of a structure. We show how a difficult mathematical expression involving complex numbers can be solved easily using a modular programming approach.

Number sorting is the extra tool that is included in this chapter. The topic itself falls under the category of data structure, which is not in the scope of our discussion. However, the problem appears quite often in various mathematical and engineering problems. In this chapter, we show how number sorting is performed with programming as its tool.

PROGRAMMING CHALLENGES

1. Code2C project computes the upper triangular matrix U from any square matrix A. Write a new program to reduce A into a lower triangular matrix L, as defined in Equation (2.10).

2. An extension to the Code2C project is the decomposition of a square matrix A into the product of its upper and lower triangular matrices, or $A = LU$. Write a new program to do this.

3. Code2F.cpp illustrates a modular approach for solving an algebraic operation involving matrices where the matrices are square and have fixed sizes. Design a C++ program to compute $Z = A^2 B^{-1} + A^{-1} B - AB$ for matrices A and B that have flexible sizes. The program must provide a check on the dimension and status of each matrix according to the fundamental mathematical properties in order to allow operations such as multiplication and computing the inverse.

4. Modify Code2F.cpp for solving the following problems:
   a. $Z = (A + 3B)^{-1}$.
   b. $Z = A^{-1} B^2 + 3A^2 B^{-1}$.
   c. $Z = (AB)^2 - 3(AB)^{-1}$.

5. An orthogonal matrix is the matrix $A = [a_{ij}]$ for $i = 1, 2, \ldots, M$, and $j = 1, 2, \ldots, N$, where $\sum_{i=1}^{M} a_{ij} = 1$ and $\sum_{j=1}^{M} a_{ij} a_{kj} = 0$. Write a program to check the orthogonality of a given matrix of any size.

6. A function $f(z)$ of a complex number z is defined as the mapping from z to $f(z)$ in such a way $f(z)$ is unique. Write a standard C++ program for the following functions:
   a. $f(z) = 1 - z$.
   b. $f(z) = 3 - 2z + 2z^2 - z^3$.
   c. $f(z) = z - \dfrac{2z - 1}{1 + z - z^2}$.
   d. $f(z) = \begin{cases} z^2 & \text{if } z > 0, \\ \dfrac{1}{1 - z} & \text{if } z \leq 0. \end{cases}$

CHAPTER 3

Numerical Interface Designs

3.1   MICROSOFT FOUNDATION CLASSES

Microsoft Corporation released its C++ compiler, bundled with a set of library functions called the Microsoft Foundation Classes (MFCs) in the late 1980s. The library functions are targeted to support program development using C++ on the Windows environment. Microsoft Foundation Classes consist of a set of more than 200 classes for exploring the resources on Windows. *Class* is a grouping whose members consist of functions and variables. Each class has several functions for things like displaying text and graphics, creating dialog windows, and managing the events in Windows. Prior to MFC, programming on Windows was a difficult task as calls to the application program interface (API) for the Windows resources involved many low-level routines using C. MFC simplified this process as it was designed to gain control over the routines using a high-level language approach.

A *window* is a rectangular region on the desktop that allows the user to view the data and to navigate using the mouse. A window on the Microsoft operating system is referred to as Windows (with the first letter in capital). There are three types of windows, namely, overlapped, pop-up, and child. An *overlapped window* is the main window where all applications originate. A *pop-up window* is a small window that appears when a certain event such as a message box is invoked. A common form of a pop-up window is an error message box when a runtime error occurs. A *child window* is a window branching from the overlapped window for representing another series of events. Dialog boxes and push buttons are some of the most common form of child windows.

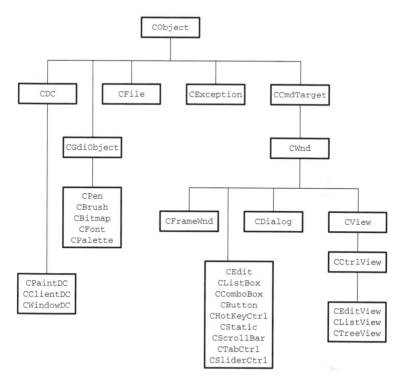

FIGURE 3.1. MFC hierarchical chart.

As a tradition, the name of a class in MFC is preceded by the letter C. Figure 3.1 shows some classes in MFC and their hierarchical structure. Top in the hierarchy is CObject, which is a base class to a host of other MFC classes. A derived class from CObject called CDC provides three important classes called CPaintDC, CClientDC, and CWindowDC, which provide the base for managing the resources in Windows. Another class called CGdiObject has functions for creating pens, brushes, and fonts.

CWnd is another important class that is derived from CObject via CCmdTarget. CWnd provides the creation of an overlapped window through CFrameWnd as well as its child windows in the form of resources on the CEdit, CListBox, and CButton classes. We will discuss these resources later in the subsequent chapters.

3.2   GRAPHICS DEVICE INTERFACE

Programming on Windows requires extensive calls to the resources provided in the hardware. Before the introduction of MFC, this task is difficult as each call to the resources involves several low-level programming skills. As an improvement, MFC makes full use of a GDI, or *graphics device interface*, which is a layer in the Windows architecture that insulates the application from direct interaction with the hardware.

TABLE 3.1. Server/client classes derived from the CDC class

Class	Description
CPaintDC	Device context for the server area on Windows.
CClientDC	Device context for the client area on Windows.
CWindowDC	Device context for the whole window.
CMetaFileDC	Device context for representing a Windows metafile or a device-independent file for reproducing an image.

This interface has an extensive set of high-level functions that can be linked from its objects for drawing and managing graphics in Windows.

A class that is commonly derived from CObject is CDC. In MFC, CDC is the base class for providing an interface with other classes, including CPaintDC, CClientDC, CWindowDC, and CMetaFileDC. An object abstraction called *device context* has been provided through these four classes, which links to the functions in CDC for supporting all the basic graphical and drawing functions on Windows. Table 3.1 describes each of these classes.

Device context is an object that is responsible for displaying text and graphics as output on Windows. In reality, a device context is a logical device that acts as an interface between a physical device (such as the monitor and printer) and the application. For achieving this task, a device context has a set of tools or attributes for putting text and drawing graphics on the screen using GDI functions. Device context is created from one of the following classes: CPaintDC, CClientDC, CWindowDC, and CMetaFileDC.

The tools in device context are represented as graphic objects such as pens, brushes, fonts, and bitmaps. Table 3.2 summarizes these objects. There are four types of device contexts in GDI: display context, memory context, information context, and printer context. A *display context* supports operations for displaying text and graphics on a video display. Before displaying text and graphics, a display context links with MFC functions for creating a pen, brush, font, color palette, and other devices. A *memory context* supports graphics operations on a bitmap and interfaces with the display context by making it compatible before displaying the image on the window. An *information context* supports the retrieval of a device data. A *printer context* provides an interface for supporting printer operations on a printer or plotter.

On Windows, everything including text is drawn as a graphics object. This is made possible as every text character and symbol is formed from pixels that may vary in

TABLE 3.2. GDI objects for text and graphics

GDI Object	Class	Description
Pen	CPen	To draw a line, rectangle, circle, polyline, etc.
Brush	CBrush	To brush a region with a color.
Color palette	CPalette	Color palettes for pens and brushes.
Font	CFont	To create a font for the text.
Bitmap	CBitmap	To store a bitmap object.

shapes and sizes. This facility allows flexibility on the shape of the text by allowing it to be displayed from a selection of dozens of different typefaces, styles, and sizes. Text and graphics are managed by GDI functions that are called every time a graphics needs to be displayed on the screen.

Color Management

The color of an object on Windows is actually the color of a segregated set of pixels that make up the object. In MFC, the color of each pixel is controlled using the function RGB() from the CDC class. RGB() consists of three arguments in the order from left to right as the red, green, and blue components. The function is declared as follows:

```
RGB(int, int, int);
```

Each component in RGB() is an integer that represents the monotone scale from 0 to 255, with 0 as the darkest value and 255 as the lightest. The monotone scales for red, green, and blue are easily obtained by blanking the other two color components, as follows:

```
Red,   r        RGB(r,0,0)
Green, g        RGB(0,g,0)
Blue,  b        RGB(0,0,b)
```

The three colors are the primitive colors for other color combinations. Figure 3.2 shows a hypercube that represents color combinations with axes at red, green, and Blue. Yellow is obtained by setting $r=g$ and $b=0$, whereas $r=0$, $g=255$, and $b=255$ produces cyan. A solid black color is obtained by setting $r=g=b=0$, whereas $r=g=b=255$

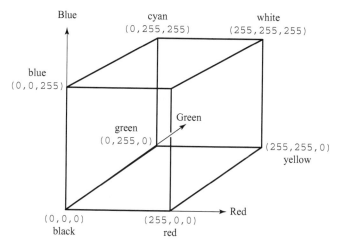

FIGURE 3.2. Red–green–blue color relationship in RGB().

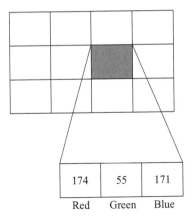

174	55	171
Red	Green	Blue

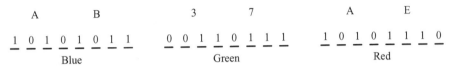

FIGURE 3.3. A pixel represented as RGB(174,55,171) in a 24-bit string.

produces pure white. It is obvious from this hypercube that a grayscale color is obtained by setting r=g=b, in RGB(r,g,b), which corresponds to a position along the diagonal line from (0,0,0) to (255,255,255) in the hypercube.

In Windows, a graphical object is formed from a rectangular composition of pixels of varying intensity. Each pixel displayed on Windows is represented by a string consisting of 24 bits of binary digits that is represented in MFC by RGB(). The first eight bits in the string starting from the right form the red component. This is followed by eight bits of green in the middle and the remaining bits make up blue. The alignment of a set of pixels in a rectangle makes up an image, a graphical object, and a text character. As one unit of the graphical object is represented by 24 bits of data, there are a total of 2^{24}, or 16,777,126, color combinations possible for supporting various graphical output requirements.

Figure 3.3 shows a pixel in the shaded square having its value defined as RGB(174,55,171) represented as a 24-bit string. The figure also shows the corresponding values in hexadecimals: AD for 171, 37 for 55, and B5 for 174. The value returned by RGB(174,55,171) is the 24-bit binary number 101010110011011110101110, or a decimal value of 11,220,910.

3.3   WRITING A BASIC WINDOWS PROGRAM

We start by discussing some basic concepts in creating an application on Windows. Microsoft Visual C++.Net provides an interface called Visual Studio for developing

TABLE 3.3. Some of the new project available options

Item	Description
Console application	Native C++ project that supports no Windows.
Win32 Project	Empty project with or without MFC.
MFC Application	Wizard approach to creating a Windows application.
Managed C++ Application	Managed C++ project with or without Windows suppor.

an application. Besides C++, this interface is shared by other languages in the family, including Visual Basic and Visual C#. To develop an application using MFC, a person must know the C++ language very well. A good knowledge in C++ is a prerequisite to developing applications on Windows. This is necessary because MFC has classes and objects defined in a manner that can only be understood if one knows the language well.

A C++ project can be created in many ways depending on the user requirements. Table 3.3 lists some of the most common ways for creating an application on the Visual Studio. In its simplest form, a standard C++ project which runs without the support of any Windows functions is a console application. This option is necessary to a beginner in C++, or a person who does not wish to use the Windows facilities. The console option is available by choosing *New Project*, *Win32 Application* and by choosing *Console Application* in *Application Type*.

A *Win32 Project* is an option for creating an empty application with or without the support of MFC. This option does not provide a guide for creating an application as the person must know all the details. One advantage in this option is the small number of code required to generate an application. The option allows the application to exist as an executable file (EXE) or as a dynamic-link library (DLL).

The *MFC Application* option is a guided approach for creating an application using a tool known as *wizard*. With this option, the details about Windows are prepared by Visual Studio through a series of menus and dialog windows in wizard. Therefore, the user can concentrate on writing the code for an application. Wizard does not provide the whole solution for the application as it only assists by generating the code related to the Windows management.

The *Managed Extension* option is a structured way of writing an application. This new option provides an opportunity to integrate the application with .Net frameworks such as Passport, Windows XP, and Tablet PC.

Code3A: The Skeleton Program

We discuss the development of a skeleton program that produces a single-line message on Windows. The program is called a skeleton because it has a minimum number of codes, and it has been designed to be as simple as possible. This simple program will become the template for all other programs in the book as it is from here that the applications will be developed.

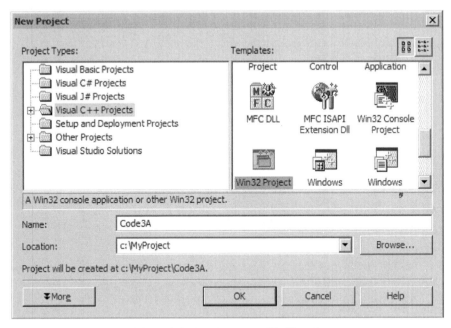

FIGURE 3.4. Creating a new Win32 project.

Step 1: Start a new project by choosing *New* from the menu, then *Project*. A window as shown in Figure 3.4 appears. Choose the Win32 Project icon, and name the project Code3A and the folder MyProject. Press the OK button to confirm.

Step 2: The window as shown in Figure 3.5 appears. Click *Application Settings*, and tick *Empty project* in the check box. Press the *Finish* button to confirm.

Step 3: From the Solution Explorer, right-click at Code3A and choose *Properties* from its menu. This is shown in Figure 3.6.

Step 4: The window in Figure 3.7 appears. In the *Use of MFC* category, choose *Use MFC in a Static Library*.

Step 5: Right-click *Source Files* in the Solution Explorer as shown in Figure 3.8, and choose *Add* from the menu followed by *Add New Item*. Name the file Code3A.h. Repeat by creating another file, Code3A.cpp.

Step 6: Enter the following codes in Code3A.h and Code3A.cpp.

```
//  Code3A.h
#include <afxwin.h>

class CCode3A : public CFrameWnd
{

public:
```

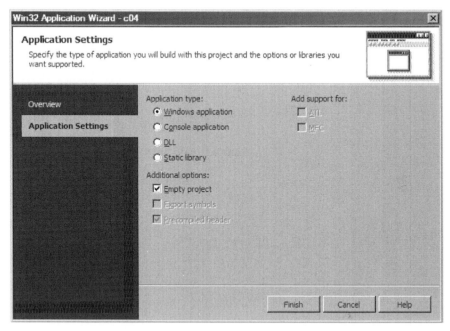

FIGURE 3.5. Empty Windows application project.

```cpp
        CCode3A();
        ~CCode3A();
        afx_msg void OnPaint();
        DECLARE_MESSAGE_MAP()
};
class CMyWinApp : public CWinApp
{
public:
        virtual BOOL InitInstance();
};

// Code3A.cpp
#include "Code3A.h"

CMyWinApp MyApplication;

BOOL CMyWinApp::InitInstance()
{
        CCode3A* pFrame = new CCode3A;
        m_pMainWnd = pFrame;
        pFrame->ShowWindow(SW_SHOW);
        pFrame->UpdateWindow();
```

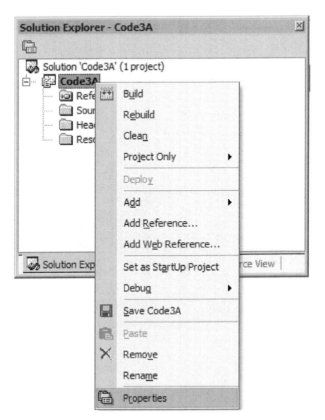

FIGURE 3.6. Choosing *Properties*.

```
      return TRUE;
}

BEGIN_MESSAGE_MAP(CCode3A,CFrameWnd)
      ON_WM_PAINT()
END_MESSAGE_MAP()

CCode3A::CCode3A()
{
Create(NULL,"Code3A: The Skeleton Program",
      WS_OVERLAPPEDWINDOW,CRect(0,0,400,200),NULL);
}

CCode3A::~CCode3A()
{
}
```

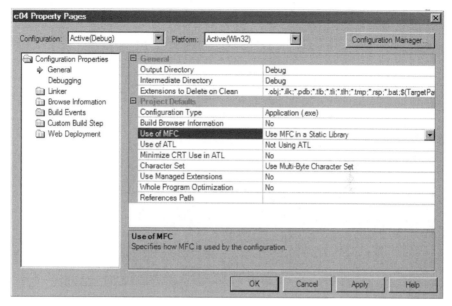

FIGURE 3.7. Choosing the MFC static library.

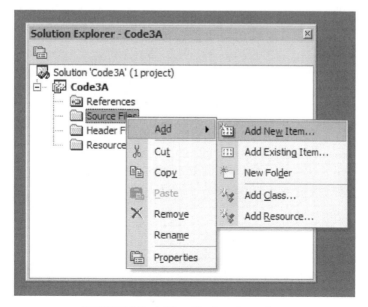

FIGURE 3.8. Adding files into the project.

FIGURE 3.9. Output from Code3A.

```
void CCode3A::OnPaint()
{
      CPaintDC dc(this);
      dc.TextOut(100,50,"Welcome to Windows...");
}
```

Step 7: The program is completed. Build the program and run to produce the output as shown in Figure 3.9.

Code3A: Discussion

Code3A is the simplest program that can be created to run on Windows. The project consists of two source files, Code3A.h and Code3A.cpp. The program displays a single line of message only involving a single application class and a single event. The application class is called CCode3A, and this class is derived from MFC's CFrameWnd. A header file called afxwin.h needs to be included as this file has all the declarations required for supporting CFrameWnd.

```
#include <afxwin.h>
```

CCode3A has a constructor function called CCode3A(), a destructor function called ~CCode3A(), and an application function called OnPaint(). The last function is a message handler that must be preceded by afx_msg in its declaration. A message handler requires a mapping through the declaration DECLARE_MESSAGE_MAP().

```
class CCode3A : public CFrameWnd
{

public:
      CCode3A();
      ~CCode3A();
```

```
        afx_msg void OnPaint();
        DECLARE_MESSAGE_MAP()
};
```

```
class CMyWinApp : public CWinApp
{
public:
        virtual BOOL InitInstance();
};
```

Any application in Windows needs to be registered, initialized, and updated. The following code segment performs these duties:

```
CMyWinApp MyApplication;
```

```
BOOL CMyWinApp::InitInstance()
{
        CCode3A* pFrame = new CCode3A;
        m_pMainWnd = pFrame;
        pFrame->ShowWindow(SW_SHOW);
        pFrame->UpdateWindow();
        return TRUE;
}
```

A message map is needed to detect events in Windows. The events to be detected should be listed in the body that begins with BEGIN_MESSAGE_MAP() and ends with END_MESSAGE_MAP(). In Code3A, the only event mapped is the display output in the main window. This event is recognized as ON_WM_PAINT, and this event is handled by its default function called OnPaint().

```
BEGIN_MESSAGE_MAP(CCode3A,CFrameWnd)
     ON_WM_PAINT()
END_MESSAGE_MAP()
```

CCode3A() is the constructor function in Code3A. The function creates memory for the class. This function is the best place to create the main window and its child windows. It is also in the constructor that most global variables and objects are initialized. In Code3A, the main window is created that occupies a rectangular area from (0,0) to (400,200).

```
CCode3A::CCode3A()
{
    Create(NULL,"Code3A: The Skeleton Program",
        WS_OVERLAPPEDWINDOW,CRect(0,0,400,200),NULL);
}
```

In Code3A, the destructor function may be optionally needed to delete the memory allocated to global arrays. This is an important step in dynamic memory allocation so that the deleted memory can be returned to the computer to be used in other applications. In our case, no global arrays have been created, and therefore, the destructor has no code.

```
CCode3A::~CCode3A()
{
}
```

OnPaint() is the default function for handling ON_WM_PAINT. The function serves as the host for displaying the output in the main window. To display the output, a device context object from CPaintDC is needed in order to call functions from MFC, and this object is called dc. The main window in this case is referred simply as this. The message is displayed as a string at position (100,50) in the window.

```
void CCode3A::OnPaint()
{
        CPaintDC dc(this);
        dc.TextOut(100,50,"Welcome to Windows...");
}
```

3.4 DISPLAYING TEXT AND GRAPHICS

All output on Windows is displayed as graphics. Windows is a form of raster graphics that displays text and graphics through tiny dots called *pixels*. A pixel is the smallest unit that makes up the screen display. A typical screen resolution of 800 × 600 is a rectangular region consisting of 800 columns and 600 rows of pixels, or a total of 420,000 pixels. A higher resolution display such as in 1,280 × 800 has 1,024,000 pixels over the same rectangular region. This means each pixel in the latter is finer than the former, and this results in a sharper display.

Text display on Windows consists of fixed alphabet characters that are formed from pixels. By default, text is normally displayed in most applications using the Times New Roman font of size 12. MFC provides a variety of other fonts for selection, including Arial, Courier, Helvetica, and Avantgarde. Each character for display can also be resized ranging from the smallest at size 6 to the largest at 72. Text can also be aligned horizontally, vertically, or to some angle. Other attributes supported by MFC include underlined text, crossed text, bold, italic, and their combinations. A variety of text output functions are available for formatting a text message and its background, as shown in Table 3.4.

A single point on Windows with its own unique coordinates is represented by a pixel. As a consequence, all text and graphics objects in Windows are constructed from a group of pixels. For example, a line is produced from a set of successive pixels aligned according to the direction of the line. A circle is obtained from pixels that

TABLE 3.4. Some common functions for displaying text

Function	Description
SetBkColor()	Sets the background color of the text.
SetTextColor()	Sets the color for the text.
TextOut()	Displays a text message at the indicated coordinates.

TABLE 3.5. Some common GDI functions for displaying graphics

Function	Description
Arc()	Draws an arc.
BitBlt()	Copies a bitmap to the current device context.
Ellipse()	Draws an ellipse (including a circle).
FillRect()	Fills a rectangular region with the indicated color.
FillSolidRect()	Creates a rectangle using the specified fill color.
GetPixel()	Gets the pixel value at the current position.
LineTo()	Draws a line to the given coordinates.
MoveTo()	Sets the current pen position to the indicated coordinates.
Polyline()	Draws a series of lines passing through the given points.
Rectangle()	Draws a rectangle according to the given coordinates.
RGB()	Creates color from the combination of red, green, and blue palettes.
SelectObject()	Selects the indicated GDI drawing object.
SetPixel()	Draws a pixel according to the chosen color.

are aligned according to its radius. It follows that a rectangle is constructed from two pairs of matching lines. A curve is obtained from the successive placement of pixels whose shape is governed by a mathematical function.

What tools are needed in graphics? They are pens, brushes, paints, and so on, just like what an artist requires. Previously, a programmer will need to apply difficult low-level routines using C or assembly languages in order to display graphics on Windows. Obviously, several ready-made tools from GDI bypass these tedious steps and cut the development time for displaying graphics on Windows. Table 3.5 summarizes some of the most common graphical GDI functions in MFC.

3.5 EVENTS AND METHODS

In object-oriented programming, an *event* is defined as a happening during the runtime that requires immediate attention and response. The response to an event is provided in the program in the form of a *method*, or a function. An event can be regarded as an interrupt where a call to a specific task in the computer is immediately performed. Some obvious examples of events are the left click of the mouse, a key stroke, a choice of an item in the menu, and a push button click.

In MFC, a message map is provided to detect the occurrence of an event during the runtime. The message map is written in the following format:

```
BEGIN_MESSAGE_MAP(CMain, CFrameWnd)
    <List of events, their ids and methods>
END_MESSAGE_MAP()
```

For example, the following code shows two events, a keyboard press and an item in the pulldown menu:

```
BEGIN_MESSAGE_MAP(CMain, CFrameWnd)
    ON_WM_KEYDOWN()
    ON_COMMAND(ID_FILESAVE, OnFileSave)
END_MESSAGE_MAP()
```

In the above example, ON_WM_KEYDOWN() is an event for detecting a keystroke, whereas ON_COMMAND(ID_FILESAVE,OnFileSave) is an event for detecting an item in the pulldown menu involving file saving. The two macros are called *message handlers*. The first event is acted on automatically by a user function called OnKeyDown(), whereas the second is acted on through the mentioned function, OnFileSave(). Since only two events are listed in the mapping, other events such as the left click of a mouse will have no effect on the program.

Table 3.6 lists some of the most common events in Windows. Of particular interest is ON_WM_PAINT, which is the most common event as it triggers the initial display of the application on the main window using a default function called OnPaint(). The display is also updated through an interrupt specified by a function called Invalidate() or InvalidateRect(). We will come across ON_WM_PAINT in almost all applications in this book later.

TABLE 3.6. Some common message handlers

Function	Description
ON_WM_PAINT	Default output drawn on the main window derived from the CFrameWnd class.
ON_WM_LBUTTONDOWN	Left button click of the mouse.
ON_WM_RBUTTONDOWN	Right button click of the mouse.
ON_WM_KEYDOWN	Key press from the keyboard.
ON_BN_CLICKED	Button click on an object from the CButton class.
ON_COMMAND	Menu items as specified in its arguments.

3.6 STANDARD CONTROL RESOURCES

MFC provides hundreds of resources or tools that can enhance the viewing quality on Windows. They include buttons, control boxes, menus, and toolbars. These tools are important in providing a user-friendly environment on Windows, and they have become a standard interface on computers these days.

Because of the limited scope in this book, it is not possible for us to discuss all Windows resources here. Instead, we will only discuss some of the most essential tools that are relevant to numerical applications.

Push Button

A push button is a rectangular object on Windows that becomes active when it is clicked with the mouse. Normally, a click at a push button indicates the user would like to see the effect when data input has been completed. This is useful, for example, in performing an analysis on a set of data, as the visual results from the analysis can help in some decision-making process.

A push button is a resource in the form of a child window created from the CButton class. An object for a push button is normally declared in the header file, as follows:

CButton *ObjectName*;

The object is created in the constructor according to a format given by

ObjectName.Create(*Title*, *DisplayOptions*, *RectangularRegion*, *HostWindow*, *Id*);

In the above format,

Title	The title in the title bar
DisplayOptions	Defines the shape of the button
RectangularRegion	Defines the coordinates and size of the button
HostWindow	The host or parent window
Id	The control id.

For example, the following statement creates a push button object called MyButton that is displayed as *Multiply*:

```
CButton MyButton;
MyButton.Create("Multiply",WS_CHILD | WS_VISIBLE |
              BS_DEFPUSHBUTTON, CRect(CPoint(100,130),
              CSize(100,25)),this, IDC_MYBUTTON);
```

A click on a push button is an event. Therefore, a message handler for this event needs to be mapped according to

```
ON_BN_CLICKED (Id, Method)
```

In the above message handler, *Id* is the control id and *Method* is the name of the function that will respond to this event. For example, a function called OnMyButton() below responds to a click on a button with id IDC_MYBUTTON:

```
ON_BN_CLICKED (IDC_MYBUTTON,OnMyButton)
```

Edit Box

An edit box collects input directly from the keyboard. A typical input is an integer or double value that is needed in a calculation. Input can also be in the form of a string, such as the name of a person or a mathematical equation for evaluation. An entry in an edit box can also be edited or deleted.

An edit box is created from an object from the CEdit class. Each edit box is recognized through a unique control id that is defined in its creating function. The entry in an edit box is collected as a string from the CString class. An entry in the form of an integer or a double is read as a string, and then it is converted to the data value using the C functions atoi() or atof(), respectively.

An edit box can be created as a global object with its declaration in the header file, according to the following format:

```
CEdit ObjectName;
```

The above example creates an object called *ObjectName* from the CEdit class. This object is created in the constructor function according to

```
ObjectName.Create(DisplayOptions, RectangularRegion,
                  HostWindow, Id);
```

The following example creates a CEdit object called eBox:

```
CEdit eBox;
eBox.Create(WS_CHILD | WS_VISIBLE | WS_BORDER,
      CRect(CPoint(250,50),CSize(80,25)),this, IDC_x);
```

Input from the user is collected as a string according to

```
ObjectName.GetWindowText(string);
```

As an example, an edit box eBox reads an input string called str. The string is converted into an integer value and stored into a variable called x using the C function, atoi().

```
CString str;
eBox.GetWindowText(str);
int x=atoi(str);
```

Static Box

The static box is a rectangular region that displays a message in the form of a string. Unlike an edit box, the content of a static box is for viewing only as no input or editing is allowed. A static box finds its usefulness in displaying the result of a calculation or in displaying a fixed message.

A static box is created from the CStatic class. As in an edit box, each static box is identified through a control id that can be defined in its creating function. The global declaration of a static box is made in the header file, as follows:

```
CStatic ObjectName ;
```

This is followed by its creation in the constructor according to

```
ObjectName.Create(Title, DisplayOptions, RectangularRegion,
             Host Window, Id ) ;
```

By default, a static box is displayed as a shaded rectangle. The shape and style of a static box can be set and modified through *DisplayOptions.* As an example, the following statements create a CStatic object called sBox:

```
sBox.Create("",WS_CHILD | WS_VISIBLE | SS_SUNKEN | SS_CENTER,
            CRect(CPoint(250,130),CSize(80,25)),this,
            IDC_STATIC) ;
```

Data in the form of a string can be displayed in a static box using the function SetWindowText() according to

```
ObjectName.SetWindowText(string ) ;
```

For example, to display the content of the variable z into a static box, sBox, the value must be formatted to a string first, as shown below:

```
double z=6.05;
CString str;
str.Format(''%lf'',z) ;
sBox.SetWindowText(str) ;
```

List View Box

List view box is an extension of the static box for displaying data usually in the form of a table. It is a more powerful tool as it has the horizontal and vertical scrolling bars for displaying data. In reality, it is not necessary to display all the data on Windows at the same time. Only a portion of the data needs to be displayed, and the scrolling bars control this portion in order to keep the overall window neat. With this wider scope, a single list view box is capable of displaying a large amount of data within its restricted window. The scrolling capability is very useful as displaying all the data from an application without this tool will be very troublesome.

A list view box can have a flexible size, and it can be placed anywhere in the main window. The object is first created from the CListCtrl class,

```
CListCtrl TableName ;
```

It is created using Create() according to

```
TableName.Create(DisplayOptions ,RectangularRegion ,
                 HostWindow, Id);
```

The following example creates a list view table from (50,50) to (400,300):

```
CListCtrl table;
CRect rcTable=CRect(50,50,400,300);
table.Create(WS_VISIBLE | WS_CHILD | WS_DLGFRAME | LVS_REPORT
           |LVS_NOSORTHEADER, rcTable, this, IDC_TABLE);
```

Further examples on the use of list view table will be discussed in the later chapters.

Code3B: Simple Multiplication Calculator

Edit and static boxes are useful resources for applications involving dialog between the user and the program. We discuss a sample application on a simple multiplication calculator that illustrates the use of edit and static boxes.

Figure 3.10 shows an output from Code3B. It consists of two edit boxes (white rectangles) for input, a static box (shaded rectangle) for the output and a push button called *Multiply*. The multiplication result from the input values of 4.5 and 1.7 in this example is displayed as 7.650000 in the static box once the push button is left clicked.

Code3B has two source files, Code3B.cpp and Code3B.h. The project extends from the skeleton program in Code3A with a few resources added. It is clear that a push button called *Multiply* represents an event that becomes active once it is left clicked with the mouse.

The application class in Code3B is called CCode3B, which is derived from MFC's CFrameWnd. The control ids are defined as macros in the header file through #define. Any integer numbers can be assigned to the ids as long as no two of them have the same value.

```
#define IDC_MYBUTTON 301
#define IDC_x 302
#define IDC_y 303
#define IDC_z 304
```

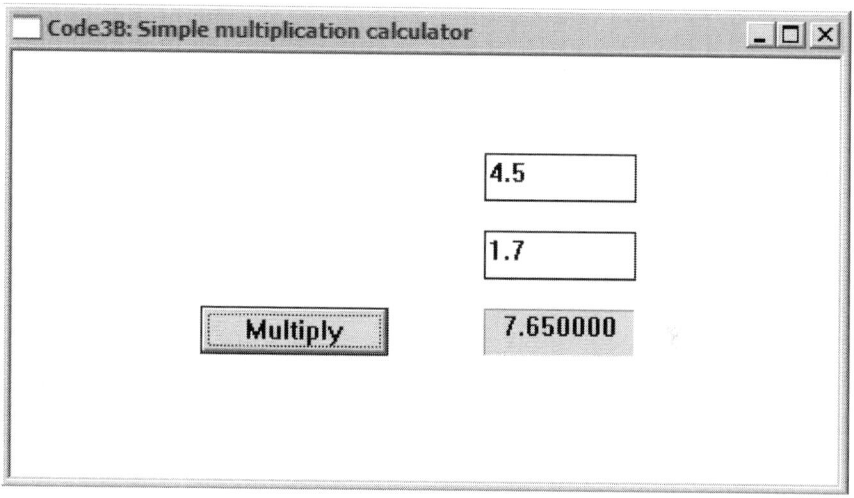

FIGURE 3.10. Output from Code3B.

The objects are declared from their respective classes. Since only one class is used, it is safe to declare the scope of these objects as private.

```
private:
    double x,y,z;
    CEdit ex,ey;
    CStatic sz;
    CButton MyButton;
```

Finally, the message mapping of the push button click requires its method called OnMyButton() to be declared according to

```
afx_msg void OnMyButton();
```

The full contents of Code3B.h are listed as follows:

```
#include <afxwin.h>
#define IDC_MYBUTTON 301
#define IDC_x 302
#define IDC_y 303
#define IDC_z 304
class CCode3B : public CFrameWnd
{
private:
        double x,y,z;
        CEdit ex,ey;
```

```
        CStatic sz;
        CButton MyButton;
public:
        CCode3B();
        ~CCode3B() {}
        afx_msg void OnMyButton();
        DECLARE_MESSAGE_MAP()
};

class CMyWinApp : public CWinApp
{
public:
        virtual BOOL InitInstance();
};
```

The coding in Code3B.cpp consists of an event and its method besides the constructor and destructor. The push button event is mapped as

```
ON_BN_CLICKED (IDC_MYBUTTON,OnMyButton)
```

In the constructor, the child windows in the form of edit boxes, a static box, and a push button are created in the main window as follows:

```
MyButton.Create("Multiply",WS_CHILD | WS_VISIBLE
        | BS_DEFPUSHBUTTON, CRect(CPoint(100,130),
        CSize(100,25)),this, IDC_MYBUTTON);
ex.Create(WS_CHILD | WS_VISIBLE | WS_BORDER,
        CRect(CPoint(250,50),CSize(80,25)),this, IDC_x);
ey.Create(WS_CHILD | WS_VISIBLE | WS_BORDER,
        CRect(CPoint(250,90),CSize(80,25)),this, IDC_y);
sz.Create("",WS_CHILD | WS_VISIBLE | SS_SUNKEN | SS_CENTER,
            CRect(CPoint(250,130),CSize(80,25)),this,
            IDC_z);
```

It is necessary to place the initial position of the caret at ex, which is the first CEdit object using

```
ex.SetFocus();
```

The method for the push button event is a function called OnMyButton(). Once activated, this function reads the input from the edit boxes through GetWindowText() and converts the strings into double variables using atof(). The calculation follows, and the result is converted into a string before it is displayed in the static box through SetWindowText(). The code for this function is shown below:

```
void CCode3B::OnMyButton()
```

```
{
    CString str;
    ex.GetWindowText(str); x=atof(str);
    ey.GetWindowText(str); y=atof(str);
    z=x*y;
    str.Format("%lf",z); sz.SetWindowText(str);
}
```

The full code listings for Code3B.cpp are shown below:

```
#include "Code3B.h"

CMyWinApp MyApplication;

BOOL CMyWinApp::InitInstance()
{
    CCode3B* pFrame = new CCode3B;
    m_pMainWnd = pFrame;
    pFrame->ShowWindow(SW_SHOW);
    pFrame->UpdateWindow();
    return TRUE;
}

BEGIN_MESSAGE_MAP(CCode3B, CFrameWnd)
    ON_BN_CLICKED (IDC_MYBUTTON,OnMyButton)
END_MESSAGE_MAP()

CCode3B::CCode3B()
{
    Create(NULL, "Code3B: Simple multiplication calculator",
    WS_OVERLAPPEDWINDOW,CRect(0,0,450,250));
    MyButton.Create("Multiply",WS_CHILD | WS_VISIBLE |
    BS_DEFPUSHBUTTON,
    CRect(CPoint(100,130),CSize(100,25)),this, IDC_MYBUTTON);
    ex.Create(WS_CHILD | WS_VISIBLE | WS_BORDER,
    CRect(CPoint(250,50),CSize(80,25)),this, IDC_x);
    ey.Create(WS_CHILD | WS_VISIBLE | WS_BORDER,
    CRect(CPoint(250,90),CSize(80,25)),this, IDC_y);
    sz.Create("",WS_CHILD | WS_VISIBLE | SS_SUNKEN | SS_CENTER,
    CRect(CPoint(250,130),CSize(80,25)),this, IDC_z);
    ex.SetFocus();
}

void CCode3B::OnMyButton()
{
    CString str;
```

```
    ex.GetWindowText(str); x=atof(str);
    ey.GetWindowText(str); y=atof(str);
    z=x*y;
    str.Format("%lf",z); sz.SetWindowText(str);
}
```

3.7 MENU AND FILE I/O

Another indispensable resource in Windows is the *menu*, which provides a choice of items for the user to choose from. The standard style of menu in Windows is the *pulldown menu*, which is a menu that appears from the Windows top panel. A pulldown menu in Windows is a resource that can be invoked at any time during the runtime.

In making a program neat and tidy, data and programs should be separated and stored in different files. This is important in order to preserve the integrity of the program by not distracting its flow. A program tells the compiler on the method for running the application, whereas data are part of the elements that make up the application. Both items are equally important, and without any one of them, the program will not be able to function properly.

Data can be stored in a file as a text file or as a binary file. The text file is a common approach, and this option is suitable in cases where the data are open for free distribution. In this case, the file does not need any security feature, and it can be opened easily using any editor or word-processing program. On the other hand, data that are restricted are normally stored as a binary file. With this approach, the file cannot be easily viewed using the normal editor or word-processing programs. The file can only be opened using some special programs, and sometimes, it can be protected with a special login and password. A more confidential data are stored as binary files with some added security features involving several encryption technologies.

In this section, we discuss the common saving and opening methods for text files using a pulldown menu. This approach is the standard technique in Windows. We start the discussion with text file input and output methods.

File Input and Output

A text file stores data in the form of a stream of bytes that represent the characters in the file. Several different methods for opening and saving files are applicable in Windows. MFC provides a special class called CFile for these purposes. However, the classic C programming approach in opening and saving a file can still be applied in both standard C++ and Visual C++. We discuss this approach.

In opening or saving a file, it is important to note that part or all the data may be lost if the program's operating procedure is not followed carefully. A common error happens when an opened file is not closed as the program ends. Both opening and saving a file involves four steps. In saving a file, the steps are as follows:

1. Create a pointer to the file. This is achieved using the built-in C structure called FILE. The following example creates a pointer called ofp:

    ```
    FILE *ofp;
    ```

2. Open the file for writing. The standard C method uses fopen(). The following example opens a file called MyFile.txt for writing:

    ```
    ofp=fopen("MyFile.txt","w");
    ```

3. Write data to the file using fprintf(). A text file stores data as strings only. Therefore, all data from other types, for example integer and double, will have to be formatted into strings. For example,

    ```
    int x=3;
    double y=6.5;
    fprintf(ofp,"x is %d, y is %lf\n",x,y);
    ```

4. Close the file using fclose().

    ```
    fclose(ofp);
    ```

Opening a file follows almost the same steps as in saving a file with some differences in the functions used. The steps are as follows:

1. Create a pointer to the file. The following example creates a pointer called ifp:

    ```
    FILE *ifp;
    ```

2. Open the file for reading, for example,

    ```
    ifp=fopen("MyFile.txt","r");
    ```

3. Read the data using fscanf(). A formatted data can be read according to their types by inserting their respective identifiers, for example,

    ```
    int x;
    double y;
    fscanf(ifp,"%d %lf\n",x,y);
    ```

4. Close the file using fclose().

    ```
    fclose(ofp);
    ```

In Windows, files are organized into folders. A folder may have many files from several different types, and it is going to be a difficult task for the user to remember the names of the required files and their location. A convenient mechanism provided in Windows is listing the names of the files for viewing in a pop-up window. This facility allows the user to select the required file for reading or saving using the mouse.

Files in a folder can be listed using a *modal window*. A modal window is a pop-up windows that does not allow access to other opened windows once it is active. Access to other opened windows can only be done by closing the modal window. We discuss the method for creating a modal window for listing and selecting files.

In listing the files, a filter is necessary to list the relevant files only based on their type. Without a filter, all files will be listed, and this makes the display a bit messy. As an example, the following statement creates a filter called `strFilter` for listing files of type `txt` only:

```
char strFilter[] = {"TXT Files (*.txt)|*.txt
               |All Files (*.*)|*.*||"};
```

The next step in creating the modal window is to create an object to MFC's `CFileDialog` class. The following example creates an object called `FileDlg`:

```
CFileDialog FileDlg(FALSE, ".txt",NULL,0,strFilter);
```

The modal window can now be invoked by linking `FileDlg` to the MFC function, `DoModal()`, as follows:

```
FileDlg.DoModal()
```

The required file can now be selected using a function called `GetFileName()`. A string is needed to store the name of this file. The following example shows how a string called `str` stores the name of the selected file:

```
char str[80];
str=FileDlg.GetFileName();
```

Code3C: Displaying Menu and File I/O

We discuss a sample project to illustrate the use of menus for file input and output. The project is called `Code3C`, which has a single menu item called *File* and five subitems, *Open, Save, Generate, Clear* and *Exit*. The submenus *Open* and *Save* read and save ten sets of (x, y) coordinates from and to files, respectively. *Generate* produces ten sets of new (x, y) coordinates at random. *Clear* erases these data and clears the screen. *Exit* terminates the program.

Figure 3.11 shows an output from `Code3C`. It consists of a pulldown menu with an opened file called `test.txt`. The file consists of the coordinates of ten points that

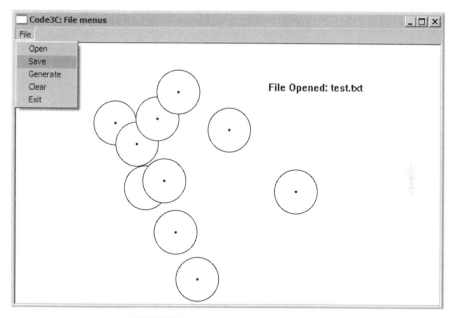

FIGURE 3.11. An output from Code3C.

serve as the centers of ten equal-sized circles. With this information, Windows draws the circles at their corresponding locations.

Code3C has a single class called CCode3C. The project consists of the source files Code3C.cpp and Code3C.h and a resource file called Code3C.rc. Nine steps are involved in creating this application. The first four are the same steps as in Code3A, whereas the rest involve the creation of Code3C.rc, which stores the information about the items in the pulldown menu.

Steps 1–4: Follow the same steps 1–4 as in the Code3A project.

Step 5: Right-click *Source Files* in the Solution Explorer, and choose *Add* from the menu followed by *Add Existing Item.* Add Code3C.h and Code3C.cpp from the supplied files.

Step 6: From the Solution Explorer, right-click *Resource Files*, and then choose *Add* and *Add Resource*, as shown in Figure 3.12. Name the resource file, Code3C.rc.

Step 7: From the Solution Explorer, double-click Code3C.rc. The Add Resource window in Figure 3.13 appears. Double-click *Menu* to start creating the menu.

Step 8: Begin with *File* as the root menu, and fill up the items in the submenu as shown in Figure 3.14. In *Properties*, name the id for each item according to ID_FILEOPEN for *Open*, ID_FILESAVE for *Save*, ID_GENERATE for *Generate*, ID_CLEAR for *Clear*, and ID_EXIT for *Exit*.

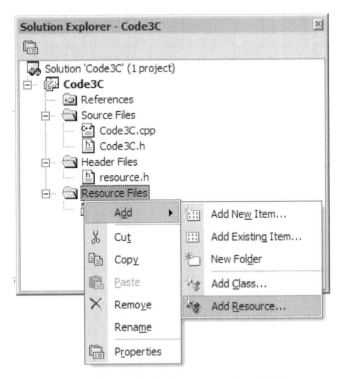

FIGURE 3.12. Adding the resources from MFC.

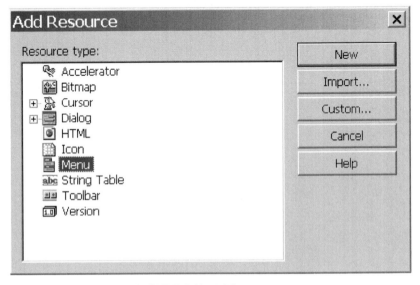

FIGURE 3.13. Adding a menu.

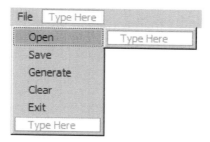

FIGURE 3.14. Adding the *File Open* submenu.

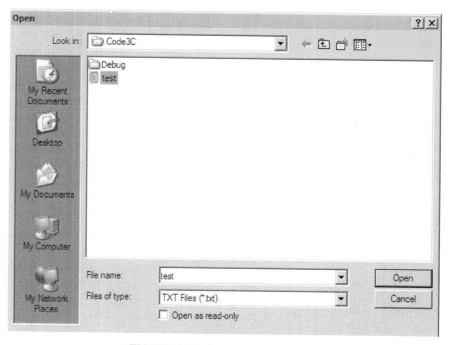

FIGURE 3.15. Selecting a file to open.

Step 9: All the steps have been completed. Build the project and run to see the results as shown in Figure 3.15.

The full listings in `Code3C.cpp` and `Code3C.h` are available in the distribution files. We will only discuss their main contents here. `Code3C.h` consists of the declarations for the class. The resource file, `Code3C.rc`, requires two header files to be included. The first file is `afxdlgs.h`, which contains the declarations for the modal window. The second file is `resource.h`, which has all the declarations on the resources in the pulldown menu. This file is create automatically when a resource file is created, and it cannot be edited by the user.

```
#include <afxwin.h>
#include <afxdlgs.h>
#include "resource.h"
```

CCode3C contains an array called pt for storing the coordinates of the centers of the circles. The class also specifies member functions for the project. Each item in the menu is considered as an event that must be attended by its respective function. Therefore, each item is mapped as an event with its declaration shown below:

```
class CCode3C : public CFrameWnd
{
private:
        CPoint *pt;
public:
        CCode3C();
        CCode3C::~CCode3C();
        afx_msg void OnFileOpen();
        afx_msg void OnFileSave();
        afx_msg void OnGenerate();
        afx_msg void OnClear();
        afx_msg void OnExit();
        DECLARE_MESSAGE_MAP()
};
```

Mapping for each event is done in Code3C.cpp, as follows:

```
BEGIN_MESSAGE_MAP(CCode3C,CFrameWnd)
    ON_COMMAND(ID_FILEOPEN,OnFileOpen)
    ON_COMMAND(ID_FILESAVE,OnFileSave)
    ON_COMMAND(ID_GENERATE,OnGenerate)
    ON_COMMAND(ID_CLEAR,OnClear)
    ON_COMMAND(ID_EXIT,OnExit)
END_MESSAGE_MAP()
```

The constructor in CCode3C creates the main window and initializes the array pt whose maximum size is n+1. In this case n is a macro whose value is predefined in Code3C.h. The contents in the constructor function are given as follows.

```
CCode3C::CCode3C()
{
        Create(NULL, "Code3C: File menus",
                WS_OVERLAPPEDWINDOW,CRect(0,0,600,400),
                NULL,MAKEINTRESOURCE(IDR_MENU1));
        pt=new CPoint [n+1];
}
```

A file is selected by invoking *Open* from the menu. This selection is mapped as ON_COMMAND(ID_FILEOPEN,OnFileOpen), which calls OnFileOpen(). The function opens the modal window, which displays all the filtered txt files in the current folder. The selected file is stored in str and opened. OnFileOpen() reads the contents of the opened file and draws circles according to their stored locations. The code for these operations in OnFileOpen() are given by

```
void CCode3C::OnFileOpen()
{
        CClientDC dc(this);
        CString str;
        CRect rc;
        FILE *ifp;
        char strFilter[] = {"TXT Files (*.txt)|*.txt
                            |All Files (*.*)|*.*||"};
        CFileDialog FileDlg(TRUE,".txt",NULL,0,strFilter);
        if (FileDlg.DoModal()==IDOK)
        {
                str=FileDlg.GetFileName();
                ifp=fopen(str,"r");
                dc.TextOut(350,50,"File Opened: "+str);
                for (int i=1;i<=n;i++)
                {
                        fscanf(ifp,"%d %d",&pt[i].x,&pt[i].y);
                        rc=CRect(pt[i].x-30,pt[i].y-30,pt[i].x+30,
                                 pt[i].y+30);
                        dc.Ellipse(rc);
                        rc=CRect(pt[i].x-1,pt[i].y-1,pt[i].x+1,
                        pt[i].y+1);
                        dc.Rectangle(rc);
                }
                fclose(ifp);
        }
}
```

OnGenerate() is a function that generates ten circles at ten random locations. Random integer numbers are produced from a C function called time() based on the clock cycle in the computer. Another function called srand() determines the scope or range of these numbers. A number from this range is then generated through rand().

```
void CCode3C::OnGenerate()
{
        CClientDC dc(this);
        CString str;
        time_t seed=time(NULL);
        srand((unsigned)seed);
```

```
OnClear();
dc.TextOut(50,50,"Generating Random Numbers");
for (int i=1;i<=n;i++)
{
        pt[i].x=100+rand()%400; pt[i].y=50+rand()%300;
        str.Format("%d %d",pt[i].x,pt[i].y);
        dc.TextOut(50,80+20*i,str);
}
}
```

OnFileSave() is doing the opposite of OnFileOpen(). The function saves the coordinates of the currently displayed circles into a string called str, which then becomes the file name.

```
void CCode3C::OnFileSave()
{
        CClientDC dc(this);
        CString str;
        FILE *ofp;
        char strFilter[] = {"TXT Files (*.txt)|*.txt
                           |All Files (*.*)|*.*||"};
        CFileDialog FileDlg(FALSE,".txt",NULL,0,strFilter);
        if( FileDlg.DoModal()==IDOK)
        {
                str=FileDlg.GetFileName();
                ofp=fopen(str,"w");
                dc.TextOut(50,20,"File Saved: "+str);
                str.Format("%d",n);
                dc.TextOut(50,50,"Contents: "+str+" randomly
                           generated numbers");
                for (int i=1;i<=n;i++)
                        fprintf(ofp,"%d %d\n",pt[i].x,pt[i].y);
                fclose(ofp);
        }
}
```

A function called OnClear() resets the coordinates of all the points to (0,0) and erases the window. The main window has its location and size stored in a rectangular object called rc, which is read through GetClientRect(). The window is cleared by spraying the background color into rc.

```
void CCode3C::OnClear()
{
        CClientDC dc(this);
```

```
        CRect rc;
        GetClientRect(&rc);
        CBrush whiteBrush(RGB(255,255,255));
        dc.FillRect(&rc,&whiteBrush);
        for (int i=1;i<=n;i++)
                pt[i]=CPoint(0,0);
}
```

OnExit() provides the formal way for exiting from the program. This is achieved by calling itself, as follows:

```
void CCode3C::OnExit()
{
        CCode3C::OnExit();
}
```

The last part of Code3C is to delete the unused array pt, and this is done in the destructor function.

```
CCode3C::~CCode3C()
{
        delete pt;
}
```

3.8 KEYBOARD CONTROL

The keys in the keyboard provide the standard input in all computers. Each key has been designed to represent a character or more in the standard American Standard Code for Information Interchange (ASCII) system. The standard keyboard in today's computers is based on the typewriter convention of key arrangement, QWERTY, which refers to the characters of the first six alphabetical keys of the standard keyboard from left.

Each key in the keyboard can also be programmed to perform things other than alphabetical input (see Table 3.7). For example, many computer games use the left, right, up, and down arrow keys for navigation in their games. Many learning-based software programs also make use of the arrow keys to perform certain functions in their course modules. This feature is interesting as personal computers have been known to have multipurpose uses, including office productivity, games, entertainment, multimedia, education, and scientific applications. Therefore, the keys contribute in making the software programs more user-friendly and interesting.

TABLE 3.7. Some common key macros

Macro	Key
VK_UP	Up arrow
VK_DOWN	Down arrow
VK_LEFT	Left arrow
VK_RIGHT	Right arrow
VK_SPACE	Spacebar
VK_RETURN	Return (Enter)

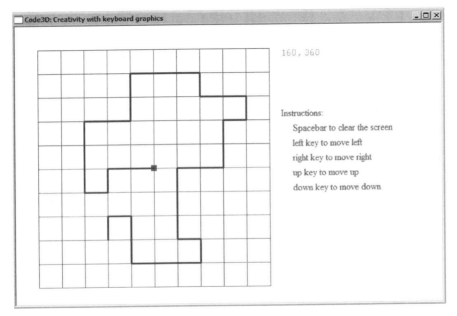

FIGURE 3.16. Keyboard creativity application.

Code3D: Creativity with Keyboard Graphics

We discuss an interesting program involving the control of keys in the keyboard. A project called Code3D has been developed to illustrate the concept. Figure 3.16 shows the output from this project, which consists of rectangular grids and the origin marked as a shaded square in the middle. A path is created by pressing on one of the following keys: left arrow, right arrow, up arrow, and down arrow. A press on the spacebar clears the path and resets the position at the origin.

Code3D has been developed based on the template in Code3A. The project has one class called CCode3D. There are two source files in the project, Code3D.cpp and Code3D.h. The header file is given as follows:

```
#include <afxwin.h>

class CCode3D : public CFrameWnd
```

```
{
protected:
      CPoint home,end,origin,pt;
      int Increment,BgColor;
public:
      CCode3D();
      ~CCode3D() {}
      afx_msg void OnPaint();
      afx_msg void OnKeyDown(UINT MyKey,UINT nRep,UINT nFlags);
      DECLARE_MESSAGE_MAP();
};

class CMyWinApp : public CWinApp
{
public:
      virtual BOOL InitInstance();
};
```

The contents of Code3D.cpp are given as

```
#include "Code3D.h"

BOOL CMyWinApp::InitInstance()
{
      CCode3D* pFrame = new CCode3D;
      m_pMainWnd = pFrame;
      pFrame->ShowWindow(SW_SHOW);
      pFrame->UpdateWindow();
      return TRUE;
}
CMyWinApp MyApplication;

BEGIN_MESSAGE_MAP(CCode3D, CFrameWnd)
      ON_WM_PAINT()
      ON_WM_KEYDOWN()
END_MESSAGE_MAP()

CCode3D::CCode3D()
{
      Create(NULL,"Code3D: Creativity with keyboard graphics",
            WS_OVERLAPPEDWINDOW,CRect(0,0,750,500));
      home=CPoint(40,40); end=CPoint(440,440);
      origin.x=(home.x+end.x)/2; origin.y=(home.y+end.y)/2;
      pt.x=origin.x; pt.y=origin.y; Increment=40;
```

```
        BgColor=RGB(255,255,255);
}

void CCode3D::OnPaint()
{
        CPaintDC dc(this);
        CString s;
        CRect rc;
        GetClientRect(&rc);
        CBrush blackBrush(BgColor);
        dc.FillRect(&rc,&blackBrush);

        CPen penGray(PS_SOLID,1,RGB(100,100,100));
        dc.SelectObject (penGray);
        for (int i=home.x;i<=end.x;i+=Increment)
            for (int j=home.y;j<=end.y;j+=Increment)
            {
                dc.MoveTo(i,j); dc.LineTo(400,j);
                dc.MoveTo(i,j); dc.LineTo(i,400);
            }
        rc=CRect(CRect(origin.x-5,origin.y-5,origin.x+5,
                origin.y+5));
        dc.FillSolidRect(&rc,RGB(200,0,0));

        CFont fontCourier,fontTimesNR;
        dc.SetBkColor(BgColor);
        fontCourier.CreatePointFont (120,"Courier");
        dc.SelectObject (fontCourier);
        dc.SetTextColor(RGB(0,200,0));
        s.Format("%d,",pt.x); dc.TextOut(end.x+20,home.y,s);
        s.Format("%d",pt.y); dc.TextOut(end.x+60,home.y,s);

        fontTimesNR.CreatePointFont (120,"Times New Roman");
        dc.SelectObject (fontTimesNR);
        dc.SetTextColor(RGB(200,0,0));
        dc.TextOut(end.x+20,home.y+100,"Instructions:");
        dc.SetTextColor(RGB(0,100,200));
        dc.TextOut(end.x+Increment,home.y+125,"Spacebar to clear
                the screen");
        dc.TextOut(end.x+Increment,home.y+150,"left key to move
                left");
        dc.TextOut(end.x+Increment,home.y+175,"right key to move
                right");
        dc.TextOut(end.x+Increment,home.y+200,"up key to move
                up");
```

```
        dc.TextOut(end.x+Increment,home.y+225,"down key to move
                down");
}


void CCode3D::OnKeyDown(UINT MyKey,UINT nRep,UINT nFlags)
{
        CClientDC dc(this);
        CRect rc;
        CPen penBlue(PS_SOLID,3,RGB(0,00,200));
        CBrush whiteBrush(RGB(255,255,255));
        dc.SelectObject(&penBlue);

        rc=CRect(end.x+10,home.y-10,end.x+100,home.y+30);
        InvalidateRect(&rc);
        dc.MoveTo(pt.x,pt.y);
        switch(MyKey)
        {
                case VK_RIGHT:
                        if (pt.x<end.x)
                                pt.x += Increment;
                        break;
                case VK_LEFT:
                        if (pt.x>home.x)
                                pt.x -= Increment;
                        break;
                case VK_UP:
                        if (pt.y>home.y)
                                pt.y -= Increment;
                        break;
                case VK_DOWN:
                        if (pt.y<end.y)
                                pt.y += Increment;
                        break;
                case VK_SPACE:
                {
                        pt.x=origin.x; pt.y=origin.y;
                        Invalidate();
                        break;
                }
        }
        dc.LineTo(pt.x,pt.y);
}
```

3.9 MFC COMPATIBILITY WITH .NET

The .Net framework was introduced by Microsoft Corporation in its Visual Studio products beginning in 2002 as a massive improvement to its programming environment. With this extension, Visual C++ was rebranded as Visual C++.Net, and this compiler supports a new style of programming environment called *Managed Extensions*, which is a new language construct that consists of additional keywords, preprocessor directives, and several compiler options, including the .Net Base Class Library (BCL). Managed Extensions also improves on the garbage-collection mechanism, reflection, and security. With this extension, all the Visual C++ environment before this date, including those that run under MFC, has been referred to as unmanaged extensions.

In general, the .Net framework was developed to enable applications to run on top of the .Net Common Language Runtime (CLR). With this framework, Visual C++.Net allows the managed and unmanaged code to be mixed freely within the same application. A new preprocessor directive such as #using in .Net imports the named metadata from a .Net executable object and library.

Because of its popularity, MFC is still supported in the .Net framework. The MFC libraries have been well established, and it will take a great amount of time for diehard followers of MFC to migrate to the new system. The reader has the option of accessing .Net from MFC, or accessing MFC from .Net. We discuss two projects called Code3E and Code3F, for accessing .Net from MFC and accessing MFC from .Net, respectively.

Accessing .Net from MFC: Code3E

Code3E shows a simple example how .Net commands can be called from MFC. The example is derived from the basic window framework in Code3A. Create a new project called Code3E by choosing *Win32 Project* in the new project template. Add two source files, Code3E.cpp and Code3E.h, into the project whose contents are exactly similar to Code3A.cpp and Code3A.h, respectively. Follow steps 1 to 6 of the Code3A project in Section 3.3 to produce the basic window.

In the *Solution Explorer*, select *Properties*. Choose *Use MFC in a Static Library* and *Yes* in *Use Managed Extensions* option, as shown in Figure 3.17. These selections make MFC as the host and .Net as its guest.

Add the following .Net code fragments into Code3E.h:

```
#using <mscorlib.dll>
#using <System.Windows.Forms.dll>
using namespace System;
using namespace System::Windows::Forms;
```

Add the following .Net messages into OnPaint() in Code3E.cpp:

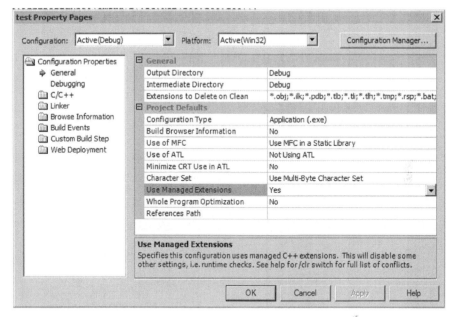

FIGURE 3.17. Accessing .Net from MFC.

```
String* str;                                 // .Net command
str=S"this message box is from .Net";        // .Net command
AfxMessageBox((CString)str);                 // .Net command,
                                                typecast to MFC
```

It works! The following code listings show the two files, Code3E.cpp and Code3E.h:

```
// Code3E.h
#include <afxwin.h>
#using <mscorlib.dll>
#using <System.Windows.Forms.dll>
using namespace System;
using namespace System::Windows::Forms;

class CMain : public CFrameWnd
{

public:
      CMain();
      ~CMain();
      afx_msg void OnPaint();
      DECLARE_MESSAGE_MAP()
};
```

```
class CMyWinApp : public CWinApp
{
public:
      virtual BOOL InitInstance();
};
```

The main program consists of the following code:

```
//  Code3E.cpp
#include "Code3E.h"

CMyWinApp MyApplication;

BOOL CMyWinApp::InitInstance()
{
  CMain* pFrame = new CMain;
  m_pMainWnd = pFrame;
  pFrame->ShowWindow(SW_SHOW);
  pFrame->UpdateWindow();
  return TRUE;
}

BEGIN_MESSAGE_MAP(CMain,CFrameWnd)
    ON_WM_PAINT()
END_MESSAGE_MAP()

CMain::CMain()
{
    Create(NULL,"Code3A: The Skeleton Program",
    WS_OVERLAPPEDWINDOW,CRect(0,0,400,300),NULL);
}

CMain::~CMain()
{
}

void CMain::OnPaint()
{
    CPaintDC dc(this);
    dc.TextOut(100,10,"Welcome to Windows...");
    String* str;                           // .Net command
    str=S"this message box is from .Net";  // .Net command
    AfxMessageBox((CString)str);           // .Net command,
                                           // typecasted to
                                           // MFC
}
```

Accessing MFC from .Net

Create a new project by choosing *Empty Project* (.Net) in the new project template. Go to *Properties* and enter the options as shown in Figure 3.17. Include the following codes into the header file:

```
#include <afxwin.h>
#using <mscorlib.dll>
#using <System.Windows.Forms.dll>
using namespace System;
using namespace System::Windows::Forms;
```

MFC routines can now be called from .Net using Managed Extension. The directive #include <afxwin.h> needs to be included to enable classes such as CString to be called from the Managed Extension. Other include files will also have to be included according to the requirement in the application.

3.10 SUMMARY

The chapter discusses the fundamental aspects of MFC from creating the simplest window, to using several basic resources, to mixing the codes from MFC and .Net. The concepts are illustrated with several simple programming projects. MFC has been widely used as a tool for proving a user-friendly interface based on Windows. Its rich resources provide all the necessary tools to produce a wide range of applications based on the friendly graphical user interface (GUI). Microsoft Corporation, which develops the new .Net platform, realizes the importance of MFC by making the platform compatible with MFC.

CHAPTER 4

Curve Visualization

4.1 TOOLS FOR VISUALIZATION

Mathematics is the queen of science. That is a well-known quote that describes the role and contribution of mathematics as the fundamental element in science and technology. In its inherent form, mathematics may be difficult to an average man on the street. However, this perception can change dramatically if the approach for solving problems in mathematics is simplified to the extent that it can be understood and appreciated by anybody.

Mathematics will become interesting if a review is made on the way it is presented. Foremost in the list is its friendliness. Mathematics should inherit a friendly look so as not to frighten people. A program or software that serves students, for example, must have friendly interfaces in the form of dialog boxes, list views, graphs, and so on. The user should be provided with ample choices or options in using a software program. The presentation must be clear and must have correctional features when the user makes an error or typing mistake.

One of the most important items in a user-friendly solution to a given problem is the graphical illustration. A curve describes a solid relationship between the variables in the given function. The visual depiction through curve drawing definitely helps in understanding the problem as well as its solution.

In this chapter, we discuss techniques for generating several types of curves for providing practical visualization to a given problem. The discussion also touches on the use of a tool called an equation parser, which allows the user to define the equation.

4.2 MYPARSER

In mathematics, a function is an operator for mapping one or more points in a domain to a single point in a range. The mapping can be one-to-one or many-to-one, which maps a single point and more than one point, respectively. When a function is assigned with a value, the whole expression is called an equation.

An equation describes the relationship or dependency between the variables in the equation. In a computer program, *equation parser* is a routine that reads a mathematical expression in the form of a string and then evaluates this expression to produce a numerical solution. For example, an input string of

```
x ^ 2*sin(x),
```

with x as a variable is interpreted as $x^2 \sin x$. If x is assigned with a value such as 2, then the expression becomes $2^2 \sin 2 = 3.637190$.

In an equation parser, a string in an expression consists of operands and an operator. An *operand* is an item that is assigned with a value, whereas an *operator* connects the operands through a mathematical operation. In the above expression, x, 2 and sin(x) are the operands, while ^ and * are the operators.

The input string for the parser is normally typed by the user directly from the keyboard. However, the standard keyboard in most computers does not support input in the form of mathematical equations. Therefore, the user will have to key in an equation as a single-line string, and the parser will interpret this string into an equation. In the above example, the parser is intelligent enough to recognize the symbol ^ as *power of*, * as *multiply*, and sin as *sine of*. The parser is also capable of processing the input items according to the priority order as required in the given expression.

There are many versions of parser in circulation today for supporting common languages such as C, C++, Basic, and Java. Some of these parsers may be purchased from vendors, whereas some others are distributed as freeware. A good parser should be robust enough for handling input items in the form of operators, functions, and tokens, capable of handling complex mathematical expressions, and have good error correction capability.

In this section, we present our parser called MyParser that will be used throughout the book. MyParser is robust, and it has most of the required features for supporting numeric-intensive calculations. The development of the parser program requires some heavy understanding of the data structure concepts and knowledge, which is beyond the scope of this book. Therefore, we will not discuss this issue. Instead, we will focus on how to incorporate the parser for solving numerical problems. In other words, we will concentrate on the usage of the parser, not on the making of a parser.

MyParser is supplied in the form of an object file called MyParser.obj, which needs to be included as one of the source files in the project. The user will find this file very useful as it can be linked to the other source files in the project for producing applications that require the use of a parser.

MyParser is easy to use as it involves only a few instructions. First, the object file MyParser.obj needs to be included in the project as a source file. Second, an external function called parse() is to be declared in the header file of the application. The third step involves the assignment of values for each argument in parse(). Finally, the function `parse()` is called, and its returned value is assigned into a double variable.

By including `MyParser.obj` in the project, all its functions can be accessed as external functions into the `.cpp` and `.h` files. The only function inside `MyParser.obj` that needs to be accessed externally is `parse()`. Since this function is external to the `.cpp` and `.h` files, it is necessary to insert the following statement in the application header file:

```
extern double parse(CString, int, double [], int []);
```

As shown, `parse()` returns a double value. The above declaration suggests `parse()` requires four arguments, a follows:

```
parse(str, nVar, array1 [], array2 []);
```

where the arguments are

str	The string expression.
nVar	The number of variables in the equation.
array1[]	The array of the input values of the operands declared as double.
array2[]	The array of the codes for the variables declared as int.

The first argument, *str*, is the string expression that represents the input equation. A variable in the expression is recognized through its code starting with 0 for *a*, 1 for *b*, and so on until the last code, 25 for *z*. Table 4.1 lists all the recognized variables for the expression. The second argument, *nVar*, represents the number of variables in the expression. The third argument is the variable array with its assigned values, whereas the last argument is its code.

We use `psv[]` as the variable array and `psi[]` as its code array. For example, the array `psi[]` defines *x* and *y* as variables in the equation whose codes are 23 and 24, respectively. They are assigned as the first two members in `psi[]`, as follows:

```
psi[1]=23; psi[2]=24;
```

In this example, *nVar* is set to 2 since only two variables are used. Another array called `psv[]` stores the assigned values of the two variables, for example, $x = 7.5$ and $y = -1.9$. They are assigned as follows:

```
psv[1]=7.5; psv[2]=-1.9;
```

The two arrays whose values have been assigned are passed to the expression *x* sin *y* in `parse()` for evaluation, and the result is stored into a variable called *z*:

```
z=parse("x*sin(y)",2,psv,psi);
```

**TABLE 4.1. Character codes
for the variables in *Str***

Variable	Code
a	0
b	1
c	2
d	3
e	4
f	5
g	6
h	7
i	8
j	9
k	10
l	11
m	12
n	13
o	14
p	15
q	16
r	17
s	18
t	19
u	20
v	21
w	22
x	23
y	24
z	25

It is not necessary to use a code for z as this variable is not inside `parse()`, and it only takes the returned value from this function.

The built-in function, `parse()`, is a powerful function that can perform most of the required mathematical operations. Tables 4.2 and 4.3 list down the mathematical functions and operators supported in `parse()`.

Data input in the domain on a function in `parse()` must adhere to the governing mathematical rules and theorems. The user must be aware of things like the domain and range of a mathematical function. For example, it is not possible to compute `asin(2)` since the argument in the function is not in the domain of $\sin^{-1}$. Therefore, `asin(2)` is a violation of the domain, and it will definitely result in an error.

An equation is read according to the standard priority rules as in all other programming languages. The order from top to bottom goes as follows:

1. Parentheses
2. Function

TABLE 4.2. Mathematical functions supported in parse()

Function	Return Value	Description
sqrt(*double*)	*double*	Square root of a number, e.g., sqrt(4)=2.
sqr(*double*)	*double*	Square of a number, e.g., sqr(4)=16.
sin(*double*)	*double*	Sine of a number, e.g., sin(2)=0.909297.
cos(*double*)	*double*	Cosine of a number, e.g., cos(2)=-0.416147.
tan(*double*)	*double*	Tangent of a number, e.g., tan(x)=-2.18504.
exp(*double*)	*double*	Exponent of a number, e.g., exp(2)=7.389056.
asin(*double*)	*double*	Inverse sine of x, $\sin^{-1} x$, e.g., asin(0.4)=0.4115.
acos(*double*)	*double*	Inverse cosine of x, $\cos^{-1} x$, e.g., acos(0.4)=1.1593.
atan(*double*)	*double*	Inverse tangent of x, $\tan^{-1} x$, e.g., atan(0.4)=0.3805.
sinh(*double*)	*double*	Hyperbolic sine of a number, e.g., sinh(0.4)=0.4107.
cosh(*double*)	*double*	Hyperbolic cosine of a number, e.g., cosh(0.4)=1.0811.
tanh(*double*)	*double*	Hyperbolic tangent of a number, e.g., tanh(0.4)=0.3800.
abs(*double*)	*double*	Absolute value of a number, e.g., abs(-0.95)=0.95.
log(*double*)	*double*	Logarithm of x, $\log x$, e.g., log(0.4)=-0.3979.

TABLE 4.3. Operators in parse()

Operator	Description
^	Power of a number, e.g., 3^2=9.
+	Addition.
−	Subtraction.
*	Multiply.
/	Divide.

3. ^
4. * and /
5. + and −

For example, (1-3*x*sin(x+y))/(x^2+3*exp(x)) is an input string that is interpreted as $\frac{1-3x\,\sin(x+y)}{x^2+3e^x}$. The priorities here are to the paired-parentheses first, followed by the functions, power, and then the rest of the items.

Code4A: Scientific Calculator

Code4A. User Manual.

1. Enter a value each for t, u, v, x, and y.
2. Enter the input string at *Expression*.
3. Click the *Compute* push button to view the results.

Development Files: Code4A.cpp, Code4A.h and MyParser.obj.

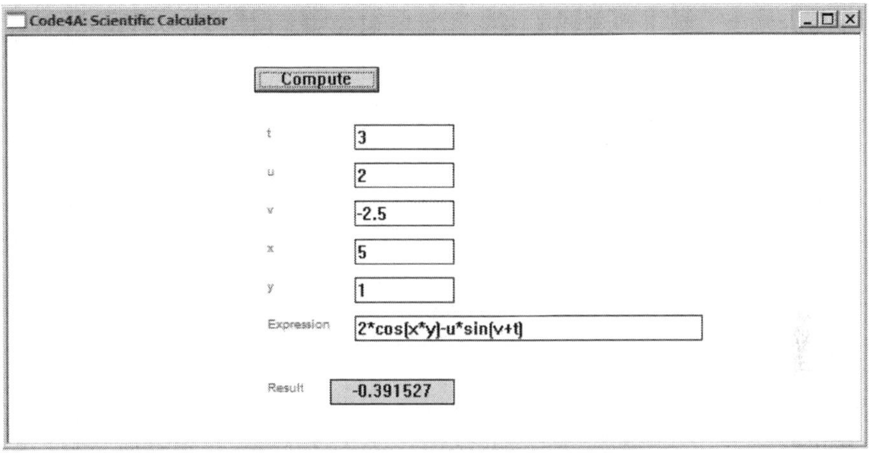

FIGURE 4.1. The scientific calculator for evaluating $2 \cos xy - u \sin(v + t)$.

An electronic calculator is a battery-powered device whose size is as small as a credit card. The calculator performs calculation on data input by the user and displays the results. This handy device is powerful enough to compute several mathematical operations at the click of some buttons to produce quick results. A basic calculator normally supports elementary operations involving addition, subtraction, multiplication, and division only. Some powerful calculators have several advanced features, including evaluating the inverse of a matrix, solving a system of linear equations, and displaying graphs.

We discuss the development of a desktop calculator that is capable of evaluating a mathematical expression. The project is called Code4A, and it consists of a class called CCode4A, and two files, Code4A.cpp and Code4A.h. The project also includes MyParser.obj for referring to an external function in this file. The calculator incorporates an equation parser for reading and evaluating an input string from the user. Figure 4.1 shows a sample runtime of Code4A whose input expression is $2 \cos xy - u \sin(v + t)$. Five variables, t, u, v, x, and y are involved whose input values are shown in the figure. An edit box called *Expression* stores the input string for the equation. The result from the calculation is displayed in the shaded rectangle in the figure.

The display from Code4A consists of several resources, including a push button called *Compute*, six edit boxes (white rectangles), and a static box (shaded rectangle). All these resources are declared in the header file, Code4A.h. Output on the shaded box is obtained once *Compute* is left-clicked.

The input in this application consists of the edit boxes that represent the variables t, u, v, x, and y and *Expression*. They are organized into a structure called INPUT whose contents are given by

TABLE 4.4. The main variables in `Code4A` and their supporting resources

Item	Content (CString)	CEdit box	Home Coordinates
t	input[1].item	input[1].ed	input[1].hm
u	input[2].item	input[2].ed	input[2].hm
v	input[3].item	input[3].ed	input[3].hm
x	input[4].item	input[4].ed	input[4].hm
y	input[5].item	input[5].ed	input[5].hm
Expression	input[6].item	input[6].ed	input[6].hm

TABLE 4.5. The main control resources in `Code4A`

Variable	Class/Type	Description
btn	CButton	Push button to start the calculation.
result	CStatic	The static box for displaying the result from the calculation.
idc	int	Control ids for the edit boxes and static boxes.

```
typedef struct
{
     CString item;
     CPoint hm;
     CEdit ed;
     CRect rc,display;
} INPUT;
INPUT input[nInput+1];
```

An array called `input[]` stores the values in `INPUT` as shown in Table 4.4. This array has the number of elements specified by `nInput+1`, which is seven in this case. The table lists the variables in this structure and their supporting resources. Input on the operands t, u, v, x, and y are provided on the edit boxes, whereas the long white rectangle collects the expression for the equation.

Table 4.5 lists other main variables and objects in `Code4A`. The push button is represented by `btn`, whereas the result from the calculation is stored as `result` in a static box. The control ids for the edit boxes and static boxes are stored as a single integer variable called `idc`. It is not necessary to use a macro for each box as what is important is that each resource must have a unique id. Hence, `idc` whose value differs by 1 for the boxes will take care of this requirement. However, a macro called `IDC_BUTTON` needs to be declared for `btn` as the resource needs a fixed reference from the message map.

The complete declaration is shown in the header file, `Code4A.h`. A single class called `CCode4A` is used in this application, and this class is derived from the MFC class, CFrameWnd. An external function called parse() is called from Code4A, and this is done using extern double parse().

```
// Code4A.h
#include <afxwin.h>
#define IDC_BUTTON 501
#define nInput 6

extern double parse(CString,int,double [],int []);

class CCode4A : public CFrameWnd
{
protected:
        typedef struct
        {
                CString label,item;
                CPoint hm;
                CEdit ed;
                CRect rc,display;
        } INPUT;
        INPUT input[nInput+1];
        CStatic result;
        CFont Arial80;
        CButton btn;
        int idc;
public:
        CCode4A();
        ~CCode4A();
        afx_msg void OnPaint(),OnButton();
        DECLARE_MESSAGE_MAP()
};

class CMyWinApp : public CWinApp
{
public:
        virtual BOOL InitInstance();
};
```

Code4A consists of two events, an output display using the standard function OnPaint() and a push button click. These two events are mapped as follows:

```
BEGIN_MESSAGE_MAP(CCode4A,CFrameWnd)
    ON_WM_PAINT()
    ON_BN_CLICKED(IDC_BUTTON,OnButton)
END_MESSAGE_MAP()
```

Basically, the constructor CCode4A() allocates memory for the class. It is also in the constructor that the main window and all the control resources are created. Besides these duties, the initial values of other main variables are assigned in the constructor. They include the location of the objects in the main window, the initial value of idc, and the labels for the edit and static boxes. CCode4A() is given as follows:

```
CCode4A::CCode4A()
{
    Create(NULL,"Code4A: Scientific Calculator",
        WS_OVERLAPPEDWINDOW,CRect(0,0,700,350),NULL);
    Arial80.CreatePointFont(80,"Arial");
    idc=301;
    input[0].hm=CPoint(200,20);
    for (int i=1;i<=nInput;i++)
        input[i].hm=CPoint(input[0].hm.x+10,
            input[0].hm.y+50+(i-1)*30);
    input[1].label="t";
    input[2].label="u";
    input[3].label="v";
    input[4].label="x";
    input[5].label="y";
    input[6].label="Expression";
    btn.Create("Compute",WS_CHILD | WS_VISIBLE | BS_DEFPUSHBUTTON,
        CRect(CPoint(input[0].hm.x,input[0].hm.y+5),
        CSize(100,20)), this,IDC_BUTTON);
    for (i=1;i<=nInput-1;i++)
        input[i].ed.Create(WS_CHILD | WS_VISIBLE | WS_BORDER,
        CRect(input[i].hm.x+70,input[i].hm.y,
        input[i].hm.x+150, input[i].hm.y+20),this,idc++);
    input[nInput].ed.Create(WS_CHILD | WS_VISIBLE | WS_BORDER,
        CRect(input[nInput].hm.x+70,input[nInput].hm.y,
        input[nInput].hm.x+350,input[nInput].hm.y+20),
        this,idc++);
    result.Create("",WS_CHILD | WS_VISIBLE | SS_CENTER | WS_BORDER,
        CRect(input[nInput].hm.x+50,input[nInput].hm.y+50,
        input[6].hm.x+150,input[6].hm.y+70),this,idc++);
}
```

OnPaint() is the standard function for displaying the initial messages triggered by an event whose id is ON_WM_PAINT. In Code4A, the initial messages consist of the labels for the edit boxes and static boxes, and they are given as

```
void CCode4A::OnPaint()
{
      CPaintDC dc(this);
      CString str;
      dc.SelectObject(Arial80);
      dc.SetBkColor(RGB(255,255,255));
      dc.SetTextColor(RGB(100,100,100));
      for (int i=1;i<=nInput;i++)
            dc.TextOut(input[i].hm.x,input[i].hm.y,
            input[i].label);
      dc.TextOut(input[6].hm.x,input[6].hm.y+50,"Result");
}
```

An event whose id is `ON_BN_CLICKED(IDC_BUTTON,OnButton)` calls the function `OnButton()` once the push button is left-clicked. `OnButton()` is the actual problem solver here. The function responds to the push button click by first reading all the input from the user, evaluates the expression, and displays the result. The input is read as strings, and they are stored in `input[i].item`. This is accomplished through the MFC function, `GetWindowText()`, as follows:

```
for (i=1;i<=nInput;i++)
      input[i].ed.GetWindowText(input[i].item);
```

The five variables in this application, t, u, v, x, and y, are recognized through their codes as defined in Table 4.1. A local array called `psi[i]` hosts these codes, which stores its assigned values as `input[i].item`. The string values in `input[i].item` are converted into doubles using the standard C++ function, `atof()`. The converted double values are stored into another array called `psv[i]`. This process is shown as

```
psi[1]=19;
psi[2]=20;
psi[3]=21;
psi[4]=23;
psi[5]=24;
for (i=1;i<=nInput;i++)
      psv[i]=atof(input[i].item);
```

The final part of `OnButton()` is to evaluate the input expression from the defined variables and their assigned values. This is done using `parse()`, and the result is stored in a double variable called `z`. The value from `z` is then formatted into a string before it is displayed in the static box, `result`. The following code segment shows how this is done:

```
z=parse(input[nInput].item,5,psv,psi);
str.Format("%lf",z);
result.SetWindowText(str);
```

4.3 DRAWING CURVES

A *curve* describes the dependency behavior of the variables in a function. A good curve can show the solid relationship between all these variables, which contributes to its understanding. Therefore, generating a curve from its given data is a very useful visualization as part of the overall solution to the problem. Readers need to be reminded that a program for drawing a curve on Windows is governed by the fundamental rules in mathematics. A curve will display properly if all the fundamental rules pertaining to its existence are obeyed.

In drawing a curve, several issues need to be addressed. Two main issues arise here. The first issue is the domain, and the second issue is singularity. As mentioned, $\sin^{-1} 2$ does not exist because the domain for $\sin^{-1} x$ is $-1 \le \sin^{-1} x \le 1$. When dealing with a mathematical function, one has to know the validity of the interval where the function is defined. A computer program will simply hang or will produce some undesirable results if a rule regarding the validity of the domain is violated.

A common mistake is to divide a number by zero. Singularity is a term that describes a point or region where the given curve is not defined. For example, the function $f(x) = 1/(x - 3)$ is singular at $x = 3$ as $f(x)$ is undefined at this point. The curve from this function will definitely produce some weird result. However, this problem can be addressed easily by adding a very small number to x, which will not affect the desired result. A very small value such as e^{-99} added to x in that function will prevent this disaster from happening.

Strategies for Drawing a Curve

Drawing a curve in a Windows environment requires a correct mapping of the points from the real coordinates, or the Cartesian system, to the pixels in Windows. As discussed, Windows is based on integer coordinates that may indirectly conflict with the Cartesian coordinate system, which is based on real numbers.

In producing high-quality curves on Windows, some strategies need to be implemented that take into account factors concerning the existence and stability of the curves. First, the rectangular region on Windows must be bounded by some fixed coordinates that make up the rectangle. This region will serve as a child window with the sides of the rectangle as the boundaries so that the curve does not explode to other region. This strategy is necessary as the window needs to display another form of output, for example, the textual explanation on the results.

The second strategy is to provide some flexibility on the curve drawing by allowing user-control facilities. For example, the generated graph $f(x)$ can be made flexible by allowing its range of x values input by the user. This means a range in $0 \leq x \leq 1$ produces an inward zooming for a coarser graph $f(x)$ that is originally in $-2 \leq x \leq 5$. This flexibility helps in viewing the same graph from a different angle.

Third, the curve to be drawn must be properly scaled so that it displays clearly on the screen. It is a good idea to use the display region fully by having the maximum and minimum values of the graph in the range shown. A good approach for this objective is to compute the graph maximum and minimum values first and then to scale the whole region according to these two values. We will illustrate this point in our case studies later. For example, in drawing a curve, the points (x_1, y_1) and (x_2, y_2) may represent the minimum and maximum values, respectively, of $y = f(x)$ in the domain. Applying this technique, the curve can then be drawn nicely with the minimum and maximum points shown clearly within the Windows viewport.

Another common error is computing a function that depends on a variable whose value has not be assigned yet. For example, computing $\sqrt{x}$ with no assigned value of x will produce a runtime error that may cause the system to halt. A runtime error may also arise in an array where its last element equals the size of the array. It is a common mistake to define, for example, a[3]=5 when the array is declared as int a[3]. In this case, only a[0], a[1], and a[2] are members of the array. Therefore, the assignment of a[3]=5 is illegal.

Cartesian-to-Windows Coordinates Conversion

Windows provides a vast opportunity for displaying the results from a mathematical modeling and simulation. The output from numeric-intensive calculations can be displayed nicely in a very structured and organized manner through the proper handling of the resources in Windows. In designing an application, a developer needs to understand these resources and their handling in order to achieve the objective of producing high-quality results. One of the most important contributions from Windows is its flexibility in producing displays. The display on Windows consists of a rectangular region that can be reconfigured according to the user requirements.

In presenting a numerical output on Windows, one has to adapt to an environment not familiar in mathematics. In mathematics, we are used to representing a point using the Cartesian coordinates. In contrast, Windows output is based on pixels whose coordinate system uses numbers from zero and positive integers only. In other words, negative numbers as well as numbers with decimal points (real numbers) are not supported in the Windows system. However, this does not mean a point with coordinates like $(-1, 3.7258)$ cannot be displayed on Windows. To display this point on Windows, we need to produce a transformation so that this point maps to a pixel on Windows. This is just another trivial mathematical mapping problem with which we should be familiar.

Figure 4.2 shows the Cartesian (left) and Windows (right) coordinates systems. We denote the coordinates of a point in the Cartesian system as (x, y) and a pixel in the Windows system as (X, Y). The Windows coordinate system starts with $(0,0)$ in the

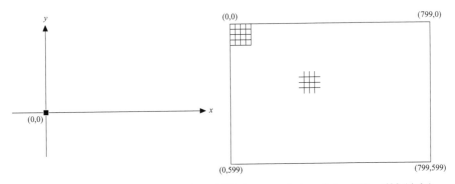

FIGURE 4.2. Cartesian system (left) and Windows resolution of size 800 × 600 (right).

top left-hand corner as its origin. The end points in Windows depend on the desktop resolution of the computer. A typical 800 × 600 screen resolution has 800 columns and 600 rows of pixels with the coordinates of (799,0), (0,599), and (799,599) in the top right-hand, bottom left-hand, and bottom right-hand corners, respectively. A higher resolution screen, such as 1280 × 1024, can be obtained by setting the properties in the desktop resolution. This setting produces finer pixels made from 1280 columns and 1024 rows for crisper display. However, very fine screen resolution also has some drawbacks. As the number of pixels increases, more memory is needed in displaying graphical objects. Some graphics-intensive applications may fail if the allocated amount of memory is not sufficient as a result from high-resolution displays.

Mapping a point (x, y) from the Cartesian system to its corresponding pixel (X, Y) on Windows is a straightforward procedure involving the linear relationships, $X = m_1 x + c_1$ and $Y = m_2 y + c_2$. In these equations, m_1 and m_2 are the gradients in the mapping $x \rightarrow X$ and $y \rightarrow Y$, respectively. The constants c_1 and c_2 are the y-axis intercepts of the lines in the Cartesian system. In solving for m_1, m_2, c_1, and c_2, a total of four equations are required that can be obtained from points: any two points in the Cartesian system and their range in Windows.

The mapping from (x, y) to (X, Y) is a type of transformation called *linear transformation*. In this transformation, all coordinates in the domain are related to their images in the range through a linear function. Figure 4.3 shows a linear transformation from a line in the Cartesian system to Windows. The line in the left is made up of the points (x_1, y_1) and (x_2, y_2), which map to the Windows coordinates, (X_1, Y_1) and (X_2, Y_2), respectively. The mapping involving $x \rightarrow X$ is linear represented as follows:

$$X_1 = m_1 x_1 + c_1, \qquad (4.1a)$$

$$X_2 = m_1 x_2 + c_1. \qquad (4.1b)$$

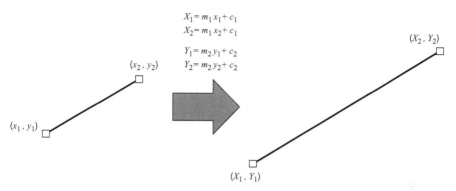

$$X_1 = m_1 x_1 + c_1$$
$$X_2 = m_1 x_2 + c_1$$
$$Y_1 = m_2 y_1 + c_2$$
$$Y_2 = m_2 y_2 + c_2$$

FIGURE 4.3. Conversion from Cartesian to Windows.

At the same time, the mapping $y \rightarrow Y$ is also linear represented as

$$Y_1 = m_2 y_1 + c_2, \tag{4.2a}$$

$$Y_2 = m_2 y_2 + c_2. \tag{4.2b}$$

Solving the first two equations, we obtain

$$m_1 = \frac{X_2 - X_1}{x_2 - x_1}, \tag{4.3a}$$

$$c_1 = X_1 - m_1 x_1. \tag{4.3b}$$

The mapping equation in $y \rightarrow Y$ is solved in the same manner to produce the following results:

$$m_2 = \frac{Y_2 - Y_1}{y_2 - y_1}, \tag{4.4a}$$

$$c_2 = Y_1 - m_2 y_1. \tag{4.4b}$$

Equations (4.1) and (4.2) provide very useful conversion criteria of coordinates from Cartesian to Windows. The contribution is obvious especially in confining certain points in the real coordinate system to be within a rectangular region in Windows.

Example 4.1. The line from $(-5, 6)$ to $(7, -4)$ in the Cartesian coordinates system maps to a line from $(50, 30)$ to $(500, 300)$ in Windows. Find a line in Windows that is mapped from the line in the Cartesian system from $(-2, 0)$ to $(4, 2)$.

Solution. In this problem, $(x_1, y_1) = (-5, 6), (x_2, y_2) = (7, -4), (X_1, Y_1) = (50, 30)$, and $(X_2, Y_2) = (500, 300)$. From Equations (4.3a), (4.3b), (4.4a), and (4.4b),

$$m_1 = \frac{500 - 50}{7 - (-5)} = 37.5,$$

$$c_1 = 50 - \frac{450}{12}(-5) = 237.5,$$

$$m_2 = \frac{300 - 30}{-4 - 6} = -27,$$

$$c_2 = 30 - (-27)6 = 192.$$

It follows that, $X_1 = 37.5(-2) + 237.5 = 162.5 \approx 163$, $Y_1 = -27(0) + 192 = 192$, $X_2 = 37.5(4) + 237.5 = 387.5 \approx 388$ and $Y_2 = -27(2) + 192 = 138$. Therefore, a line from $(-2, 0)$ to $(4, 2)$ in the Cartesian system maps directly to $(163, 192)$ to $(388, 138)$ in Windows.

Code4B: Drawing a Polynomial

Code4B. User Manual.

The program is a demonstration, and, therefore, no input is required.

Development Files: Code4B.cpp and Code4B.h.

We discuss a small project called Code4B for drawing a polynomial on Windows. Figure 4.4 shows the polynomial $f(x) = 2x^4 + 2x^3 - 9x^2 - 4x + 1$, which is the output of this project. The curve is drawn in the range of $-1 \le x \le 1.8$ with the maximum and minimum points as well as the left and right points of the curve shown. The coordinates shown are the real coordinates from the Cartesian system.

It is no secret that a curve is drawn on Windows by placing pixels successively through iterations from left to right according to its given function. In Code4B, the polynomial is produced by iterating the points (x_i, y_i) using the function $y_i = f(x_i)$, for $i = 0$ to m, where x_0 and x_m are the left and right values, respectively, and m is the number of subintervals in the given range. Pixels are placed on the display using the MFC functions, SetPixel() or LineTo(). With large m, the successive pixels lie very close to each other so that they appear like a curve.

The main advantage from Code4B is the ease in which the polynomial is drawn according to scale on Windows. The drawing area in the window is bounded by a rectangle (not shown in the output) with the CPoint objects called home in the upper left-hand corner and end in the lower right-hand corner. With these defined boundaries, the curve is scaled in such a way that the maximum and minimum values touch the upper and lower boundaries of the rectangle. Therefore, the curve fits in nicely in the drawing area as it does not appear to be too big or too small this way.

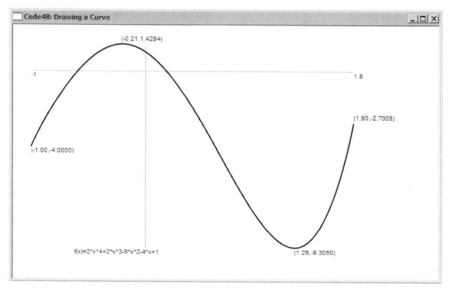

FIGURE 4.4. The polynomial $f(x) = 2x^4 + 2x^3 - 9x^2 - 4x + 1$ generated from Code4B.

Code4B is a brief project consisting of two files, Code4B.cpp and Code4B.h. Code4B.cpp is a relatively compact file consisting of about 100 lines of code. It has one class called CCode4B, which is derived from the MFC class, CFrameWnd. There is only one event in this project, namely, the main display, which is handled by the function OnPaint(). This implies Code4B does not take input from the user, and that most variables and objects are locally based.

In Code4B, the points in the Cartesian system with (x, y) real coordinates are declared in a structure called PT. The points along the curve are represented by an array called pt, which is linked to this structure, as follows:

```
typedef struct
{
        double x,y;
} PT;
PT *pt,max,min,left,right;
pt=new PT [m+1];
```

From this structure, the ith point in the graph is denoted as pt[i], with $i = 0, 1, \ldots, m$. The coordinates x_i and y_i of the point (x_i, y_i) are represented as pt[i].x and pt[i].y, respectively. Other members of this structure are left, right, min and max, which represent the left, right, minimum, and maximum points of the curve, respectively.

Figure 4.5 shows the drawing area of the curve on Windows showing the CPoint objects, home, end, left, right, max, and min. The drawing area is a rectangle

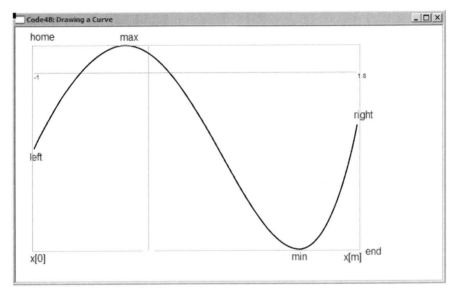

FIGURE 4.5. Windows objects in the drawing area.

designated by the CPoint objects, home (top left-hand corner) and end (bottom right-hand corner). With this designation, the drawing area for the curve is restricted to within the boundaries defined by the rectangle. A pixel in Windows is denoted px whose components, px.x and px.y, are the results from mapping pt[i].x and pt[i].y from the real coordinates according to Equations (4.1) and (4.2). This is illustrated in Figure 4.6. The left and right points of the curve in the given range are the CPoint objects, left and right. The maximum value of the curve is max, whereas its minimum is min.

Table 4.6 lists all the main variables in Code4B. There are m uniform subintervals in the x range whose width is given as h. Equations (4.1) and (4.2) convert the points from the real coordinates in the Cartesian system to their corresponding pixels in Windows with the constants determined by m1, c1, m2, and c2. The main objects in the project are listed in Table 4.7.

$$px.x = m1*pt[i].x + c1$$

$$px.y = m2*pt[i].y + c2$$

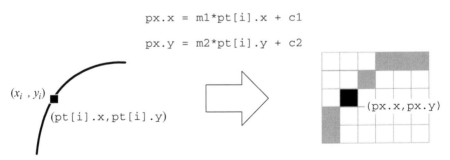

FIGURE 4.6. Mapping from (x_i, y_i) to (px.x,px.y).

TABLE 4.6. The main variables in Code4B

Variable	Type	Description
pt[i].x, pt[i].y	double	(x_i, y_i) coordinates of a point in Cartesian.
left.x, left.y	double	Left point of the curve in the interval.
right.x, right.y	double	Right point of the curve in the interval.
min.x, min.y	double	Minimum point of the curve in the interval.
max.x, max.y	double	Maximum point of the curve in the interval.
m1, c1	double	Constants m_1 and c_1 from Equation (4.1).
m2, c2	double	Constants m_2 and c_2 from Equation (4.2).
h	double	Width of each subinterval.
m	macro	Total number of subintervals.

TABLE 4.7. Main objects in Code4B

Object	Class	Description
px	CPoint	Pixel on Windows with (px.x,px.y) as its coordinates.
home	CPoint	Home (top left-hand corner) coordinates of the graphical area.
end	CPoint	End (bottom right-hand corner) coordinates of the graphical area.

In drawing the curve, the width of each subinterval from x_0 to x_m is first computed, and this is given by

```
h=(pt[m].x-pt[0].x)/(double)m;
```

The curve is drawn by iterating on x_i from $i = 0$ to $i = m$, incrementing the value of x_i by h at each step. To determine the maximum and minimum values in the curve, comparisons are made through the expressions max.y<pt[i].y and min.y>pt[i].y, respectively. The following code segment shows how this is done:

```
max.y=0; min.y=0;
for (int i=0;i<=m;i++)
{
        pt[i].y=2*pow(pt[i].x,4)+2*pow(pt[i].x,3)
                    -9*pow(pt[i].x,2)-4*pt[i].x+1;
        if (max.y<pt[i].y)
              max=pt[i];
        if (min.y>pt[i].y)
              min=pt[i];
        if (i<m)
              pt[i+1].x=pt[i].x+h;
}
```

For mapping points from the real coordinates (pt[i].x,pt[i].y) into Windows coordinates (px.x,px.y), Equations (4.1) and (4.2) need to be solved. From Figure

4.5, we assign the variables in Equations (4.1) and (4.2) as follows:

x_1 is pt [o] .x, y_1 is min.y,
x_2 is pt [m] .x, y_2 is max.y,
X_1 is home.x, Y_1 is end.y,
X_2 is end.x, Y_2 is home.y.

Equations (4.4a) and (4.4b) then become

```
m1=(double)(end.x-home.x)/(pt[m].x-pt[0].x);
c1=(double)home.x-pt[0].x*m1;
m2=(double)(home.y-end.y)/(max.y-min.y);
c2=(double)end.y-min.y*m2;
```

In drawing a curve, it is important to have the x- and y-axes in the real coordinates shown for reference. These axes are drawn as follows:

```
CPen pGray(PS_SOLID,1,RGB(200,200,200));
dc.SelectObject(pGray);
px=CPoint(m1*0+c1,m2*min.y+c2); dc.MoveTo(px);
px=CPoint(m1*0+c1,m2*max.y+c2); dc.LineTo(px);
px=CPoint(m1*pt[0].x+c1,m2*0+c2); dc.MoveTo(px);
str.Format("%.01f",pt[0].x); dc.TextOut(px.x,px.y,str);
px=CPoint(m1*pt[m].x+c1,m2*0+c2); dc.LineTo(px);
str.Format("%.11f",pt[m].x); dc.TextOut(px.x,px.y,str);
```

The last step is to draw the curve by drawing points through iterations from left to right. The curve is drawn by moving the pen to pt [0], then drawing a line to the next point in the iteration, and so on until pt [m] .x. Each point in the real coordinates pt [i] is mapped to its corresponding point on Windows px [i] using the conversion formula in Equations (4.1) and (4.2). At each iteration, comparisons are made to determine the maximum and minimum points of the curve. The following code shows how this is done:

```
CPen pDark(PS_SOLID,2,RGB(50,50,50));
dc.SelectObject(pDark);
for (int i=0;i<=m;i++)
{
        px=CPoint((int)(m1*pt[i].x+c1), (int)(m2*pt[i].y+c2));
        if (i==0)
        {
                dc.MoveTo(px);
                str.Format("(%.21f,%.41f)",pt[0].x,pt[0].y);
                dc.TextOut(px.x,px.y,str);
        }
        else
                dc.LineTo(px);
        if (pt[i].y==max.y)
```

```
        {
                str.Format("(%.2lf,%.4lf)",max.x,max.y);
                dc.TextOut(px.x,px.y-15,str);
        }
        if (pt[i].y==min.y)
        {
                str.Format("(%.2lf,%.4lf)",min.x,min.y);
                dc.TextOut(px.x,px.y,str);
        }
        if (i<m)
                pt[i+1].x=pt[i].x+h;
}
str.Format("(%.2lf,%.4lf)",pt[m].x,pt[m].y);
dc.TextOut(px.x,px.y-15,str);
dc.TextOut(100,350,"f(x)=2*x 4+2*x 3-9*x 2-4*x+1");
delete pt;
```

The above code draws the curve perfectly in the given interval. Once this is done, it is important for us to delete the array pt from the structure as the allocated memory is still active although the program has ended. This is achieved through delete pt.

4.4 GENERATING CURVES USING MYPARSER

Code4C. User Manual.

1. Enter the input strings of the functions.
2. Enter the range of values from left to right.
3. Click the corresponding method to view the results.

Development Files: Code4C.cpp, Code4C.h, and MyParser.obj.

We have discussed Code4B for drawing a simple polynomial. The project illustrates some very important fundamental concepts in drawing a curve. However, the generated curve is static as the input is determined solely by the programmer. There is no flexibility as the end user will not get a chance to key in his/her own equations directly without the need to recompile the code. The program is also not user-friendly as there are no events associated with its interaction with the user.

A practical curve drawing program is one that allows interaction and has the user-friendliness features through MyParser. Curve drawing is considered complete if it incorporates an equation parser as one of its interactive tools. One useful application of an equation parser is in drawing a curve whose input equation is defined by the user. MyParser reads the equation keyed in by the user as a string, processes this string, and generates values for drawing the curve. We discuss two types of curves, namely, $y = f(x)$, which is a curve that is dependent on a single variable x, and the parametric curve given by $(x, y) = (f(t), g(t))$.

The function $y = f(x)$ is the most fundamental form of curve in mathematics. This function is a direct mapping from x to y through the relationship defined in $f(x)$. This curve allows a single point in x to map to a point in y. It also allows several values of x to map to a single value in y. However, mapping a single point in x to multiple points of y is not allowed.

A *parametric curve* is a curve made up of points that are dependent on one or more parameters. A two-dimensional parametric curve represented by (x, y), where x and y are terms or functions that are dependent on the parameter t in the interval given by $t_0 \le t \le t_m$, is expressed as follows:

$$(x, y) = (f(t), g(t)). \tag{4.5}$$

In a simple case like $x = t$ and $y = t^2$ for $0 \le t \le 2$, the parametric equations are equivalent to a single equation given by $y = x^2$, for $0 \le x \le 2$. In some other cases, the relationship given by Equation (4.5) may not be expressed implicitly as a single function of x. For example, $x = 2 \cos t$ and $y = 2 \sin t$ are parametric equations representing a circle whose equation is $x^2 + y^2 = 4$.

We discuss Code4C, which provides flexibility by allowing the user to write an equation directly into an edit box. Figure 4.7 shows a sample output from quite a

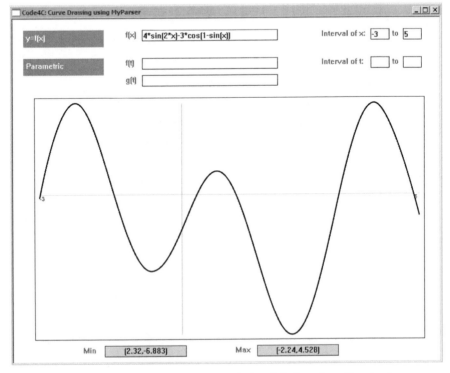

FIGURE 4.7. An exotic curve from $f(x) = 4 \sin 2x - 3 \cos(1 - \sin x)$.

sophisticated function given by $f(x) = 4 \sin 2x - 3 \cos(1 - \sin x)$ for $-3 \le x \le 5$. The parser in the program reads the expression keyed in by the user and computes the equation for $i = 0$ to m, where m is the number of intervals (which is also the number of iterations).

Code4C supports two types of graphs, the function $y = f(x)$ and the parametric curve, $x = f(t)$ and $y = g(t)$. The two graphs can be selected using a menu in the form of shaded rectangular boxes. Input for the first type of graph is provided in the form of CEdit boxes that read the string for $f(x)$ and range of values of x in the domain. For the second graph, CEdit input boxes are provided to read the strings $x = f(t)$ and $y = g(t)$ and the start and end values of t. The curve is shown in the big rectangular region, whereas its maximum and minimum points are displayed in the static boxes at the bottom of the window.

Figure 4.8 shows another example of a curve from a set of parametric equations given by $x(t) = 1 + t \cos t$ and $y(t) = t \sin(1 - t)$ for $-30 \le t \le 50$. In this example, the program determines the left and right values of x, as well as the maximum and minimum values of y from the given range of t. Creativity continues from here. The user should try several equations in mind and should view the results on the window.

Figure 4.9 illustrates the development steps in Code4C. A global variable called fMenu has been created to monitor the progress in the execution, with fMenu=0

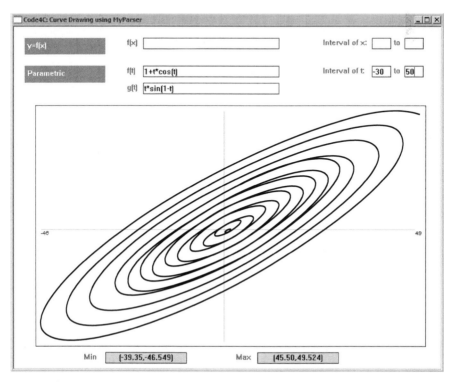

FIGURE 4.8. The parametric curve $x(t) = 1 + t \cos(t)$ and $y(t) = t \sin(1 - t)$.

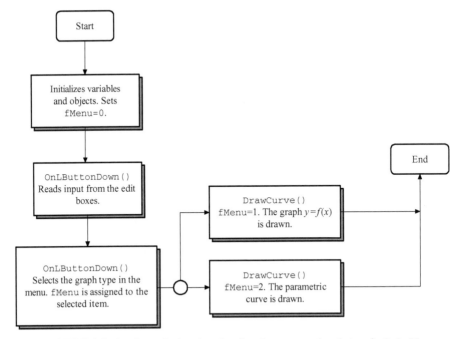

FIGURE 4.9. A schematic drawing showing the computational steps in Code4C.

as its initial value, and fMenu=1 and fMenu=2 indicating a selection of the first or second item in the menu, respectively. Input from the user in the edit boxes is read in OnLButtonDown(). Once input has been completed, a click at an item in the menu results in its assignment of value to fMenu. Another function called DrawCurve() draws the selected graph according to the input values.

Code4C includes three source files, Code4C.cpp, Code4C.h, and MyParser.obj. A single class called CCode4C is used, and this class is derived from CFrameWnd. The same structure called PT as in Code4B is used, but this time t is included as its additional member to support the parametric equations, $x(t)$ and $y(t)$. The structure is declared as follows:

```
typedef struct
{
        double t,x,y;                    // x,y and t coordinates
} PT;
PT *pt,max,min,left,right;
```

The objects for the menu are organized into a structure called MENU. The structure consists of members that represent the title, home coordinates, and rectangular objects of the items. An array called menu[] stores these values.

```
typedef struct
{
      CString item;              // title
      CPoint hm;                 // home coordinates
      CRect rc;                  // rectangular object
} MENU;
MENU menu[nMenuItems+1];
```

Input is organized into a structure called INPUT.

```
typedef struct
{
      CString item;              // title
      CPoint hm;                 // home coordinates
      CEdit ed;                  // edit box
      CRect rc;                  // rectangular object
} INPUT;
INPUT input[nInput+1];
```

Another structure is OUTPUT, which organizes the objects for displaying the output. The structure organizes objects for the static boxes and their home coordinates and rectangular objects through an array called output[].

```
typedef struct
{
      CPoint hm;                 // home coordinates
      CStatic st;                // static box
      CRect rc;                  // rectangular object
} OUTPUT;
OUTPUT output[nOutput+1];
```

The selected curve is displayed in the big rectangular region. A structure called CURVE organizes the objects comprising the starting and end coordinates of the rectangle and the rectangular object. This structure is linked to a variable called curve.

```
typedef struct
{
      CPoint hm,end;             // start and end coordinates
      CRect rc;                  // rectangular object
} CURVE;
CURVE curve;
```

There are five main functions in Code4C, and they are listed in Table 4.8. OnPaint() responds to ON_WM_PAINT, which updates the output in the main window. A function called OnLButtonDown() responds to ON_WM_LBUTTONDOWN, which

TABLE 4.8. Functions in Code4C

Function	Type	Description
CCode4C()	Constructor	Creates the main window and initializes the variables and objects.
~CCode4C()	Destructor	Destroy the class and the array pt[].
DrawCurve()	void	Draw the curve selected from the menu.
OnLButtonDown()	void	Message handler for the left-click of the mouse.
OnPaint()	void	Output display in the main window.

is immediately activated when an item in the menu is left-clicked. Another function, called DrawCurve(), draws and label the selected curve once input has been completed.

Two events are mapped in this application, and they are the default display on the main window and the left-click of the mouse.

```
BEGIN_MESSAGE_MAP(CCode4C,CFrameWnd)
    ON_WM_PAINT()
    ON_WM_LBUTTONDOWN()
END_MESSAGE_MAP()
```

The code for the constructor function is listed below. Initial values are assigned to the global variables and objects. They include the home coordinates of all the menu items, edit and static boxes. The menu items are identified through a flag variable called fMenu, with fMenu=0 as its initial value to denote no selection has been made. The id for each edit and static boxes is uniquely assigned with the first box read given with an initial value of idc=301. To guarantee a unique id for each box, this value is incremented every time an edit box or static box is created.

```
CCode4C::CCode4C()
{
    int i;
    Create(NULL,"Code4C: Curve Drawing using MyParser",
        WS_OVERLAPPEDWINDOW,CRect(0,0,800,640),NULL);
    pt=new PT [m+1];
    fMenu=0; idc=301;
    menu[1].item="y=f(x)";
    menu[2].item="Parametric";
    curve.hm=CPoint(50,150);
    curve.end=CPoint(750,560);
    curve.rc=CRect(curve.hm.x,curve.hm.y,curve.end.x,
        curve.end.y);
    input[1].hm=CPoint(240,20);
    input[2].hm=CPoint(660,20);
    input[3].hm=input[2].hm+CPoint(60,0);
```

```
input[4].hm=CPoint(660,70);
input[5].hm=input[4].hm+CPoint(60,0);
input[6].hm=input[1].hm+CPoint(0,50);
input[7].hm=input[6].hm+CPoint(0,30);
output[1].hm=CPoint(170,580);
output[2].hm=output[1].hm+CPoint(280,0);
for (i=1;i<=nMenuItems;i++)
{
      menu[i].hm=CPoint(20,20+(i-1)*50);
      menu[i].rc=CRect(menu[i].hm.x,menu[i].hm.y,
            menu[i].hm.x+150,menu[i].hm.y+30);
}
for (i=1;i<=nOutput;i++)
      output[i].st.Create("",WS_CHILD | WS_VISIBLE
            | SS_CENTER | WS_BORDER,
            CRect(output[i].hm.x,output[i].hm.y,
            output[i].hm.x+150, output[i].hm.y+20),this,
            idc++);
for (i=1;i<=nInput;i++)
      input[i].ed.Create(WS_CHILD | WS_VISIBLE | WS_BORDER,
            CRect(input[i].hm.x,input[i].hm.y,
            input[i].hm.x+((i>1 && i<6)?35:250),
            input[i].hm.y+20), this,idc++);
Arial80.CreatePointFont (80,"Arial");
}
```

OnPaint() displays the menu items and the labels for the edit boxes and static boxes. The function also updates the display in the main window and draws curve through DrawCurve() when it receives a call through InvalidateRect().

```
void CCode4C::OnPaint()
{
      CPaintDC dc(this);
      CString str;
      dc.SetBkColor(RGB(150,150,150));
      dc.SetTextColor(RGB(255,255,255));
      dc.SelectObject(Arial80);
      for (int i=1;i<=2;i++)
      {
            dc.FillSolidRect(&menu[i].rc,RGB(150,150,150));
            dc.TextOut(menu[i].hm.x+5,menu[i].hm.y+5,
                  menu[i].item);
      }
      CRect rc=CRect(curve.hm.x-10,curve.hm.y-10,curve.end.x+10,
            curve.end.y+10);
```

```
        dc.Rectangle(rc);
        dc.SetBkColor(RGB(255,255,255));
        dc.SetTextColor(RGB(100,100,100));
        dc.TextOut(input[1].hm.x-30,input[1].hm.y,"f(x)");
        dc.TextOut(input[2].hm.x-90,input[2].hm.y,
                "Interval of x:");
        dc.TextOut(input[3].hm.x-20,input[3].hm.y,"to");
        dc.TextOut(input[4].hm.x-90,input[4].hm.y,
                "Interval of t:");
        dc.TextOut(input[5].hm.x-20,input[5].hm.y,"to");
        dc.TextOut(input[6].hm.x-30,input[6].hm.y,"f(t)");
        dc.TextOut(input[7].hm.x-30,input[7].hm.y,"g(t)");
        dc.TextOut(output[1].hm.x-40,output[1].hm.y,"Min");
        dc.TextOut(output[2].hm.x-40,output[2].hm.y,"Max");
        if (fMenu==1 || fMenu==2)
                DrawCurve();
}
```

The above code updates the display on certain parts of the window whenever the function `InvalidateRect()` is invoked. This is observed in the last two lines when fMenu is assigned with the value of 1 or 2 whenever a menu item is selected. The selection causes a curve to be drawn from the function `DrawCurve()`.

`OnLButtonDown()` responds to `ON_WM_LBUTTONDOWN`. This happens when an item in the menu is left-clicked. This function has two parameters, nFlags and pxClick. We are only concerned with pxClick here as it represents an object for recording the pixel location where the click is made. `OnLButtonDown()` first reads the input made on the edit boxes through the MFC function, `GetWindowText()`. Each input is read as a string recognized as input[i].item. Inputs such as the left and right values of x are recognized as real values. Therefore, their input strings, input[2].item and input[3].item, respectively, are converted to double using the C++ function, atof(). Other inputs, input[1].item, input[6].item, and input[7].item, are global strings that represent the input expressions that will be read and evaluated by the parser.

```
void CCode4C::OnLButtonDown(UINT nFlags,CPoint pxClick)
{
        input[1].ed.GetWindowText(input[1].item);
        input[2].ed.GetWindowText(input[2].item);
        input[3].ed.GetWindowText(input[3].item);
        input[4].ed.GetWindowText(input[4].item);
        input[5].ed.GetWindowText(input[5].item);
        input[6].ed.GetWindowText(input[6].item);
        input[7].ed.GetWindowText(input[7].item);
        pt[0].x=atof(input[2].item);
```

```
pt[m].x=atof(input[3].item);
pt[0].t=atof(input[4].item);
pt[m].t=atof(input[5].item);
if (menu[1].rc.PtInRect(pxClick))
        fMenu=1;
if (menu[2].rc.PtInRect(pxClick))
        fMenu=2;
InvalidateRect(curve.rc);
}
```

When a shaded rectangle in the window is left-clicked, pxClick stores the pixel coordinates at the location. A MFC function called PtInRect() checks whether the click is within the named rectangle. For example, the following expression checks whether if the first box has been clicked:

```
menu[1].rc.PtInRect(pxClick)
```

where menu[1].rc is the rectangle and pxClick is the location of the click. The function returns TRUE (1) if the click is inside the rectangle, and FALSE (0) if it is outside of the rectangle. A return value of 1 validates the selection, and the program immediately assigns fMenu to indicate the type of function chosen.

The last statement in OnLButtonDown() updates the display in the main window. This is done through InvalidateRect(), which calls OnPaint() for updating a rectangular region in the display denoted by curve.rc.

The most important content in Code4C is the function for drawing a curve, DrawCurve(). The code of this function is derived mostly from a function of the same name in Code4B with some enhancement from the parser. The parser incorporates a function called parse(), which is derived from the external object file, MyParser.obj. This external function is called through the statement

```
extern double parse(CString,int,double [],int []);
```

in the header file.

DrawCurve() draws a curve based on the assigned value assigned to fMenu, where fMenu=1 indicates the curve $y = f(x)$ and fMenu=2 for the parametric curve. The first curve is drawn according to the following code segment:

```
if (fMenu==1)
{
        h=(pt[m].x-pt[0].x)/((double)(m));
        for (i=0;i<=m;i++)
        {
                psi[1]=23; psv[1]=pt[i].x;
                pt[i].y=parse(input[1].item,1,psv,psi);
```

```
if (i==0)
{
        left.x=0; right.x=0;
        min=pt[0]; max=pt[0];
}
left.x=((left.x>pt[i].x)?pt[i].x:left.x);
right.x=((right.x<pt[i].x)?pt[i].x:right.x);
if (min.y>pt[i].y)
{
        min.x=pt[i].x; min.y=pt[i].y;
}
if (max.y<pt[i].y)
{
        max.x=pt[i].x; max.y=pt[i].y;
}
if (i<m)
        pt[i+1].x=pt[i].x+h;
    }
}
```

The above code computes $y_i = f(x_i)$ for $i = 0, 1, \ldots, m$. The code also determines the maximum and minimum points in the curve as well as the left and right intervals for x. The x variable is identified as a code in psi[1]=23. The value assigned to this variable is stored as psv[1]. These two values make up the arguments in parse(), which reads the string expression from input[1].item, as follows:

```
pt[i].y=parse(input[1].item,1,psv,psi);
```

The entry of 1 in the above statement denotes only one variable is involved in the expression. The above statement computes the expression in input[1].item and returns the value as pt[i].y.

The second type of curve is generated from fMenu=2. Two variables are involved, t and x, and they are identified as codes 19 and 23, respectively. The first function in this parametric equation, $x = f(t)$, is evaluated from the global string input[6].item through

```
psi[1]=19; psv[1]=pt[i].t;
pt[i].x=parse(input[6].item,1,psv,psi);
```

The second function $y = g(t)$ is read as input[7].item and evaluated through

```
psi[1]=23; psv[1]=pt[i].x;
pt[i].y=parse(input[7].item,1,psv,psi);
```

The following code segment evaluates $x_i = f(t_i)$ and $y_i = g(t_i)$ for $i = 0, 1, \ldots, m$ besides determining the maximum, minimum, left, and right points in the curve:

```
if (fMenu==2)
{
    h=(pt[m].t-pt[0].t)/((double)(m));
    for (int i=0;i<=m;i++)
    {
        psi[1]=19; psv[1]=pt[i].t;
        pt[i].x=parse(input[6].item,1,psv,psi);
        psi[1]=23; psv[1]=pt[i].x;
        pt[i].y=parse(input[7].item,1,psv,psi);
        if (i==0)
        {
            left.x=0; right.x=0;
            min=pt[0]; max=pt[0];
        }
        left.x=((left.x>pt[i].x)?pt[i].x:left.x);
        right.x=((right.x<pt[i].x)?pt[i].x:right.x);
        if (min.y>pt[i].y)
        {
            min.x=pt[i].x; min.y=pt[i].y;
        }
        if (max.y<pt[i].y)
        {
            max.x=pt[i].x; max.y=pt[i].y;
        }
        if (i<m)
            pt[i+1].t=pt[i].t+h;
    }
}
```

With all computed values for (x_i, y_i) determined from the parser, the last step in the project is to display the curve in the designated graphical region. This is done through

```
CPen pDark(PS_SOLID,2,RGB(50,50,50));
dc.SelectObject(pDark);
for (i=0;i<=m;i++)
{
    px=CPoint((int)(m1*pt[i].x+c1),  (int)(m2*pt[i].y+c2));
    if (i==0)
        dc.MoveTo(px);
    else
        dc.LineTo(px);
}
```

4.5 SUMMARY

In this chapter, we discussed in detail the techniques for drawing several types of curves on Windows using the resources in MFC. As the platforms on Windows and the real coordinates in mathematics differ, some simple transformations are required to make them compatible. In addition, we also discussed MyParser, which is a very useful tool in mathematics for developing user-friendly mathematical applications. MyParser incorporates an equation parser that reads and evaluates a mathematical equation from the user in the form of a string. The user should find the supplied object file in MyParser very useful as this file can be linked to other source files in a project for solving many numeric-intensive applications. Some useful applications include the final year undergraduate, M.Sc., and Ph.D. projects, and modeling and simulation in research.

PROGRAMMING CHALLENGES

1. Test on the following curves using Code4B:

   a. $y = 1 - 3x + 5x^2$.
   b. $y = 3x \sin x - 5x^2 \cos(1 - x)$.
   c. $y = \frac{3x-1}{1+4x^2}$.
   d. $x(t) = 2\cos(2t - 1)$ and $y(t) = t \sin(t^2 - 1)$.
   e. $x(t) = 3 \sin 2t$ and $y = 3 \cos^2 t$.

2. The polar coordinate system is represented as $r = f(\theta)$, where r is the radius from the origin and θ is the angle measured from the x-axis in the counter-clockwise direction. The conversion from (r, θ) to (x, y) is done according to the following rules:

$$x = r \cos \theta \text{ and } y = r \sin \theta.$$

   Improve on the program in Code4A to include a polar curve given by $r = \sin 5\theta$.

3. Improve on the program in Code4C to include equations from the polar coordinates.

4. The curve drawing project in Code4C does not include mechanisms for checking the domain of the function and for testing for its singularity at the points in the given interval. These issues are important to consider as they may cause the program to crash if not handled carefully. Study these issues, and incorporate them into Code4C.

Systems of Linear Equations

5.1   INTRODUCTION

A linear system consists of a set of equations with some governing parameters or variables that control the system. The variables that make up the system can be stated in the form of differential equations or autonomous variables. The system works in a dependent manner where changes in one or more variables affect the performance of the whole system in general.

Definition 5.1. An equation in the following form with the unknowns x_i and constants a_i, for $i = 1, 2, \ldots, n$, is said to be a *linear equation*:

$$a_1 x_1 + a_2 x_2 + \cdots + a_n x_n = b. \tag{5.1}$$

In this definition, each unknown x_i must have an index power of 1. Also, a linear equation cannot have a variable inside sine, cosine, tangent, exponent, logarithm, and other operators. It follows that any equation in the form other than Equation (5.1) is called a *nonlinear equation*.

The following equations are examples of linear equations:

$$3x - 5y + z = -2,$$

$$4x_1 + 2x_2 - 3x_3 = 0.$$

A *system of linear equations* consists of a set of linear equations and a set of unknowns. In general, a system of linear equations with m equations and n unknowns is expressed as follows:

$$a_{11}x_1 + a_{12}x_2 + \cdots + a_{1n}x_n = b_1,$$
$$a_{21}x_1 + a_{22}x_2 + \cdots + a_{2n}x_n = b_2, \qquad (5.2)$$
$$\cdots$$
$$a_{m1}x_1 + a_{m2}x_2 + \cdots + a_{mn}x_n = b_n.$$

The above system is represented in matrix form as $Ax = b$, as follows:

$$
\begin{bmatrix}
a_{11} & a_{12} & \cdots & a_{1n} \\
a_{21} & a_{22} & \cdots & a_{2n} \\
\cdots & \cdots & \cdots & \cdots \\
a_{m1} & a_{m2} & \cdots & a_{mn}
\end{bmatrix}
\begin{bmatrix}
x_1 \\
x_2 \\
\cdots \\
x_n
\end{bmatrix}
=
\begin{bmatrix}
b_1 \\
b_2 \\
\cdots \\
b_n
\end{bmatrix},
\qquad (5.3)
$$

where $A = [a_{ij}]$ is the coefficient matrix of the system, for $i = 1, 2, \ldots, m$ and $j = 1, 2, \ldots, n$. In this system, m is the number of rows in the coefficient matrix, whereas n is the number of columns (unknowns). The vector $b = [b_i]$ is the given constants, whereas $x = [x_i]$ is the vector of the unknowns. If $b_i = 0$ for all i, then the linear system is called a *homogeneous system*; otherwise, if at least one of $b_i \neq 0$, then the system is *nonhomogeneous*.

In designing computer programs for solving the problem involving a system of linear equations, it is important to incorporate good programming habits. These habits include making the program easy to modify in order to have the problem scalable and flexible. A program is said to be *scalable* if the code for a program with small data applies to one with large data as well. A flexible program is one that adapts well to changes in the parameters governing the problem with minimum modification in the program. It is our duty to discuss these approaches in this chapter.

5.2 EXISTENCE OF SOLUTIONS

The solution to a system of linear equations may or may not exist, depending on the given equations and several other factors. To complicate matters, a solution that exists may not be unique, and there can be an infinite number of solutions for the system. We study all cases that lead to the existence of the solution for a system of linear equations.

Definition 5.2. A system of linear equations is said to have *consistent* solutions if all its solutions satisfy each equation in the system. If one or more of the solutions do not satisfy at least one equation in the system, then the system is said to be *inconsistent*.

The above definition describes the state of the solutions in a system of linear equations. Referring to Equation (5.3), if the solution to the first equation also satisfies all other equations in the system, then this solution is consistent. It follows that the system normally produces consistent solutions if $m \leq n$ and inconsistent solutions if $m > n$. A consistent system may have either unique or many solutions. A set of inconsistent solutions consists of solutions that are valid in some but not in all equations in the system.

Ideally, good and consistent solutions are obtained if the coefficient matrix is square; that is, the number of equations in the system equals the number of unknowns, or $m = n$. With this assumption, Equation (5.3) can now be rewritten as follows:

$$
\begin{bmatrix}
a_{11} & a_{12} & \cdots & a_{1n} \\
a_{21} & a_{22} & \cdots & a_{2n} \\
\cdots & \cdots & \cdots & \cdots \\
a_{n1} & a_{n2} & \cdots & a_{nn}
\end{bmatrix}
\begin{bmatrix}
x_1 \\
x_2 \\
\cdots \\
x_n
\end{bmatrix}
=
\begin{bmatrix}
b_1 \\
b_2 \\
\cdots \\
b_n
\end{bmatrix}.
\tag{5.4}
$$

Equation (5.4) is also expressed compactly in matrix form as $A\,x = b$, where

$$
A =
\begin{bmatrix}
a_{11} & a_{12} & \cdots & a_{1n} \\
a_{21} & a_{22} & \cdots & a_{2n} \\
\cdots & \cdots & \cdots & \cdots \\
a_{n1} & a_{n2} & \cdots & a_{nn}
\end{bmatrix},
\quad
b =
\begin{bmatrix}
b_1 \\
b_2 \\
\cdots \\
b_n
\end{bmatrix}
\text{ and } x =
\begin{bmatrix}
x_1 \\
x_2 \\
\cdots \\
x_n
\end{bmatrix}.
\tag{5.5}
$$

In this representation, the square matrix A is the coefficient matrix of the system, whereas x is the solution vector.

Theorem 5.1. If the coefficient matrix A of a system of linear equations is square and if $|A| \neq 0$, then the system has consistent and unique solutions.

The above theorem states that a system whose coefficient matrix is not singular has consistent and unique solutions. Unique solutions imply each unknown in the system has only one value as its solution, and this value satisfies all equations in the system.

Example 5.1. The following system is consistent with unique solutions:

$$
\begin{aligned}
x + y &= 3, \\
x - 4y &= -2.
\end{aligned}
$$

Solution. It can be shown that $|A| = \begin{vmatrix} 1 & 1 \\ 1 & -4 \end{vmatrix} = -5 \neq 0$. Solving the above system produces $(x, y) = (2, 1)$, which is unique and applicable to both equations.

Theorem 5.2. If the augmented matrix $A \mid b$ reduces to $U \mid v$ after row operations with one or more zero rows in U and the corresponding row(s) is v is also zero, then the system is consistent with an infinite number of solutions. If one or more corresponding row(s) in v is not zero, then the solution does not exist.

Theorem 5.2 says an infinite solution exists if a row (or more) in the augmented matrix has zero contents in all its (their) elements. The theorem is best illustrated using two examples, one with an infinite number of solutions and another with no solution.

Example 5.2. The following system is consistent with many solutions:

$$x + 2y = 1,$$
$$2x + 4y = 2.$$

Solution. Reduce the augmented matrix, as follows:

$$\begin{matrix} 1 & 2 \\ 2 & 4 \end{matrix} \begin{vmatrix} 1 \\ 2 \end{vmatrix} \sim \begin{matrix} 1 & 2 \\ 0 & 0 \end{matrix} \begin{vmatrix} 1 \\ 0 \end{vmatrix} \qquad \text{through } R_2 \leftarrow R_2 - 2R_1.$$

Since all the elements in the second row of the augmented matrix have zero values, the system has infinite solutions. $(x, y) = (1, 0)$ and $(x, y) = (-1, 1)$ are some of the possible solutions to the system.

Example 5.3. The following system is inconsistent:

$$x + 2y = 1,$$
$$3x - y = -2,$$
$$2x + y = 3.$$

Solution. Reduce the augmented matrix, as follows:

$$\begin{matrix} 1 & 2 \\ 3 & -1 \\ 2 & 1 \end{matrix} \begin{vmatrix} 1 \\ -2 \\ 3 \end{vmatrix}$$

$$\sim \begin{matrix} 1 & 2 \\ 0 & -7 \\ 0 & -5 \end{matrix} \begin{vmatrix} 1 \\ -5 \\ 1 \end{vmatrix} \qquad \text{through } R_2 \leftarrow R_2 - 3R_1 \text{ and } R_2 \leftarrow R_2 - 2R_1$$

$$\sim \begin{matrix} 1 & 2 \\ 0 & -7 \\ 0 & 0 \end{matrix} \begin{vmatrix} 1 \\ -5 \\ -32/7 \end{vmatrix} \qquad \text{through } R_3 \leftarrow R_3 - \tfrac{5}{7}R_2.$$

The last row of the final reduced augmented matrix has zero elements on the left-hand side and nonzero on the right-hand side. This contradicts the equation, and by Theorem 5.2, no solution exists for this system.

TABLE 5.1. Common methods for solving a system of linear equations

Method	Type	Condition
Gaussian elimination	Elimination	Nonzero pivot elements and nonsingular coefficient matrix.
Gaussian elimination with partial pivoting	Elimination	Nonsingular rearranged matrix.
Gauss–Jordan	Elimination	Nonzero pivot elements and nonsingular coefficient matrix.
Crout	LU factorization	Nonzero diagonal elements in the L matrix.
Doolittle	LU factorization	Nonzero diagonal elements in the U matrix.
Cholesky	LU factorization	Positive-definite coefficient matrix.
Thomas	LU factorization	Tridiagonal coefficient matrix.
Jacobi	Iterative	Diagonally dominant coefficient matrix.
Gauss-Seidel	Iterative	Diagonally dominant coefficient matrix.

In this chapter, we will concentrate on discussing the system of linear equations in the form of Equation (5.4) that is consistent with a unique solution. In general, there are many techniques for solving this system, and they can be categorized as elimination, LU factorization, and iterative methods. An *elimination* technique reduces the augmented matrix in the system to its triangular form before applying backward substitutions for the solution. The *LU factorization* technique factorizes the coefficient matrix as a product of its upper and lower triangular matrices before generating the solution through substitutions. Lastly, an *iterative* technique performs iterations on all unknowns until their values converge to within a tolerable range.

Table 5.1 shows the common methods for solving a system of linear equations that can be categorized under the elimination, LU factorization, and iterative techniques. Note that some methods are very sensitive to the conditions imposed on the coefficient matrix, and we will address them later in the chapter.

5.3 GAUSSIAN ELIMINATION TECHNIQUES

The elimination technique reduces the rows of a matrix into a simpler form through a series of row operations, as discussed earlier in Chapter 2. We discuss three methods commonly used in solving a system of linear equations, namely, the Gaussian elimination method, Gaussian elimination method with partial pivoting, and Gauss–Jordan method.

Gaussian Elimination Method

The Gaussian elimination method is an elimination technique for solving a system of linear equations through two series of steps. The method works best when the

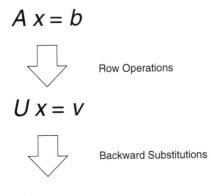

$$A x = b$$

Row Operations

$$U x = v$$

Backward Substitutions

solution **x**

FIGURE 5.1. Main steps in the Gaussian elimination method.

coefficient matrix is not singular; that is, $|A| \neq 0$ so that the solution obtained is unique according to Theorem 5.1. The Gaussian elimination method consists of two main steps, as depicted in Figure 5.1. The first step is the row operations for eliminating the elements in the coefficient matrix so as to reduce the matrix into its upper triangular form. If all elements in this triangular matrix have nonzero values, then the matrix definitely has a unique solution. The second step, which is backward substitutions, follows directly from the triangular matrix relationship for producing the solution.

We discuss a system of linear equations $Ax = b$ having n linear equations and n unknowns. The row operations technique in the Gaussian elimination method is similar to the technique for matrix reduction discussed in Section 2.4. This time row operations are applied to the augmented matrix $A \mid b$ from the system of linear equations according to the following relationships:

$$a_{ij} \leftarrow a_{ij} - p^* a_{kj} \quad \text{and} \quad b_i \leftarrow b_i - p^* b_k \quad \text{where} \quad p = a_{ik}/a_{kk},$$

for $k = 1, 2, \ldots, n - 1$, followed by $i = k + 1, k + 2, \ldots, n$ and $j = 1, 2, \ldots, n$.

The above step reduces $A \mid b$ to $U \mid v$. Backward substitutions are performed on $U \mid v$ to produce the solution x according to the following:

$$x_n = \frac{v_n}{u_{nn}} \quad \text{and} \quad x_i = \frac{v_i - \sum_{k=i+1}^{n} u_{ik} x_k}{u_{ii}}, \quad \text{for } i = n - 1, n - 2, \ldots, 1.$$

Example 5.4. Solve the following system of linear equations using the Gaussian elimination method:

$$2x_1 - x_2 + 7x_3 - x_4 = -1,$$
$$-2x_1 + 5x_2 - 3x_3 - 8x_4 = 3,$$
$$8x_1 + 2x_2 - x_3 + 2x_4 = -2,$$
$$x_1 - 8x_2 + x_3 - 2x_4 = 4.$$

Solution. The above system is represented as the following augmented matrix:

$$
\begin{array}{cccc|c}
2.000 & -1.000 & 7.000 & -1.000 & -1.000 \\
-2.000 & 5.000 & -3.000 & -8.000 & 3.000 \\
8.000 & 2.000 & -1.000 & 2.000 & -2.000 \\
1.000 & -8.000 & 1.000 & -2.000 & 5.000
\end{array}
$$

Operations with Respect to the First Row. The row operation is $R_i \leftarrow R_i - pR_1$ for the rows $i = 2, 3, 4$. The pivot element of the first row is $a_{11} = 2$, and, therefore, $p = a_{i1}/a_{11}$. All elements in the second, third, and fourth rows are reduced to their corresponding values using the relationships $a_{ij} \leftarrow a_{ij} - pa_{1j}$ and $b_i \leftarrow b_i - pb_1$ for the columns $j = 1, 2, 3, 4$.

$$
\begin{array}{cccc|c}
2.000 & -1.000 & 7.000 & -1.000 & -1.000 \\
0.000 & 4.000 & 4.000 & -9.000 & 2.000 \\
0.000 & 6.000 & -29.000 & 6.000 & 2.000 \\
0.000 & -7.500 & -2.500 & -1.500 & 5.500
\end{array}
$$

Operations with Respect to the Second Row. The row operation is $R_i \leftarrow R_i - pR_2$ for the rows $i = 3, 4$. The pivot element of the second row is $a_{22} = 4$, and, therefore, $p = a_{i2}/a_{22}$. All elements in the third and fourth rows are reduced to their corresponding values using the relationships $a_{ij} \leftarrow a_{ij} - pa_{2j}$ and $b_i \leftarrow b_i - pb_2$ for the columns $j = 1, 2, 3, 4$.

$$
\begin{array}{cccc|c}
2.000 & -1.000 & 7.000 & -1.000 & -1.000 \\
0.000 & 4.000 & 4.000 & -9.000 & 2.000 \\
0.000 & 0.000 & -35.000 & 19.500 & -1.000 \\
0.000 & 0.000 & 5.000 & -18.375 & 8.250
\end{array}
$$

Operations with Respect to the Third Row. The row operation is $R_i \leftarrow R_i - pR_3$ for the rows $i = 4$. The pivot element of the third row is $a_{33} = -35$, and, therefore, $p = a_{i3}/a_{33}$. All elements in the fourth row are reduced to their corresponding values using the relationships $a_{ij} \leftarrow a_{ij} - pa_{3j}$ and $b_i \leftarrow b_i - pb_3$ for the columns $j = 1, 2, 3, 4$.

$$
\begin{array}{cccc|c}
2.000 & -1.000 & 7.000 & -1.000 & -1.000 \\
0.000 & 5.000 & 5.000 & -9.000 & 2.000 \\
0.000 & 0.000 & -35.000 & 19.500 & -1.000 \\
0.000 & 0.000 & 0.000 & -15.589 & 8.107
\end{array}
$$

Row operations on A and b produce U and v, respectively, where

$$
U = \begin{bmatrix}
2 & -1 & 7 & -1 \\
0 & 4 & 4 & -9 \\
0 & 0 & -35 & 19.5 \\
0 & 0 & 0 & -15.589
\end{bmatrix} \quad \text{and} \quad v = \begin{bmatrix}
-1 \\
2 \\
-1 \\
8.107
\end{bmatrix}.
$$

Backward Substitutions. The reduced system of linear equations after row operations is

$$\begin{bmatrix} 2 & -1 & 7 & -1 \\ 0 & 4 & 4 & -9 \\ 0 & 0 & -35 & 19.5 \\ 0 & 0 & 0 & -15.589 \end{bmatrix} \begin{bmatrix} x_1 \\ x_2 \\ x_3 \\ x_4 \end{bmatrix} = \begin{bmatrix} -1 \\ 2 \\ -1 \\ 8.107 \end{bmatrix}.$$

This system is solved by applying backward substitutions to get

$$x_4 = v_4/u_{44} = -0.520,$$
$$x_3 = [v_3 - (u_{34}*x_4)]/u_{33} = -0.261,$$
$$x_2 = [v_2 - (u_{24}*x_4 + u_{23}*x_3)]/u_{22} = -0.409,$$
$$x_1 = [v_1 - (u_{14}*x_4 + u_{13}*x_3 + u_{12}*x_2)]/u_{11} = -0.050.$$

The C++ code for this method follows from the same coding for reducing a square matrix to its upper triangular form and for finding the inverse of a matrix in Sections 3.3 and 3.4, respectively. Row operations reduce $A \mid b$ to $U \mid v$, as shown below:

```
double Sum,m;
for (k=1; k<=n-1; k++)
     for (i=k+1; i<=n; i++)
     {
          p=a[i][k]/a[k][k];
          for (j=1; j<=n; j++)
               a[i][j] -= p*a[k][j];
          b[i] -= p*b[k];
     }
```

This is followed by backward substitutions:

```
for (i=n; i>=1; i--)
{
     Sum=0;
     x[i]=0;
     for (j=i ;j<=n; j++)
          Sum += a[i][j]*x[j];
     x[i]=(b[i]-Sum)/a[i][i];
}
```

Code5A.cpp solves a system of linear equations using the Gaussian elimination method. The data for the system is obtained from an external file called Code5A.in.

Code5A.cpp: Gaussian elimination method

```
#include <fstream.h>
#include <iostream.h>
#define n 4
```

```
void main()
{
    int i,j,k;
    double *x,*b,**a;
    for (i=0;i<=n;i++)
    {
        x=new double [n+1];
        b=new double [n+1];
        a=new double *[n+1];
        for (j=0;j<=n;j++)
            a[j]=new double [n+1];
    }
    ifstream InFile("Code5A.in");
    cout.setf(ios::fixed);
    cout.precision(5);
    cout << "Our input data: " << endl;
    for (i=1;i<=n;i++)
    {
        cout << "a: ";
        for (j=1;j<=n;j++)
        {
            InFile >> a[i][j];
            cout << a[i][j] << " ";
        }
        InFile >> b[i];
        cout << "b=" << b[i] << " " << endl;
    }
    InFile.close();

    // row operations
    double Sum,p;
    for (k=1;k<=n-1;k++)
        for (i=k+1;i<=n;i++)
        {
            p=a[i][k]/a[k][k];
            for (j=1;j<=n;j++)
                a[i][j]-=p*a[k][j];
            b[i] -= p*b[k];
        }

    // backward substitutions
    for (i=n;i>=1;i--)
    {
        Sum=0;
        x[i]=0;
```

```
        for (j=i;j<=n;j++)
              Sum += a[i][j]*x[j];
        x[i]=(b[i]-Sum)/a[i][i];
    }
        cout << endl;

    // display results
    cout << "results after row operations:" << endl;
    for (i=1; i<=n; i++)
    {
        cout << "u: ";
        for (j=1;j<=n;j++)
              cout << a[i][j] << " ";
        cout << "v=" << b[i] << " " << endl;
    }
    cout << endl << "results after backward substitutions: x=";
    for (i=1; i<=n; i++)
        cout << x[i] << " ";
    cout << endl;
    for (i=0;i<=n;I++)
        delete a[i];
    delete a,x,b;
}
```

Gaussian Elimination Method with Partial Pivoting

Very often, the diagonal elements of a matrix have values equal or close to zero. A matrix with this feature is called an *ill-conditioned matrix*. Division on a zero value produces an undefined number. Division on a small number whose value is very close to zero is the inverse of the division process that produces a very large number. Such a number does not fit to be a pivot element in the row operations as the result affects the accuracy of the division significantly.

An ill-conditioned matrix is treated differently when the Gaussian elimination method is applied. To avoid division by zero or a number close to this value, *partial pivoting* on the diagonal element is performed. Partial pivoting is a technique of interchanging a row whose pivot element has a value equal to or close to zero with another row whose element in the same column as the pivot element has the largest modulus value. Therefore, partial pivoting requires an additional step of interchanging two rows before row operations are performed.

Consider the following system:

$$3x_1 + 5x_2 + 1 = 5,$$
$$3x_1 + 5x_2 - x_3 = 2,$$
$$6x_1 - 2x_2 - 3x_3 = -1.$$

The augmented matrix $A \mid b$ is

$$
\begin{array}{ccc|c}
3 & 5 & 1 & 5 \\
3 & 5 & -1 & 2 \\
-6 & -2 & -3 & -1
\end{array}
$$

With the Gaussian elimination method, this augmented matrix reduces to the following form with $R_2 \leftarrow R_2 - R_1$:

$$
\begin{array}{ccc|c}
3 & 5 & 1 & 5 \\
0 & 0 & -2 & -3 \\
-6 & -2 & -3 & -1
\end{array}
$$

Continuing the next row operation with a_{22} as the pivot element will be disastrous as this element is zero, and $p = \frac{a_{32}}{a_{22}}$ is undefined. Therefore, Gaussian elimination fails to solve this system. However, rearranging the rows by exchanging rows 1 and 3 manages to avoid this problem, as follows:

$$
\begin{array}{ccc|c}
-6 & -2 & -3 & -1 \\
3 & 5 & -1 & 2 \\
3 & 5 & 1 & 5
\end{array}
$$

Reduction with $R_2 \leftarrow R_2 + 2R_1$ produces

$$
\begin{array}{ccc|c}
-6 & -2 & -3 & -1 \\
0 & 1 & -7 & 0 \\
3 & 5 & 1 & 5
\end{array}
$$

which rectifies the above problem. We discuss an example here.

Example 5.5. Solve the following system of linear equations using the Gaussian elimination method with partial pivoting:

$$
\begin{aligned}
-x_2 + 7x_3 - x_4 &= -1, \\
-2x_1 - 3x_3 - 8x_4 &= 3, \\
8x_1 + 2x_2 - x_3 + 2x_4 &= -2, \\
x_1 - 8x_2 + x_3 - 2x_4 &= 5.
\end{aligned}
$$

Solution. The above matrix is ill-conditioned as some pivot elements have zero values, namely a_{11} and a_{22}. Applying the Gaussian elimination method directly to solve the system will be disastrous as m is undefined due to division by these zero values. Let's try on the partial pivoting technique. The augmented matrix for this system is shown as follows:

$$
\begin{array}{cccc|c}
\multicolumn{4}{c}{A} & b \\
0.000 & -1.000 & 7.000 & -1.000 & -1.000 \\
-2.000 & 0.000 & -3.000 & -8.000 & 3.000 \\
8.000 & 2.000 & -1.000 & 2.000 & -2.000 \\
1.000 & -8.000 & 1.000 & -2.000 & 5.000
\end{array}
$$

Test for the Dominance of the Pivot Element in Row 1. The modulus of the pivot element in the first row is not the largest as $|a_{11}| < |a_{31}|$. Therefore, an interchange between rows 1 and 3 is performed.

$$
\begin{array}{cccc|c}
8.000 & 2.000 & -1.000 & 2.000 & -2.000 \\
-2.000 & 0.000 & -3.000 & -8.000 & 3.000 \\
0.000 & -1.000 & 7.000 & -1.000 & -1.000 \\
1.000 & -8.000 & 1.000 & -2.000 & 5.000
\end{array}
$$

Operations with Respect to the First Row. The row operation is $R_i \leftarrow R_i - pR_1$, where $a_{11} = 8$ is the pivot element of the first row and $m = a_{i1}/a_{11}$ for the rows $i = 2, 3, 4$. All elements in the second, third, and fourth rows are reduced to their corresponding values using the relationships $a_{ij} \leftarrow a_{ij} - pa_{1j}$ and $b_i \leftarrow b_i - pb_1$ for the columns $j = 1, 2, 3, 4$.

$$
\begin{array}{cccc|c}
8.000 & 2.000 & -1.000 & 2.000 & -2.000 \\
0.000 & 0.500 & -3.250 & -7.500 & 2.500 \\
0.000 & -1.000 & 7.000 & -1.000 & -1.000 \\
0.000 & -8.250 & 1.125 & -2.250 & 5.250
\end{array}
$$

Test for the Dominance of the Pivot Element in Row 2. The modulus of the pivot element in the second row is not the largest as $|a_{22}| < |a_{42}|$. Therefore, an interchange between rows 2 and 4 is performed.

$$
\begin{array}{cccc|c}
8.000 & 2.000 & -1.000 & 2.000 & -2.000 \\
0.000 & -8.250 & 1.125 & -2.250 & 5.250 \\
0.000 & -1.000 & 7.000 & -1.000 & -1.000 \\
0.000 & 0.500 & -3.250 & -7.500 & 2.500
\end{array}
$$

Operations with Respect to the Second Row. The row operation is $R_i \leftarrow R_i - pR_2$, where $a_{22} = -8.250$ is the pivot element of the second row, and therefore, $p = a_{i2}/a_{22}$

for the rows $i = 3, 4$. All elements in the third and fourth rows are reduced to their corresponding values using the relationships $a_{ij} \leftarrow a_{ij} - pa_{2j}$ and $b_i \leftarrow b_i - pb_2$ for the columns $j = 1, 2, 3, 4$.

$$
\begin{array}{cccc|c}
8.000 & 2.000 & -1.000 & 2.000 & -2.000 \\
0.000 & -8.250 & 1.125 & -2.250 & 5.250 \\
0.000 & 0.000 & 6.864 & -0.727 & -1.515 \\
0.000 & 0.000 & -3.182 & -7.636 & 2.758
\end{array}
$$

Test for the Dominance of the Pivot Element in Row 3. No interchange of rows is performed as the modulus of the pivot element in the third row is larger than the elements below it.

Operations with Respect to the Third Row. The row operation is $R_i \leftarrow R_i - pR_3$, where $a_{33} = 6.864$ is the pivot element of the third row. Therefore, $p = a_{i3}/a_{33}$ for the row $i = 4$. All elements in the fourth row are reduced to their corresponding values using the relationships $a_{ij} \leftarrow a_{ij} - pa_{3j}$ and $b_i \leftarrow b_i - pb_3$ for the columns $j = 1, 2, 3, 4$.

$$
\begin{array}{cccc|c}
8.000 & 2.000 & -1.000 & 2.000 & -2.000 \\
0.000 & -8.250 & 1.125 & -2.250 & 5.250 \\
0.000 & 0.000 & 6.864 & -0.727 & -1.515 \\
0.000 & 0.000 & 0.000 & -7.974 & 2.055
\end{array}
$$

Row operations with partial pivoting reduce $A \mid b$. to $U \mid v$. with the following values:

$$
U = \begin{bmatrix} 8 & 2 & -1 & 2 \\ 0 & -8.25 & 1.125 & -2.25 \\ 0 & 0 & 6.864 & -0.727 \\ 0 & 0 & 0 & -7.974 \end{bmatrix} \text{ and } v = \begin{bmatrix} -2 \\ 4.25 \\ -1.515 \\ 2.055 \end{bmatrix}.
$$

Backward Substitutions. The same backward substitution technique as in the Gaussian elimination method is applied to produce the solutions, $x_4 = -0.258$, $x_3 = -0.248$, $x_2 = -0.479$, and $x_1 = -0.097$.

Gauss–Jordan Method

The Gauss–Jordan method is another elimination technique for solving the $Ax = b$ problem involving row operations. Unlike the Gaussian elimination method, the Gauss–Jordan method is wholly based on row operations and does not involve backward substitutions. The method assumes the coefficient matrix to be nonsingular and fails to produce the desired result if the matrix is singular.

The Gauss–Jordan method is implemented as follows: Reduce the coefficient matrix A into an identity matrix I by eliminating the elements in the rows and columns

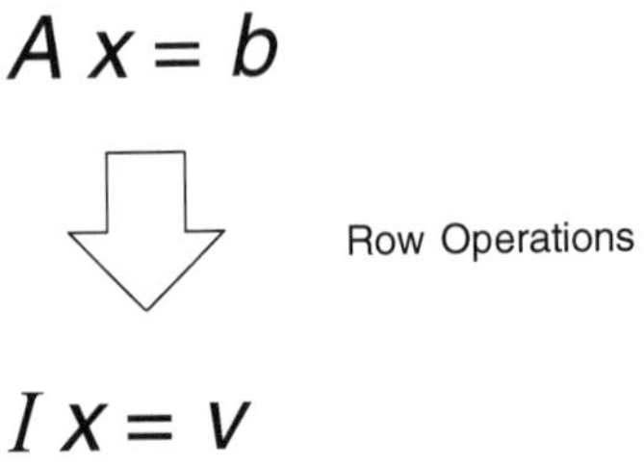

$$A x = b$$

Row Operations

$$I x = v$$

FIGURE 5.2. The single main step in the Gauss–Jordan method.

of the other diagonals. The augmented matrix $A \mid b$. reduces to $I \mid v$. through row operations where v is the final solution. This is described in Figure 5.2.

The Gauss–Jordan method is illustrated through Example 5.6.

Example 5.6. Solve the problem from Example 5.4 using the Gauss–Jordan method.

Solution. Form the augmented matrix $A \mid b$:

$$
\begin{array}{rrrr|r}
2.000 & -1.000 & 7.000 & -1.000 & -1.000 \\
-2.000 & 5.000 & -3.000 & -8.000 & 3.000 \\
8.000 & 2.000 & -1.000 & 2.000 & -2.000 \\
1.000 & -8.000 & 1.000 & -2.000 & 5.000
\end{array}
$$

Reduce a_{11} value to 1 through $R_1 \leftarrow R_1/a_{11}$.

$$
\begin{array}{rrrr|r}
1.000 & -0.500 & 3.500 & -0.500 & -0.500 \\
-2.000 & 5.000 & -3.000 & -8.000 & 3.000 \\
8.000 & 2.000 & -1.000 & 2.000 & -2.000 \\
1.000 & -8.000 & 1.000 & -2.000 & 5.000
\end{array}
$$

Eliminate a_{21}, a_{31}, and a_{41} through $R_2 \leftarrow R_2 - a_{21}R_1$, $R_3 \leftarrow R_3 - a_{31}R_1$, and $R_4 \leftarrow R_4 - a_{41}R_1$, respectively.

$$
\begin{array}{rrrr|r}
1.000 & -0.500 & 3.500 & -0.500 & -0.500 \\
0.000 & 5.000 & 5.000 & -9.000 & 2.000 \\
0.000 & 6.000 & -29.000 & 6.000 & 2.000 \\
0.000 & -7.500 & -2.500 & -1.500 & 5.500
\end{array}
$$

Reduce a_{22} to 1 through $R_2 \leftarrow R_2/a_{22}$.

$$
\begin{array}{rrrr|r}
1.000 & -0.500 & 3.500 & -0.500 & -0.500 \\
0.000 & 1.000 & 1.000 & -2.250 & 0.500 \\
0.000 & 6.000 & -29.000 & 6.000 & 2.000 \\
0.000 & -7.500 & -2.500 & -1.500 & 5.500
\end{array}
$$

Eliminate a_{12}, a_{32}, and a_{42} through $R_1 \leftarrow R_1 - a_{12}R_2$, $R_3 \leftarrow R_3 - a_{32}R_2$, and $R_4 \leftarrow R_4 - a_{42}R_2$, respectively.

1.000	0.000	5.000	−1.625	−0.250
0.000	1.000	1.000	−2.250	0.500
0.000	0.000	−35.000	19.500	−1.000
0.000	0.000	5.000	−18.375	8.250

Reduce a_{33} to 1 through $R_3 \leftarrow R_3/a_{33}$.

1.000	0.000	5.000	−1.625	−0.250
0.000	1.000	1.000	−2.250	0.500
−0.000	−0.000	1.000	−0.557	0.029
0.000	0.000	5.000	−18.375	8.250

Eliminate a_{13}, a_{23}, and a_{43} through $R_1 \leftarrow R_1 - a_{13}R_3$, $R_2 \leftarrow R_2 - a_{23}R_3$, and $R_4 \leftarrow R_4 - a_{43}R_3$, respectively.

1.000	0.000	0.000	0.604	−0.364
0.000	1.000	0.000	−1.693	0.471
0.000	0.000	1.000	−0.557	0.029
0.000	0.000	0.000	−15.589	8.107

Reduce a_{43} to 1 through $R_4 \leftarrow R_4/a_{44}$.

1.000	0.000	0.000	0.604	−0.364
0.000	1.000	0.000	−1.693	0.471
0.000	0.000	1.000	−0.557	0.029
0.000	0.000	0.000	1.000	−0.520

Eliminate a_{14}, a_{24}, and a_{34} through $R_1 \leftarrow R_1 - a_{14}R_4$, $R_2 \leftarrow R_2 - a_{24}R_4$, and $R_4 \leftarrow R_4 - a_{34}R_4$, respectively.

1.000	0.000	0.000	0.000	−0.050
0.000	1.000	0.000	0.000	−0.409
0.000	0.000	1.000	0.000	−0.261
0.000	0.000	0.000	1.000	−0.520

The final step gives the solutions as $x_1 = -0.050$, $x_2 = -0.409$, $x_3 = -0.261$, and $x_4 = -0.520$, which are the same as in Example 5.4.

5.4 LU FACTORIZATION METHODS

LU Factorization Concepts

An important feature of a nonsingular square matrix A is it can be factorized into a pair of upper and lower triangular matrices, U and L, as follows:

$$
\begin{bmatrix}
a_{11} & a_{12} & \cdots & a_{1n} \\
a_{21} & a_{22} & \cdots & a_{2n} \\
\cdots & \cdots & \cdots & \cdots \\
a_{n1} & a_{n2} & \cdots & a_{nn}
\end{bmatrix}
=
\begin{bmatrix}
l_{11} & 0 & 0 & 0 \\
l_{21} & l_{22} & 0 & 0 \\
\cdots & \cdots & \cdots & 0 \\
l_{n1} & l_{n2} & \cdots & l_{nn}
\end{bmatrix}
\begin{bmatrix}
u_{11} & u_{12} & \cdots & u_{1n} \\
0 & u_{22} & \cdots & u_{2n} \\
\cdots & \cdots & \cdots & \cdots \\
0 & 0 & \cdots & u_{nn}
\end{bmatrix}.
\tag{5.6}
$$

The factorization into triangular matrices helps in solving the linear equations as the unknowns can be directly determined through forward or backward substitutions. It can be generalized that for $A_{n \times n}$, the number of unknowns in the matrices L and U produced from the factorization of A is n^2.

This is illustrated using an example on a 4×4 matrix A, as follows:

$$A = LU, \text{ where } L = \left[l_{ij} \right], \ l_{ij} = 0 \text{ for } i < j, \text{ and } u_{ij} = 0 \text{ for } i > j.$$

$$
\begin{bmatrix}
a_{11} & a_{12} & a_{13} & a_{14} \\
a_{21} & a_{22} & a_{23} & a_{24} \\
a_{31} & a_{32} & a_{33} & a_{34} \\
a_{41} & a_{42} & a_{43} & a_{44}
\end{bmatrix}
=
\begin{bmatrix}
l_{11} & 0 & 0 & 0 \\
l_{21} & l_{22} & 0 & 0 \\
l_{31} & l_{32} & l_{33} & 0 \\
l_{41} & l_{42} & l_{43} & l_{44}
\end{bmatrix}
\begin{bmatrix}
u_{11} & u_{12} & u_{13} & u_{14} \\
0 & u_{22} & u_{23} & u_{24} \\
0 & 0 & u_{33} & u_{34} \\
0 & 0 & 0 & u_{44}
\end{bmatrix}
$$

$$
=
\begin{bmatrix}
l_{11}u_{11} & l_{11}u_{12} & l_{11}u_{13} & l_{11}u_{14} \\
l_{21}u_{11} & l_{21}u_{12} + l_{22}u_{22} & l_{21}u_{13} + l_{22}u_{23} & l_{21}u_{14} + l_{22}u_{24} \\
l_{31}u_{11} & l_{31}u_{12} + l_{32}u_{22} & l_{31}u_{13} + l_{32}u_{23} + l_{33}u_{33} & l_{31}u_{14} + l_{32}u_{24} + l_{33}u_{34} \\
l_{41}u_{11} & l_{41}u_{12} + l_{42}u_{22} & l_{41}u_{13} + l_{42}u_{23} + l_{43}u_{33} & l_{41}u_{14} + l_{42}u_{24} + l_{43}u_{34} + l_{44}u_{44}
\end{bmatrix}.
$$

The values u_{ij} and l_{ij} for $i, j = 1, 2, 3, 4$ can be obtained by comparing the two matrices element by element. Comparing the terms one by one, we have $a_{11} = l_{11}u_{11}$, $a_{12} = l_{11}u_{12}$, and so on. From these relationships, the values of all elements in L and U can be determined.

We now discuss the solution to the LU factorization technique. The objective is to solve the system of linear equations given by $Ax = b$. Factorize $A = LU$ to get

$$(LU)x = b.$$

This simplifies into

$$L(Ux) = b.$$

Let $Ux = w$, where $w = [w_i]$ for $i = 1, 2, \ldots, n$ is a temporary vector. This gives $Lw = b$, where the values in w can be determined through forward substitutions. The values of x are then obtained from backward substitutions in $Ux = w$.

The general factorization technique in $A = LU$ is simplified through four main methods for solving several different cases of the system of linear equations. Basically, these four methods differ in their representation in the triangular matrices, as follows:

1. If $u_{ii} = 1$ for $i = 1, 2, \ldots, n$ in matrix U, then the method is called the Crout method.

2. If $l_{ii} = 1$ for $i = 1, 2, \ldots, n$ in matrix L, then the method is called the Doolittle method.

3. If $U = L^T$ for the positive-definite matrix A, then the method is called the Cholesky method.

4. The Thomas algorithm is a special adaptation of the Crout method when A is a tridiagonal matrix.

We discuss each method and its C++ implementation.

Crout Method

The Crout method simplifies the whole factorization process by setting the values of all diagonal elements of matrix U equal to 1; that is, $u_{ii} = 1$ for $i = 1, 2, \ldots, n$. This technique had the advantage over the Gaussian elimination method where the number of unknowns in L and U is $n^2 - n$, which is lower.

Figure 5.3 shows a schematic flowchart of the Crout method. The values of the elements in L and U are determined by comparing the two matrices, A and LU, element by element starting from the top row downward from left to right. These values can be formulated as follows:

$$l_{ij} = a_{ij} - \sum_{k=1}^{j-1} l_{ik} u_{kj}, \qquad (5.7a)$$

$$u_{ij} = \left(a_{ij} - \sum_{k=1}^{i-1} l_{ik} u_{kj} \right) \Big/ l_{ii}. \qquad (5.7b)$$

Example 5.7. Solve the system from Example 5.4 using the Crout method.

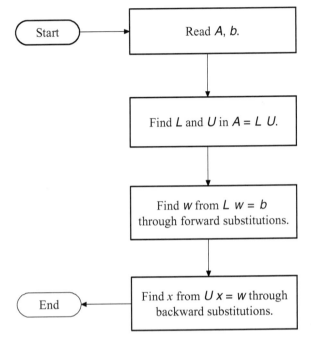

FIGURE 5.3. Schematic flowchart of the Crout and Doolittle methods.

Solution. Let $A = LU$, or

$$
\begin{bmatrix}
2 & -1 & 7 & -1 \\
-2 & 5 & -3 & -8 \\
8 & 2 & -1 & 2 \\
1 & -8 & 1 & -2
\end{bmatrix}
=
\begin{bmatrix}
l_{11} & 0 & 0 & 0 \\
l_{21} & l_{22} & 0 & 0 \\
l_{31} & l_{32} & l_{33} & 0 \\
l_{41} & l_{42} & l_{43} & l_{44}
\end{bmatrix}
\begin{bmatrix}
1 & u_{12} & u_{13} & u_{14} \\
0 & 1 & u_{23} & u_{24} \\
0 & 0 & 1 & u_{34} \\
0 & 0 & 0 & 1
\end{bmatrix}
$$

$$
=
\begin{bmatrix}
l_{11} & l_{11}u_{12} & l_{11}u_{13} & l_{11}u_{14} \\
l_{21} & l_{21}u_{12}+l_{22} & l_{21}u_{13}+l_{22}u_{23} & l_{21}u_{14}+l_{22}u_{24} \\
l_{31} & l_{31}u_{12}+l_{32} & l_{31}u_{13}+l_{32}u_{23}+l_{33} & l_{31}u_{14}+l_{32}u_{24}+l_{33}u_{34} \\
l_{41} & l_{41}u_{12}+l_{42} & l_{41}u_{13}+l_{42}u_{23}+l_{43} & l_{41}u_{14}+l_{42}u_{24}+l_{43}u_{34}+l_{44}
\end{bmatrix}.
$$

By matching the resultant matrix from the multiplication with the original A matrix element by element, we get the following L and U matrices:

$$
L =
\begin{bmatrix}
2 & 0 & 0 & 0 \\
-2 & 4 & 0 & 0 \\
8 & 6 & -35 & 0 \\
1 & -7.5 & 5 & -15.589
\end{bmatrix}
\quad \text{and} \quad
U =
\begin{bmatrix}
1 & -0.5 & 3.5 & -0.5 \\
0 & 1 & 1 & -2.25 \\
0 & 0 & 1 & -0.557 \\
0 & 0 & 0 & 1
\end{bmatrix}.
$$

The vector w is obtained from $Lw = b$ through forward substitutions:

$$\begin{bmatrix} 2 & 0 & 0 & 0 \\ -2 & 4 & 0 & 0 \\ 8 & 6 & -35 & 0 \\ 1 & -7.5 & 5 & -15.589 \end{bmatrix} \begin{bmatrix} w_1 \\ w_2 \\ w_3 \\ w_4 \end{bmatrix} = \begin{bmatrix} -1 \\ 3 \\ -2 \\ 4 \end{bmatrix},$$

which is the same as

$$2w_1 = -1,$$
$$-2w_1 + 4w_2 = 3,$$
$$8w_1 + 6w_2 - 35w_3 = -2,$$
$$w_1 - 7.5w_2 + 5w_3 - 15.589w_4 = 4,$$

to produce $w_1 = -0.5, w_2 = 0.5, w_3 = 0.029$, and $w_4 = -0.52$. Finally, x is obtained from $Ux = w$ through backward substitutions, as follows:

$$\begin{bmatrix} 1 & -0.5 & 3.5 & -0.5 \\ 0 & 1 & 1 & -2.25 \\ 0 & 0 & 1 & -0.557 \\ 0 & 0 & 0 & 1 \end{bmatrix} \begin{bmatrix} x_1 \\ x_2 \\ x_3 \\ x_4 \end{bmatrix} = \begin{bmatrix} -0.5 \\ 0.5 \\ 0.029 \\ -0.520 \end{bmatrix}.$$

This produces the solutions, $x_4 = -0.520$, $x_3 = -0.261$, $x_2 = -0.409$, and $x_1 = -0.050$.

We discuss the C++ implementation of the Crout method. The main routine in this method is in factorizing $A = LU$. The code is obtained by comparing the left-hand and right-hand sides of the equations, element by element, to produce:

```
for (i=1;i<=n;i++)
      l[i][1]=a[i][1];
for (j=2;j<=n;j++)
      u[1][j]=a[1][j]/l[1][1];
for (j=2; j<=n-1;j++)
{
      for (i=j; i<=n; i++)
      {
            z=0;
            for (k=1;k<=j-1;k++)
                  z += l[i][k]*u[k][j];
                  l[i][j]=a[i][j]- z;
      }
      for (k=j+1;k<=n;k++)
      {
            z=0;
            for (r=1;r<=j-1;r++)
                  z += l[j][r]*u[r][k];
```

```
                    u[j][k]=(a[j][k]-z)/l[j][j];
        }
}
z=0;
for (k=1;k<=n-1;k++)
        z += l[n][k]*u[k][n];
l[n][n]=a[n][n]-z;
```

The next step is to apply forward substitutions into the equation $Lw = b$ using the given values in b and the computed values in L. The code is shown as follows:

```
w[1]=b[1]/l[1][1];
for (j=1; j<=n; j++)
{
        z=0;
        for (k=1; k<=j-1;k++)
                z += l[j][k]*w[k];
        w[j] =(b[j]-z)/l[j][j];
}
```

Finally, we obtain the solution x through backward substitutions from $Ux = w$, as follows:

```
for (i=n; i>=1; i--)
{
        z=0;
        for (k=i+1; k<=n; k++)
                z += u[i][k]*x[k];
        x[i]=w[i]-z;
        cout << "x[" << i << "]=" << x[i] << endl;
}
```

Code5B.cpp shows a complete C++ code for the Crout method, which reads input from Code5B.in.

```
Code5B.cpp: Crout Method
#include <iostream.h>
#include <fstream.h>
#define n 4

void main()
{
        int i,j,k,r;
        double z,*x,*w,*b,**a,**l,**u;
        for (i=0;i<=n;i++)
        {
                x=new double [n+1];
                w=new double [n+1];
```

```
       b=new double [n+1];
       a=new double *[n+1];
       l=new double *[n+1];
       u=new double *[n+1];
       for (j=0;j<=n;j++)
       {
              a[j]=new double [n+1];
              l[j]=new double [n+1];
            . u[j]=new double [n+1];
       }
}
cout.setf(ios::fixed);
cout.precision(5);

// read the input data
ifstream InFile("Code5B.in");
for (i=1;i<=n;i++)
{
       for (j=1;j<=n;j++)
       {
              InFile >> a[i][j];
              l[i][j]=u[i][j]=0;
              if (i==j)
                     u[i][j]=1;
       }
       InFile >> b[i];
}
InFile.close();

// compute L and U
for (i=1;i<=n;i++)
       l[i][1]=a[i][1];
for (j=2;j<=n;j++)
       u[1][j]=a[1][j]/l[1][1];

for (j=2; j<=n-1;j++)
{
       for (i=j; i<=n; i++)
       {
              z=0;
              for (k=1;k<=j-1;k++)
                     z += l[i][k]*u[k][j];
              l[i][j]=a[i][j]- z;
       }
       for (k=j+1;k<=n;k++)
       {
```

```
                z=0;
                for (r=1;r<=j-1;r++)
                        z += l[j][r]*u[r][k];
                u[j][k]=(a[j][k]-z)/l[j][j];
        }
}
z=0;
for (k=1;k<=n-1;k++)
        z += l[n][k]*u[k][n];
l[n][n]=a[n][n]-z;

cout << endl << "L matrix:" << endl;
for (i=1;i<=n;i++)
{
        for (j=1;j<=n;j++)
                cout << l[i][j] << " ";
        cout << endl;
}
cout << endl;

cout << endl << "U matrix:" << endl;
for (i=1;i<=n;i++)
{
        for (j=1;j<=n;j++)
                cout << u[i][j] << " ";
        cout << endl;
}
cout << endl;

// forward substitutions for finding w
w[1]=b[1]/l[1][1];
for (j=1;j<=n;j++)
{
        z=0;
        for (k=1; k<=j-1;k++)
                z += l[j][k]*w[k];
        w[j] =(b[j]-z)/l[j][j];
}
cout << endl << "w vector:" << endl;
for (i=1;i<=n;i++)
{
        cout << w[i] << endl;
}
cout << endl;

// find x through backward substitutions
```

```
for (i=n;i>=1;i--)
{
        z=0;
        for (k=i+1; k<=n; k++)
                z += u[i][k]*x[k];
        x[i]=w[i]-z;
        cout << "x[" << i << "]=" << x[i] << endl;
}

// deallocate all used arrays
for (i=0;i<=n;i++)
        delete a[i],l[i],u[i];
delete x,w,b,a,l,u;
}
```

Doolittle Method

The Doolittle method is the other side of Crout, where the values of the diagonal elements in the lower triangular matrix are set to 0, or $l_{ii} = 1$, for $i = 1, 2, \ldots, n$ in factorizing $A = LU$. Doolittle shares the same flowchart as Crout in terms of execution, as shown in Figure 5.3.

As in Crout, the values of the elements in L and U are evaluated by comparing matrix A with the product LU, element by element starting from the top row, from left to right. The rest of the steps in Doolittle are exactly similar to that of Crout: find the values in w through forward substitutions from the equation $Lw = b$, and then get the solution x from $Ux = w$ through backward substitutions.

Example 5.8. Solve the system from Example 5.4 using the Doolittle method.

Solution.

$$
\begin{bmatrix} 2 & -1 & 7 & -1 \\ -2 & 5 & -3 & -8 \\ 8 & 2 & -1 & 2 \\ 1 & -8 & 1 & -2 \end{bmatrix} = \begin{bmatrix} 1 & 0 & 0 & 0 \\ l_{21} & 1 & 0 & 0 \\ l_{31} & l_{32} & 1 & 0 \\ l_{41} & l_{42} & l_{43} & 1 \end{bmatrix} \begin{bmatrix} u_{11} & u_{12} & u_{13} & u_{14} \\ 0 & u_{22} & u_{23} & u_{24} \\ 0 & 0 & u_{33} & u_{34} \\ 0 & 0 & 0 & u_{44} \end{bmatrix}
$$

$$
= \begin{bmatrix} u_{11} & u_{12} & u_{13} & u_{14} \\ l_{21}u_{11} & l_{21}u_{12} + u_{22} & l_{21}u_{13} + u_{23} & l_{21}u_{14} + u_{24} \\ l_{31}u_{11} & l_{31}u_{12} + l_{32}u_{22} & l_{31}u_{13} + l_{32}u_{23} + u_{33} & l_{31}u_{14} + l_{32}u_{24} + u_{34} \\ l_{41}u_{11} & l_{41}u_{12} + l_{42}u_{22} & l_{41}u_{13} + l_{42}u_{23} + l_{43}u_{33} & l_{41}u_{14} + l_{42}u_{24} + l_{43}u_{34} + u_{44} \end{bmatrix}.
$$

Direct comparison between the two matrices produces L and U, as follows:

$$
L = \begin{bmatrix} 1 & 0 & 0 & 0 \\ -1 & 1 & 0 & 0 \\ 4 & 1.5 & 1 & 0 \\ 0.5 & -1.875 & 0.143 & 1 \end{bmatrix} \text{ and } U = \begin{bmatrix} 2 & -1 & 7 & -1 \\ 0 & 4 & 4 & -9 \\ 0 & 0 & -35 & 19.5 \\ 0 & 0 & 0 & 15.589 \end{bmatrix}.
$$

From $Lw = b$, the temporary matrix $w = [-1 \quad 2 \quad -1 \quad 8.107]^T$ is obtained through forward substitutions. Finally, applying backward substitutions into $Ux = w$. we get the solution $x = [-0.050 \quad 0.409 \quad 0.261 \quad 0.520]^T$.

Cholesky Method

The Cholesky method is a method that can be applied to a system of linear equations whose coefficient matrix is a special type of matrix called *positive-definite*. In most cases, the method will not work if this requirement is not fulfilled. However, the method may work in some exceptional cases where the coefficient matrix is symmetric but not positive definite.

A square matrix $A = [a_{ij}]$ is said to be symmetric if $A = A^T$, or $a_{ij} = a_{ji}$. A symmetric matrix is one where the elements in its ith column have equal entries as the ith row in the matrix. This type of matrix is common in many applications.

Definition 5.3. A square symmetric matrix A of size $n \times n$ is said to be positive definite if it is symmetric, and the matrix satisfies the following requirement:

$$x^T Ax > 0, \text{ where } x = [x_1 \, x_2 \cdots x_n]^T \text{ is any } n \times 1 \text{ vector.}$$

Definition 5.3 is an analytical method for determining whether a given square matrix is positive definite. The expression $x^T Ax$ in the definition produces a *characteristic polynomial*, $P(x) > 0$.

Example 5.9. Determine whether the following matrix is positive definite:

$$A = \begin{bmatrix} 2 & -1 & 0 \\ -1 & 2 & -1 \\ 0 & -1 & 2 \end{bmatrix}.$$

Solution. Let $x = [x_1 \quad x_2 \quad x_3]^T$, and then

$$x^T Ax = \begin{bmatrix} x_1 & x_2 & x_3 \end{bmatrix} \begin{bmatrix} 2 & -1 & 0 \\ -1 & 2 & -1 \\ 0 & -1 & 2 \end{bmatrix} \begin{bmatrix} x_1 \\ x_2 \\ x_3 \end{bmatrix}$$

$$= \begin{bmatrix} x_1 & x_2 & x_3 \end{bmatrix} \begin{bmatrix} 2x_1 - x_2 \\ -x_1 + 2x_2 - x_3 \\ -x_2 + 2x_3 \end{bmatrix}$$

$$= 2x_1^2 - 2x_1x_2 + 2x_2^2 - 2x_2x_3 + 2x_3^2$$

$$= x_1^2 + (x_1 - x_2)^2 + (x_2 - x_3)^2 + x_3^2 > 0.$$

Therefore, A is positive definite.

In most cases, the characteristic polynomial method of determining the positive-definiteness of a matrix may not be easy to implement especially when the size of the matrix is large. The difficulty arises in converting the given equation into the form of sum of squares. Alternative computational approaches using the properties of a positive-definite matrix can also be applied, as follows:

Method 1. A square matrix A is positive definite if all conditions as stated below are satisfied:

1. A is symmetric; that is, $a_{ij} = a_{ji}$ for $i, j = 1, 2, \ldots, n$.
2. All the diagonal elements in A are positive, or $a_{ii} > 0$ for $i = 1, 2, \ldots, n$.
3. The largest absolute value in each row of A is the diagonal element.
4. $a_{ii}a_{jj} > a_{ij}^2$ for $i, j = 1, 2, \ldots, n$.

It can be shown that, from Example 5.10, all four conditions above are satisfied to prove that A is positive definite.

Method 2. All diagonal elements of the matrix U reduced from the symmetric matrix A through row operations are positive.

A reduces to U through row operations, as follows:

$$
\begin{bmatrix} 2 & -1 & 0 \\ -1 & 2 & -1 \\ 0 & -1 & 2 \end{bmatrix} \sim \begin{bmatrix} 2 & -1 & 0 \\ 0 & 3 & -2 \\ 0 & 0 & 4 \end{bmatrix}.
$$

A is positive definite since $u_{ii} > 0$ for $i = 1, 2, 3$.

Method 3. All eigenvalues of the square matrix A are real and positive.

Figure 5.4 shows the schematic flowchart of the method. The method starts by assuming $U = L^T$ in the equation $A = LU$. Hence, the Cholesky method requires finding the unknowns in L only. This assumption reduces the number of unknowns to $1 + 2 + \cdots + n$, which is lower than that in Crout and Doolittle. For example, the number of variables when A is a 4×4 matrix is only 10, against 16 in the previous two methods.

As in Crout and Doolittle, the elements in L and U are found by directly comparing A with LU. In summary, these values are obtained using the following equations:

$$
l_{kk} = \sqrt{a_{kk} - \sum_{j=1}^{k-1} l_{kj}^2} \text{ for } k = 1, 2, \ldots, n, \tag{5.8a}
$$

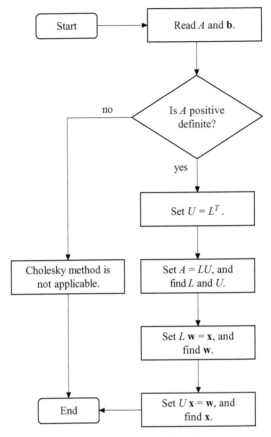

FIGURE 5.4. Schematic flowchart of the Cholesky method.

and

$$l_{ki} = \frac{a_{ki} - \sum_{j=1}^{i-1} l_{ij} l_{kj}}{l_{ii}} \text{ for } i = 1, 2, \dots, k-1. \tag{5.8b}$$

Once L and U have been found, the same forward and backward substitutions steps as in Crout and Doolittle are applied before arriving at the solution.

Example 5.10. Solve the following system of linear equations using the Cholesky method:

$$\begin{bmatrix} 6 & 2 & 1 & -1 \\ 2 & 4 & 1 & 0 \\ 1 & 1 & 4 & -1 \\ -1 & 0 & -1 & 3 \end{bmatrix} \begin{bmatrix} x_1 \\ x_2 \\ x_3 \\ x_4 \end{bmatrix} = \begin{bmatrix} 2 \\ -3 \\ 1 \\ -1 \end{bmatrix}.$$

Solution. Set $A = LU$, and let $U = L^T$. This produces

$$
\begin{bmatrix}
6 & 2 & 1 & -1 \\
2 & 4 & 1 & 0 \\
1 & 1 & 4 & -1 \\
-1 & 0 & -1 & 3
\end{bmatrix}
=
\begin{bmatrix}
l_{11} & 0 & 0 & 0 \\
l_{21} & l_{22} & 0 & 0 \\
l_{31} & l_{32} & l_{33} & 0 \\
l_{41} & l_{42} & l_{43} & l_{44}
\end{bmatrix}
\begin{bmatrix}
l_{11} & l_{21} & l_{31} & l_{41} \\
0 & l_{22} & l_{32} & l_{42} \\
0 & 0 & l_{33} & l_{43} \\
0 & 0 & 0 & l_{44}
\end{bmatrix}
$$

$$
=
\begin{bmatrix}
l_{11}^2 & l_{11}l_{21} & l_{11}l_{31} & l_{11}l_{41} \\
l_{21}l_{11} & l_{21}^2 + l_{22}^2 & l_{21}l_{31} + l_{22}l_{32} & l_{21}l_{41} + l_{22}l_{42} \\
l_{31}l_{11} & l_{31}l_{21} + l_{32}l_{22} & l_{31}^2 + l_{32}^2 + l_{33}^2 & l_{31}l_{41} + l_{32}l_{42} + l_{33}l_{43} \\
l_{41}l_{11} & l_{41}l_{21} + l_{42}l_{22} & l_{41}l_{31} + l_{42}l_{32} + l_{43}l_{33} & l_{41}^2 + l_{42}^2 + l_{43}^2 + l_{44}^2
\end{bmatrix}.
$$

All the elements in the left and right matrices are compared to produce

$$
L =
\begin{bmatrix}
2.449 & 0 & 0 & 0 \\
0.816 & 1.826 & 0 & 0 \\
0.408 & 0.365 & 1.924 & 0 \\
-0.408 & 0.183 & -0.468 & 1.607
\end{bmatrix},
$$

$$
U = L^T =
\begin{bmatrix}
2.449 & 0.816 & 0.408 & -0.408 \\
0 & 1.826 & 0.365 & 0.183 \\
0 & 0 & 1.924 & -0.468 \\
0 & 0 & 0 & 1.607
\end{bmatrix}.
$$

Applying forward substitutions on $Lw = b$:

$$
\begin{bmatrix}
2.449 & 0 & 0 & 0 \\
0.816 & 1.826 & 0 & 0 \\
0.408 & 0.365 & 1.924 & 0 \\
-0.408 & 0.183 & -0.468 & 1.607
\end{bmatrix}
\begin{bmatrix}
w_1 \\
w_2 \\
w_3 \\
w_4
\end{bmatrix}
=
\begin{bmatrix}
2 \\
-3 \\
1 \\
-1
\end{bmatrix},
$$

we get $w = [0.816 \quad -2.008 \quad 0.728 \quad 0.025]^T$. Finally, the last step is to perform backward substitutions on $Ux = w$:

$$
\begin{bmatrix}
2.449 & 0.816 & 0.408 & -0.408 \\
0 & 1.826 & 0.365 & 0.183 \\
0 & 0 & 1.924 & -0.468 \\
0 & 0 & 0 & 1.607
\end{bmatrix}
\begin{bmatrix}
x_1 \\
x_2 \\
x_3 \\
x_4
\end{bmatrix}
=
\begin{bmatrix}
0.816 \\
-2.008 \\
0.728 \\
0.025
\end{bmatrix},
$$

and this produces the solution $w = [0.665 \quad -1.178 \quad 0.382 \quad 0.016]^T$.

We now discuss the program design for the Cholesky method. The difficulty in this method lies on finding the values of matrix L according to Equation (5.8a) and (5.8b). The code is given as follows:

```
for (k=1;k<=n;k++)
{
      Sum=0;
      for (j=1;j<=k-1;j++)
            Sum += pow(l[k][j],2);
      l[k][k]=sqrt(a[k][k]-Sum);
      for (i=1;i<=k-1;i++)
      {
            Sum=0;
            for (j=1;j<=i-1;j++)
                  Sum += l[i][j]*l[k][j];
            l[k][i]=(a[k][i]-Sum)/l[i][i];
      }
}
```

Once L is found, the other triangular matrix U is simply its transpose, or $U = [u_{ij}] = [l_{ji}]$. This is achieved by setting u[i][j]=l[j][i] in the program for all i and j. The code for the rest of the steps involving forward and backward substitutions is similar to those in Crout and Doolittle, and it will not be discussed here. Code5C.cpp shows the complete program of the Cholesky method, which reads A and b values from an external file called Code5C.in.

```
Code5C.cpp: Cholesky method.
#include <iostream.h>
#include <fstream.h>
#include <math.h>
#define n 4

void main()
{
      int i,j,k;
      double *x,*w,*b,**a,**l,**u;
      double Sum;
      for (i=0;i<=n;i++)
      {
            x=new double [n+1];
            w=new double [n+1];
            b=new double [n+1];
            a=new double *[n+1];
            l=new double *[n+1];
            u=new double *[n+1];
            for (j=0;j<=n;j++)
            {
```

```
            a[j]=new double [n+1];
            l[j]=new double [n+1];
            u[j]=new double [n+1];
        }
    }
    cout.setf(ios::fixed);
    cout.precision(5);

    // read the input data
    ifstream InFile("Code5C.in");
    for (i=1;i<=n;i++)
    {
        for (j=1;j<=n;j++)
        {
            InFile >> a[i][j];
            l[i][j]=u[i][j]=0;
        }
        InFile >> b[i];
    }
    InFile.close();

    for (k=1;k<=n;k++)
    {
        Sum=0;
        for (j=1;j<=k-1;j++)
            Sum += pow(l[k][j],2);
        l[k][k]=sqrt(a[k][k]-Sum);
        for (i=1;i<=k-1;i++)
        {
            Sum=0;
            for (j=1;j<=i-1;j++)
                Sum += l[i][j]*l[k][j];
            l[k][i]=(a[k][i]-Sum)/l[i][i];
        }
    }
    for (i=1;i<=n;i++)
        for (j=1;j<=n;j++)
            u[j][i]=l[i][j];
    cout << endl << "L matrix:" << endl;
    for (i=1;i<=n;i++)
    {
        for (j=1;j<=n;j++)
            cout << l[i][j] << " ";
        cout << endl;
    }
    cout << endl;
```

```
cout << endl << "U matrix:" << endl;
for (i=1;i<=n;i++)
{
      for (j=1;j<=n;j++)
            cout << u[i][j] << " ";
      cout << endl;
}
cout << endl;

// forward substitution for finding w
for (j=1;j<=n;j++)
{
      Sum=0;
      for (k=1; k<=j-1;k++)
            Sum += l[j][k]*w[k];
      w[j] =(b[j]-Sum)/l[j][j];
}
cout << endl << "w vector:" << endl;
for (i=1;i<=n;i++)
{
      cout << w[i] << endl;
}
cout << endl;

// find x through backward substitutions
for (i=n;i>=1;i--)
{
      Sum=0;
      for (k=i+1; k<=n; k++)
            Sum += u[i][k]*x[k];
      x[i]=w[i]-Sum;
      cout << "x[" << i << "]=" << x[i] << endl;
}

for (i=0;i<=n;i++)
      delete a[i],l[i],u[i];
delete x,w,b,a,l,u;
}
```

Thomas Algorithm

Another *LU* factorization method is the Thomas algorithm, which is a derivative of the Crout factorization method. The algorithm can only be applied in solving a system of linear equation whose coefficient matrix is a special type of matrix called a *tridiagonal matrix*. An advantage in applying the Thomas algorithm is its computational time at $O(n)$ is lower than the Gaussian elimination approach, which stands at $O(n^3)$.

Definition 5.4. A *tridiagonal matrix* is a three-band matrix whose elements other than in the main diagonal and upper and lower diagonals have the value of zero, or

$$a_{ij} = \begin{cases} 0 & \text{if } j \geq i + 2 \text{ for } i = 1, 2, \ldots, n - 2, \\ 0 & \text{if } j \leq i - 2 \text{ for } i = 3, 4, \ldots, n, \\ * & \text{for } i - 1 < j < i + 2. \end{cases}$$

In the above definition, * stands for any value including zero. An example of a tridiagonal matrix of size 5×5 is

$$\begin{bmatrix} 3 & -2 & 0 & 0 & 0 \\ 2 & 7 & -1 & 0 & 0 \\ 0 & -1 & 4 & 1 & 0 \\ 0 & 0 & 5 & -2 & 3 \\ 0 & 0 & 0 & 1 & 4 \end{bmatrix}.$$

The tridiagonal matrix is often encountered in engineering problems. A typical application is the computational fluid dynamics (CFD) problem, which involves breaking down the big problem into smaller ones by converting the continuous differential equations in the system into ones that are discrete. The discrete equations are then reduced further into one or more systems of linear equations, including some that may have the triangular form. Therefore, the task of computing a triadiagonal system of linear equations has become a challenging but rewarding exercise in numerical computing.

The Thomas algorithm starts with a three-band coefficient matrix A. This matrix is factorized and reduced into L and U, both of which are two-band. The next step is to get w from the equation $Lw = b$ through forward substitutions. Finally, the solution x is obtained from $Ux = w$ through backward substitutions.

In implementing the Thomas algorithm approach, the process is very much simplified by setting the values of all diagonal elements in U to 1; that is, $u_{11} = 1$. This same approach is deployed in the Crout method. We discuss this approach through an example.

Example 5.11. Solve the following system using the Thomas Algorithm:

$$\begin{bmatrix} 4 & -2 & 0 & 0 \\ 1 & -5 & 1 & 0 \\ 0 & 2 & 4 & 1 \\ 0 & 0 & -1 & -3 \end{bmatrix} \begin{bmatrix} x_1 \\ x_2 \\ x_3 \\ x_4 \end{bmatrix} = \begin{bmatrix} 1 \\ 2 \\ -1 \\ 3 \end{bmatrix}.$$

Solution. The coefficient matrix A is tridiagonal. Let A and its factors, L and U, be the following unknowns with α_i and β_i:

$$A = \begin{bmatrix} d_1 & e_1 & 0 & 0 \\ c_2 & d_2 & e_2 & 0 \\ 0 & c_3 & d_3 & e_3 \\ 0 & 0 & c_4 & d_4 \end{bmatrix}, \quad U = \begin{bmatrix} 1 & \beta_1 & 0 & 0 \\ 0 & 1 & \beta_2 & 0 \\ 0 & 0 & 1 & \beta_3 \\ 0 & 0 & 0 & 1 \end{bmatrix}, \quad \text{and } L = \begin{bmatrix} \alpha_1 & 0 & 0 & 0 \\ c_2 & \alpha_2 & 0 & 0 \\ 0 & c_3 & \alpha_3 & 0 \\ 0 & 0 & c_4 & \alpha_4 \end{bmatrix}.$$

In U, all diagonal elements have values equal to 1. Next, factorize $A = LU$:

$$\begin{bmatrix} d_1 & e_1 & 0 & 0 \\ c_2 & d_2 & e_2 & 0 \\ 0 & c_3 & d_3 & e_3 \\ 0 & 0 & c_4 & d_4 \end{bmatrix} = \begin{bmatrix} \alpha_1 & 0 & 0 & 0 \\ c_2 & \alpha_2 & 0 & 0 \\ 0 & c_3 & \alpha_3 & 0 \\ 0 & 0 & c_4 & \alpha_4 \end{bmatrix} \begin{bmatrix} 1 & \beta_1 & 0 & 0 \\ 0 & 1 & \beta_2 & 0 \\ 0 & 0 & 1 & \beta_3 \\ 0 & 0 & 0 & 1 \end{bmatrix},$$

$$\begin{bmatrix} 4 & -2 & 0 & 0 \\ 1 & -5 & 1 & 0 \\ 0 & 2 & 4 & 1 \\ 0 & 0 & -1 & -3 \end{bmatrix} = \begin{bmatrix} \alpha_1 & \alpha_1\beta_1 & 0 & 0 \\ c_2 & c_2\beta_1 + \alpha_2 & \alpha_2\beta_2 & 0 \\ 0 & c_3 & c_3\beta_2 + \alpha_3 & \alpha_3\beta_3 \\ 0 & 0 & c_4 & c_4\beta_3 + \alpha_4 \end{bmatrix}.$$

Comparing the two matrices element by element produces the following matrices:

$$L = \begin{bmatrix} 4 & 0 & 0 & 0 \\ 1 & -4.5 & 0 & 0 \\ 0 & 2 & 4.444 & 0 \\ 0 & 0 & -1 & -2.775 \end{bmatrix} \quad \text{and } U = \begin{bmatrix} 1 & -0.5 & 0 & 0 \\ 0 & 1 & -0.222 & 0 \\ 0 & 0 & 1 & 0.225 \\ 0 & 0 & 0 & 1 \end{bmatrix}.$$

The next step is to compute w from $Lw = b$ through forward substitutions:

$$\begin{bmatrix} 4 & 0 & 0 & 0 \\ 1 & -4.5 & 0 & 0 \\ 0 & 2 & 4.444 & 0 \\ 0 & 0 & -1 & -2.775 \end{bmatrix} \begin{bmatrix} w_1 \\ w_2 \\ w_3 \\ w_4 \end{bmatrix} = \begin{bmatrix} 1 \\ 2 \\ -1 \\ 3 \end{bmatrix}.$$

We get $w = [0.25 \quad -0.389 \quad -0.05 \quad -1.063]^T$. Finally, applying backward substitutions to $Ux = w$:

$$\begin{bmatrix} 1 & -0.5 & 0 & 0 \\ 0 & 1 & -0.222 & 0 \\ 0 & 0 & 1 & 0.225 \\ 0 & 0 & 0 & 1 \end{bmatrix} \begin{bmatrix} x_1 \\ x_2 \\ x_3 \\ x_4 \end{bmatrix} = \begin{bmatrix} 0.25 \\ -0.389 \\ -0.05 \\ -1.063 \end{bmatrix},$$

produces the solution $x = [0.077 \quad -0.347 \quad 0.189 \quad -1.063]^T$.

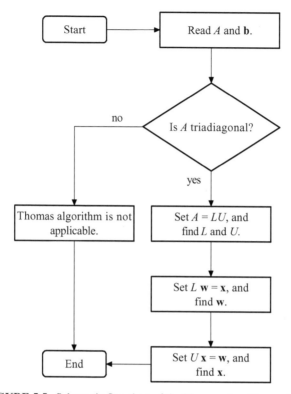

FIGURE 5.5. Schematic flowchart of the Thomas algorithm method.

The implementation of Thomas algorithm follows the same path as the previous LU factorization methods. This is shown in the schematic flowchart of Figure 5.5. The task of finding L and U, and solving for x through forward and backward substitutions, is shown below:

```
for (i=1;i<=n;i++)
{
      d[i]=a[i][i];
      if (i<=n-1)
            e[i]=a[i][i+1];
      if (i>=2)
            c[i]=a[i][i-1];
}

for (i=2;i<=n;i++)
{
      c[i] /= d[i-1];
      d[i] -= c[i]*e[i-1];
}
```

```
for (i=2;i<=n;i++)
     b[i] -= c[i]*b[i-1];

cout << "x values: " << endl;
x[n]=b[n]/d[n];
cout << x[i] << " ";
for (i=n-1;i>=1;i--)
{
     x[i]=(b[i]-e[i]*x[i+1])/d[i];
     cout << x[i] << " ";
}
```

Code5D.cpp shows the full source code for Thomas algorithm, which reads A and b values from an input file called Code5D.in.

Code5D.cpp: Thomas algorithm

```
#include <iostream.h>
#include <fstream.h>
#define n 5

void main()
{
     int i,j,k;
     double *x,*w,*b,*c,*d,*e,**a;
     for (i=0;i<=n;i++)
     {
          x=new double [n+1];
          w=new double [n+1];
          b=new double [n+1];
          c=new double [n+1];
          d=new double [n+1];
          e=new double [n+1];
          a=new double *[n+1];
          for (j=0;j<=n;j++)
               a[j]=new double [n+1];
     }
     cout.setf(ios::fixed);
     cout.precision(5);

     // read the input data
     ifstream InFile("Code5D.in");
     for (i=1;i<=n;i++)
     {
          for (j=1;j<=n;j++)
               InFile >> a[i][j];
```

```
        InFile >> b[i];
    }
    InFile.close();

    for (i=1;i<=n;i++)
    {
        d[i]=a[i][i];
        if (i<=n-1)
            e[i]=a[i][i+1];
        if (i>=2)
            c[i]=a[i][i-1];
    }

    for (i=2;i<=n;i++)
    {
        c[i] /= d[i-1];
        d[i] -= c[i]*e[i-1];
    }
    for (i=2;i<=n;i++)
        b[i] -= c[i]*b[i-1];

    cout << "x values: " << endl;
    x[n]=b[n]/d[n];
    cout << x[i] << " ";
    for (i=n-1;i>=1;i--)
    {
        x[i]=(b[i]-e[i]*x[i+1])/d[i];
        cout << x[i] << " ";
    }
    cout << endl;
    for (i=0;i<=n;i++)
        delete a[i];
    delete x,w,b,c,d,e,a;
}
```

5.5 ITERATIVE TECHNIQUES

An iterative method is easy to implement as it is a repetition of the same set of variables that does not require too much computer memory. An iteration starts with some given initial values to the variables. These values are constantly updated at each iteration, and eventually they converge to the desired values after some successful iterations. The iterations only stop once the values have converged to some numbers where stopping criteria have been met.

The stopping point for the iterations is normally based on a comparison between an error in the iteration with some small value. A common error for an iterative process is the maximum from the difference in vector magnitude norm based on Equation (3.8), or

$$\|\boldsymbol{x}^{(k+1)} - \boldsymbol{x}^{(k)}\|_\infty = \max\left(\left|x_1^{(k+1)} - x_1^{(k)}\right|, \left|x_2^{(k+1)} - x_2^{(k)}\right|, \ldots, \left|x_n^{(k+1)} - x_n^{(k)}\right|\right).$$
(5.9)

In the above equation, iteration on $\boldsymbol{x} = [x_1 \quad x_2 \quad \cdots \quad x_n]^T$ is indicated by the superscript k. Equation (5.9) specifies the error as the difference between the computed values of $\boldsymbol{x}$ at iteration $k + 1$ with the ones at iteration k. The difference is compared with some small number ε. If $\|\boldsymbol{x}^{(k+1)} - \boldsymbol{x}^{(k)}\|_\infty > \varepsilon$, the iteration continues by computing new values of $\boldsymbol{x}$. Otherwise, if $\|\boldsymbol{x}^{(k+1)} - \boldsymbol{x}^{(k)}\|_\infty \le \varepsilon$, the iteration stops immediately, and the iterative technique is said to have converged to the solution $\boldsymbol{x}$.

We discuss two iterative methods that are commonly applied in solving a system of linear equations. The first system is the Jacobi method, which updates the values of the unknowns when an iteration has completed. The second system is the Gauss–Seidel method, which performs an update on the values immediately without waiting for an iteration to complete. These two methods are basically similar, having the same formula for updating the values. Both methods require the coefficient matrix of the system to be a special type of matrix called a *diagonally-dominant matrix*. The difference between them lies in the way updates are made: Gauss–Seidel uses the latest values of the variables, whereas Jacobi is not using the latest values.

Definition 5.5. A square matrix A is said to be *diagonally-dominant* if the absolute value of its diagonal element in each row is greater than the sum of the absolute value of all other elements in the same row. This condition is expressed as follows:

$$|a_{ii}| > \sum_{\substack{j=1 \\ j \ne i}}^{N} |a_{ij}|, \text{ for } i = 1, 2, \ldots, n.$$
(5.10)

In the above definition, a diagonally-dominant matrix must always have the largest absolute values at its diagonal elements. As an illustration, the following matrix is diagonally-dominant:

$$\begin{bmatrix} 5 & -1 & 3 \\ 2 & -7 & -1 \\ 4 & -2 & 8 \end{bmatrix}$$

$$\text{row 1}: |5| > |-1| + |3|,$$
$$\text{row 2}: |-7| > |2| + |-1|,$$
$$\text{row 3}: |8| > |4| + |-2|.$$

The following matrix is not diagonally-dominant but becomes diagonally-dominant by exchanging rows 2 and 3:

$$\begin{bmatrix} 5 & -1 & 3 \\ 2 & 3 & -6 \\ 3 & -7 & 2 \end{bmatrix} \sim \begin{bmatrix} 5 & -1 & 3 \\ 3 & -7 & 2 \\ 2 & 3 & -6 \end{bmatrix}.$$

The diagonally-dominance criteria test on a coefficient matrix is implemented in C++, as follows:

```
for (i=1;i<=n;i++)
{
        Sum=0;
        for (j=1;j<=n;j++)
                if (i!=j)
                        Sum += fabs(a[i][j]);
        if (fabs(a[i][i])<=Sum)
        {
                cout <<"Unsuccessful, the matrix is not
                        diagonally-dominant" << endl;
                return;
        }
}
```

Jacobi Method

The Jacobi method is an iterative method that performs updates based on the values of its variables from the previous iteration. The method requires the coefficient matrix A of the linear system to be positive definite to guarantee convergence to its solutions. Otherwise, the computed values of the variables will not converge and will cause a very significant error to the solution.

The Jacobi method starts with a formulation of the variables from the given equations in the system. The rth equation in the $n \times n$ system of linear equations is given as follows:

$$\sum_{j=1}^{n} a_{rj}x_j = b_r, \text{ or}$$

$$a_{r1}x_1 + a_{r2}x_2 + \cdots + a_{rr}x_r + \cdots + a_{rn}x_n = b_r.$$

From this equation, the variable x_r is formulated as a subject of the rest of the variables:

$$x_r = \frac{b_r - \sum_{j=1, j \neq r}^{n} a_{rj}x_j}{a_{rr}}$$

$$= \frac{b_r - a_{r1}x_1 - a_{r2}x_2 - \cdots - a_{r,r-1}x_{r-1} - a_{r,r+1}x_{r+1} - \cdots - a_{r,n-1}x_{n-1} - a_{rn}x_n}{a_{rr}}.$$

At iteration k, an update is made on x_r based on the current values of the variables $\mathbf{x}^{(k)} = [x_1^{(k)} \quad x_2^{(k)} \quad \cdots \quad x_n^{(k)}]^T$ to produce new values at the one-step ahead iteration or iteration $k + 1$, which is $\mathbf{x}^{(k+1)} = [x_1^{(k+1)} \quad x_2^{(k+1)} \quad \cdots \quad x_n^{(k+1)}]^T$. This is performed as follows:

$$x_r^{(k+1)} = \frac{b_r - a_{r1}x_1^{(k)} - a_{r2}x_2^{(k)} - \cdots - a_{r,r-1}x_{r-1}^{(k)} - a_{r,r+1}x_{r+1}^{(k)} - \cdots - a_{rn}x_n^{(k)}}{a_{rr}}.$$

(5.11)

The complete formulation of all variables in the system of linear equations based on Equation (5.11) is shown below:

$$x_1^{(k+1)} = \frac{b_1 - a_{12}x_2^{(k)} - a_{13}x_3^{(k)} - \cdots - a_{1r}x_r^{(k)} - \cdots - a_{1n}x_n^{(k)}}{a_{11}},$$

$$x_2^{(k+1)} = \frac{b_2 - a_{22}x_1^{(k)} - a_{23}x_3^{(k)} - \cdots - a_{2r}x_r^{(k)} - \cdots - a_{2n}x_n^{(k)}}{a_{22}},$$

$$\cdots$$

$$x_r^{(k+1)} = \frac{b_r - a_{r1}x_1^{(k)} - a_{r2}x_2^{(k)} - \cdots - a_{r,r-1}x_{r-1}^{(k)} - a_{r,r+1}x_{r+1}^{(k)} - \cdots - a_{rn}x_n^{(k)}}{a_{rr}},$$

$$\cdots$$

$$x_n^{(k+1)} = \frac{b_n - a_{n1}x_1^{(k)} - a_{n2}x_2^{(k)} - \cdots - a_{nr}x_r^{(k)} - \cdots - a_{n,n-1}x_{n-1}^{(k)}}{a_{nn}}.$$

Figure 5.6 shows the schematic flowchart of the Jacobi method. The iterations start by assigning initial values to the variables at iteration 0, which is denoted as $x^{(0)} = [x_1^{(0)} \quad x_2^{(0)} \quad \cdots \quad x_n^{(0)}]^T$. The normal initial values for these variables are 0, or $x_i^{(0)} = 0$ for $i = 1, 2, \ldots, n$, although they can be any small numbers. From these initial values, the values of the variables are updated according to Equation (5.13) to produce $x^{(1)} = [x_1^{(1)} \quad x_2^{(1)} \quad \cdots \quad x_n^{(1)}]^T$.

The stopping criteria for the iterations are normally set according to the maximum magnitude of the vector difference of Equation (5.9) or $\|x^{(k+1)} - x^{(k)}\|_\infty \leq \varepsilon$. In this case, the error is $\|x^{(1)} - x^{(0)}\|_\infty$. If this value is equal to or smaller than the predefined value ε, the iteration stops immediately. Otherwise, the process continues at iteration 1 with the same step explained earlier repeated.

The same process as above is repeated for the subsequent iterations until the stopping criteria of Equation (5.9) is fulfilled. The final solution is x, which consists of the values in the last iteration where the stopping criteria are met.

Example 5.12. Solve the following system using the Jacobi method with the initial value of $x^{(0)} = [0 \quad 0 \quad 0 \quad 0]^T$ until $\|x^{(k+1)} - x^{(k)}\|_\infty \leq \varepsilon$, where $\varepsilon = 0.005$:

$$x_1 - 5x_2 + x_3 - x_4 = 3,$$
$$-2x_1 + x_2 - 5x_3 + x_4 = -2,$$
$$4x_1 - 2x_2 + x_3 - 8x_4 = 2,$$
$$6x_1 - x_2 + 2x_3 + 2x_4 = -1.$$

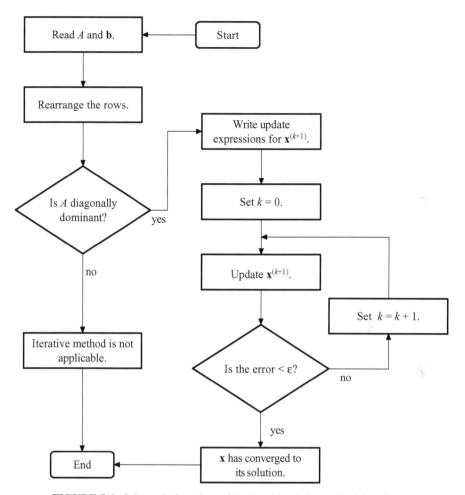

FIGURE 5.6. Schematic flowchart of the Jacobi and Gauss–Seidel methods.

Solution. It can be shown that the coefficient matrix in the above system is not positive definite. Therefore, the rows are interchanged in such a way that the dominant coefficient of each row forms the diagonal in the new coefficient matrix. This results in the following arrangement where the coefficient matrix is positive definite:

$$6x_1 - x_2 + 2x_3 + 2x_4 = -1,$$
$$x_1 - 5x_2 + x_3 - x_4 = 3,$$
$$-2x_1 + x_2 - 5x_3 + x_4 = -2,$$
$$4x_1 - 2x_2 + x_3 - 8x_4 = 2.$$

We get expressions for x_1, x_2, x_3, and x_4, as follows:

$$x_1 = \frac{-1 + x_2 - 2x_3 - 2x_4}{6},$$

$$x_2 = \frac{3 - x_1 - x_3 + x_4}{-5},$$

$$x_3 = \frac{-2 + 2x_1 - x_2 - x_4}{-5},$$

$$x_4 = \frac{2 - 4x_1 + 2x_2 - x_3}{-8}.$$

From Equation (5.11), the values of x_1, x_2, x_3, and x_4 are updated at each iteration k as follows:

$$x_1^{(k+1)} = \frac{-1 + x_2^{(k)} - 2x_3^{(k)} - 2x_4^{(k)}}{6},$$

$$x_2^{(k+1)} = \frac{3 - x_1^{(k)} - x_3^{(k)} + x_4^{(k)}}{-5},$$

$$x_3^{(k+1)} = \frac{-2 + 2x_1^{(k)} - x_2^{(k)} - x_4^{(k)}}{-5},$$

$$x_4^{(k+1)} = \frac{2 - 4x_1^{(k)} + 2x_2^{(k)} - x_3^{(k)}}{-8}.$$

With $k = 0$:

$$x_1^{(1)} = \frac{-1 + x_2^{(0)} - 2x_3^{(0)} - 2x_4^{(0)}}{6} = \frac{-1 + (0) - 2(0) - 2(0)}{6} = -0.167,$$

$$x_2^{(1)} = \frac{3 - x_1^{(0)} - x_3^{(0)} + x_4^{(0)}}{-5} = \frac{3 - 0 - 0 + 0}{-5} = -0.6,$$

$$x_3^{(1)} = \frac{-2 + 2x_1^{(0)} - x_2^{(0)} - x_4^{(0)}}{-5} = \frac{-2 + 2(0) - 0 - 0}{-5} = 0.4,$$

$$x_4^{(1)} = \frac{2 - 4x_1^{(0)} + 2x_2^{(0)} - x_3^{(0)}}{-8} = \frac{2 - 4(0) + 2(0) - 0}{-8} = -0.25,$$

$$\text{error} = \left\| x^{(1)} - x^{(0)} \right\|_\infty = \max(|-0.167|, |-0.6|, |-0.4|, |-0.25|) = 0.6 > \varepsilon.$$

TABLE 5.2. Results from Jacobi iterations in Example 5.12

k	$x_1^{(k)}$	$x_2^{(k)}$	$x_3^{(k)}$	$x_4^{(k)}$	$\|x^{(k+1)} - x^{(k)}\|_\infty$
0	0.000	0.000	0.000	0.000	0.600
1	−0.167	−0.600	0.400	−0.250	0.150
2	−0.317	−0.503	0.297	−0.133	0.112
3	−0.305	−0.577	0.399	−0.245	0.045
4	−0.314	−0.532	0.357	−0.208	0.021
5	−0.305	−0.550	0.378	−0.229	0.011
6	−0.308	−0.540	0.366	−0.218	0.005
7	−0.306	−0.545	0.372	−0.223	

The error is greater than the tolerance. Therefore, the iterations continue with $k = 1$:

$$x_1^{(2)} = \frac{-1 + x_2^{(1)} - 2x_3^{(1)} - 2x_4^{(1)}}{6} = \frac{-1 + (-0.6) - 2(-0.4) - 2(-0.25)}{6} = -0.317,$$

$$x_2^{(2)} = \frac{3 - x_1^{(1)} - x_3^{(1)} + x_4^{(1)}}{-5} = \frac{3 - (-0.167) - (-0.4) + (-0.25)}{-5} = -0.503,$$

$$x_3^{(2)} = \frac{-2 + 2x_1^{(1)} - x_2^{(1)} - x_4^{(1)}}{-5} = \frac{-2 + 2(-0.167) - (-0.6) - (-0.25)}{-5} = 0.297,$$

$$x_4^{(2)} = \frac{2 - 4x_1^{(1)} + 2x_2^{(1)} - x_3^{(1)}}{-8} = \frac{2 - 4(-0.167) + 2(-0.6) - (-0.4)}{-8} = -0.133,$$

$$\text{error} = \left\|x^{(2)} - x^{(1)}\right\|_\infty = \max(|-0.317|, |-0.503|, |0.297|, |-0.133|) = 0.15 > \varepsilon.$$

The error is reduced significantly in this iteration, which means convergence to the solution is beginning to take its shape. Continuing the process gives the results as shown in Table 5.2.

The final solution is $x^{(7)} = [-0.306 \quad -0.545 \quad 0.372 \quad -0.223]^T$, which is achieved at $k = 6$ when $\|x^{(7)} - x^{(6)}\|_\infty \leq \varepsilon$.

Gauss–Seidel Method

The Gauss–Seidel method is an iterative method that updates the values of a variable based on the latest values of other variables in the system. Because of these latest values, convergence to the solution takes place faster than the Jacobi method.

The implementation of the Gauss–Seidel method is very similar to the Jacobi method, as shown in the flowchart in Figure 5.6. For an equation given by $\sum_{i=1}^{n} a_{ij}x_i = b_i$, where $i = 1, 2, \ldots, n$, the rth unknown, or x_r, is expressed as follows:

$$x_r = \frac{b_r - \sum_{j=1, j \neq r}^{n} a_{rj}x_j}{a_{rr}}.$$

The update on x_r at iteration k is the one-step ahead value given as follows:

$$x_r^{(k+1)} = \frac{b_r - a_{r1}x_1^{(k+1)} - a_{r2}x_2^{(k+1)} - \cdots - a_{r,r-1}x_{r-1}^{(k+1)} - a_{r,r+1}x_{r+1}^{(k)} - \cdots - a_{rn}x_n^{(k)}}{a_{rr}}.$$

$$(5.12)$$

In the above equation, the update on x_r involves the latest values of x_i for $i = 1, 2, \ldots, r-1$ at iteration $k+1$ and the values of x_i for $i = r+1, r+2, \ldots, n$ at iteration k. The complete update on the variables is shown as follows:

$$x_1^{(k+1)} = \frac{b_1 - a_{12}x_2^{(k)} - a_{13}x_3^{(k)} - \cdots - a_{1r}x_r^{(k)} - \cdots - a_{1n}x_n^{(k)}}{a_{11}},$$

$$x_2^{(k+1)} = \frac{b_2 - a_{22}x_1^{(k+1)} - a_{23}x_3^{(k)} - \cdots - a_{2r}x_r^{(k)} - \cdots - a_{2n}x_n^{(k)}}{a_{22}},$$

$$\cdots$$

$$x_r^{(k+1)} = \frac{b_r - a_{r1}x_1^{(k+1)} - a_{r2}x_2^{(k+1)} - \cdots - a_{r,r-1}x_{r-1}^{(k+1)} - a_{r,r+1}x_{r+1}^{(k)} - \cdots - a_{rn}x_n^{(k)}}{a_{rr}},$$

$$\cdots$$

$$x_n^{(k+1)} = \frac{b_n - a_{n1}x_1^{(k+1)} - a_{n2}x_2^{(k+1)} - \cdots - a_{nr}x_r^{(k+1)} - \cdots - a_{n,n-1}x_{n-1}^{(k+1)}}{a_{nn}}.$$

The standard error in the iterations is the maximum magnitude norm as expressed in Equation (5.9). This error is compared with the preset threshold value ε, where the iterations stop immediately if the error is smaller than or equal to this value.

Example 5.13. Solve the system of linear equations in Example 5.12 using the Gauss–Seidel method with the initial value of $x^{(0)} = [0 \ \ 0 \ \ 0 \ \ 0]^T$ until $\|x^{(k+1)} - x^{(k)}\|_\infty < \varepsilon$, where $\varepsilon = 0.005$.

Solution. The coefficient matrix of the above system is not diagonally-dominant. Rearrange the rows so that the diagonal elements become dominant, as follows:

$$6x_1 - x_2 + 2x_3 + 2x_4 = -1,$$
$$x_1 - 5x_2 + x_3 - x_4 = 3,$$
$$-2x_1 + x_2 - 5x_3 + x_4 = -2,$$
$$4x_1 - 2x_2 + x_3 - 8x_4 = 2,$$

From Equation (5.15), updates are performed according to

$$x_1^{(k+1)} = \frac{-1 + x_2^{(k)} - 2x_3^{(k)} - 2x_4^{(k)}}{6},$$

$$x_2^{(k+1)} = \frac{3 - x_1^{(k+1)} - x_3^{(k)} + x_4^{(k)}}{-5},$$

$$x_3^{(k+1)} = \frac{-2 + 2x_1^{(k+1)} - x_2^{(k+1)} - x_4^{(k)}}{-5},$$

$$x_4^{(k+1)} = \frac{2 - 4x_1^{(k+1)} + 2x_2^{(k+1)} - x_3^{(k+1)}}{-8}.$$

Starting with $k = 0$:

$$x_1^{(1)} = \frac{-1 + x_2^{(0)} - 2x_3^{(0)} - 2x_4^{(0)}}{6} = \frac{-1 + 0 - 0 - 0}{6} = -0.167,$$

$$x_2^{(1)} = \frac{3 - x_1^{(1)} - x_3^{(0)} + x_4^{(0)}}{-5} = \frac{3 - (-0.167) - 0 - 0}{-5} = -0.633,$$

$$x_3^{(1)} = \frac{-2 + 2x_1^{(1)} - x_2^{(1)} - x_4^{(0)}}{-5} = \frac{-2 + 2(-0.167) - (-0.633) - 0}{-5} = 0.340,$$

$$x_4^{(1)} = \frac{2 - 4x_1^{(1)} + 2x_2^{(1)} - x_3^{(1)}}{-8} = \frac{2 - 4(-0.167) + 2(-0.633) - (0.340)}{-8} = -0.133,$$

$$\text{error} = \left\| x^{(1)} - x^{(0)} \right\|_\infty = \max(0.167, 0.633, 0.340, 0.133) = 0.633 > \varepsilon.$$

The iteration continues with $k = 1$ since the error is greater than ε:

$$x_1^{(2)} = \frac{-1 + x_2^{(1)} - 2x_3^{(1)} - 2x_4^{(1)}}{6} = \frac{-1 + 0.167 - 2(0.633) - 2(0.340)}{6} = -0.341,$$

$$x_2^{(2)} = \frac{3 - x_1^{(2)} - x_3^{(1)} + x_4^{(1)}}{-5} = \frac{3 + 0.341 - 0.340 - 0.133}{-5} = -0.574,$$

$$x_3^{(2)} = \frac{-2 + 2x_1^{(2)} - x_2^{(2)} - x_4^{(1)}}{-5} = \frac{-2 + 2(-0.341) - (-0.574) - (-0.133)}{-5} = 0.395,$$

$$x_4^{(2)} = \frac{2 - 4x_1^{(2)} + 2x_2^{(2)} - x_3^{(2)}}{-8} = \frac{2 - 4(-0.341) + 2(-0.574) - 0.395}{-8} = -0.228,$$

$$\text{error} = \left\| x^{(2)} - x^{(1)} \right\|_\infty = \max(0.341, 0.574, 0.395, 0.228) = 0.175 > \varepsilon.$$

Additional iterations produce results as shown in Table 5.3.

TABLE 5.3. Results from Example 5.12

k	$x_1^{(k)}$	$x_2^{(k)}$	$x_3^{(k)}$	$x_4^{(k)}$	$\|x^{(k+1)} - x^{(k)}\|_\infty$
0	0.000	0.000	0.000	0.000	0.633
1	−0.167	−0.633	0.340	−0.133	0.175
2	−0.341	−0.574	0.395	−0.228	0.035
3	−0.318	−0.539	0.374	−0.228	0.013
4	−0.305	−0.541	0.368	−0.221	0.002
5	−0.306	−0.543	0.369	−0.221	

The iterations stop at $k = 4$ to produce $x^{(5)}$ as the solution where the error given by $\|x^{(5)} - x^{(4)}\|_\infty = 0.002$ is less than ε.

Code5E. Gauss-Seidel method.

```
#include <iostream.h>
#include <fstream.h>
#include <math.h>
#define n 4                    // array size
#define MAX 10                 // maximum number of iterations

void main()
{
        int i,j,k;
        double *x,*xOld,*b,**a;
        double Sum,error;

        // allocate memory
        x=new double [n+1];
        xOld=new double [n+1];
        b=new double [n+1];
        a=new double *[n+1];
        for (i=0;i<=n;i++)
                a[i]=new double [n+1];
        cout.setf(ios::fixed);
        cout.precision(3);

        // read the input data
        ifstream InFile("Code5E.in");
        for (i=1;i<=n;i++)
        {
                x[i]=xOld[i]=0;
                for (j=1;j<=n;j++)
                        InFile >> a[i][j];
                InFile >> b[i];
        }
```

```
InFile.close();

// test for diagonally-dominance
for (i=1;i<=n;i++)
{
     Sum=0;
     for (j=1;j<=n;j++)
          if (i!=j)
               Sum += fabs(a[i][j]);
     if (fabs(a[i][i])<=Sum)
     {
          cout <<"Unsuccessful, the matrix is not
               diagonally-dominant" << endl;
          return;
     }
}

// perform iterations & update x
for (k=0;k<=MAX;k++)
{
     error=0;
     cout << "k=" << k << ": ";
     for (i=1;i<=n;i++)
     {
          Sum=0;
          for (j=1;j<=n;j++)
               if (i!=j)
                    Sum += a[i][j]*x[j];
          xOld[i]=x[i];
          x[i]=(b[i]-Sum)/a[i][i];
          cout << x[i] << " ";
          error=((error>fabs(x[i]-xOld[i]))?error:
               fabs(x[i]-xOld[i]));
     }
     cout << error << endl;
}

// delete memory
for (i=0;i<=n;i++)
     delete a[i];
delete x,xOld,b,a;
}
```

5.6 VISUALIZING THE SOLUTION: CODE5

Code5. User Manual.

1. Enter the matrix values starting from the top left-hand corner. The size of the matrix terminates at the first unfilled diagonal element.
2. Click the corresponding method to view the results.

Development Files: Code5.cpp, Code5.h and Code5.rc.

We now discuss the design and development of several modules for solving a system of linear equations. The modules are integrated into a single system based on Windows using the resources in the Microsoft Foundation Classes library. The interface is user-friendly with dialog boxes for input, and buttons for updates, which serves as a black box for solving the system of linear equations problem to its user. A user-friendly interface must have features that allow easy input of data, modification of the input data, a choice of several methods for the same data, and an error-proof mechanism to detect problems in data entry. The last factor is important since problems like singular matrix as input can cause the system to crash instead of producing a good solution.

Figure 5.7 shows the Windows interface for solving a system of up to five linear equations problem. The interface consists of a modal window with input in the form of edit boxes (white), the output in static boxes (gray), buttons for activating the methods, and text messages. The buttons are labeled *Gauss, Crout, Cholesky, Thomas,* and *Gauss–Seidel,* which activate the named methods for solving the problem. The *Reset* button clears all entries and refreshes the display.

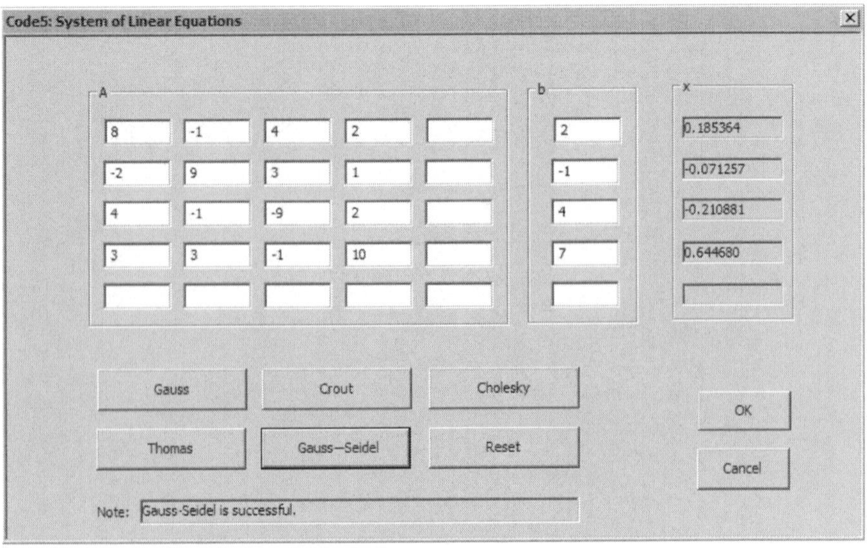

FIGURE 5.7. Windows interface for solving a system of up to five linear equations.

The figure also shows a sample run of the Gauss–Seidel method with A and b as input in the edit boxes, to produce the output x in the static boxes once the *Gauss–Seidel* button is clicked. The program is intelligent enough to recognize the coefficient matrix A as diagonally dominant. Otherwise, if A is not diagonally dominant, then a click on the *Gauss–Seidel* button will not produce the desired results. The same input produces the same results when the *Gauss* and *Crout* buttons are clicked, but it will respond with some error messages if the *Cholesky* and *Thomas* buttons are clicked. In this case, the program can verify correctly that A is not positive definite and tridiagonal, respectively.

Any size of matrix, from two to five, can be entered. Entries can be made beginning from the top left-hand corner. The size of the matrix is then automatically determined from the first unfilled entry in the diagonal elements.

We discuss the steps for producing this interface. They consist of creating a new Win32 project codenamed Code5 using the MFC static library, creating the resource file Code5.rc, and inserting codes for the Gauss, Crout, Cholesky, Thomas, and Gauss–Seidel methods. Only three files need to be created, namely Code5.cpp, Code5.h, and Code5.rc. A single class called CCode5 is used, and this class is derived from MFC's CDialog.

Figure 5.8 shows a schematic drawing of the computational steps in Code5. The process starts with the resource file Code5.rc for creating the dialog window. The dialog window becomes the main window that hosts several child windows consisting of edit boxes for collecting the input, static boxes for displaying the output, and buttons for selecting a method for solving the problem.

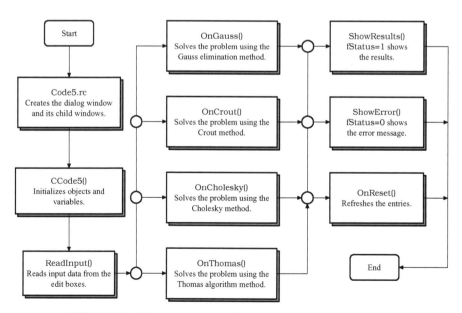

FIGURE 5.8. Schematic drawing of the computational steps in Code5.

Global objects and variables are initialized in the constructor, CCode5(). They include the matrices and vectors for the system of linear equations. ReadInput() reads the input data in the edit boxes, and the data are then passed for processing to a function when a push button that points to the function is clicked. The activated function returns either fStatus=1 or fStatus=0, which indicates whether the supplied data fit into the named method or not, respectively. With fStatus=1, the results from the function are displayed in the static boxes through ShowResults(), and the method is flagged as successful. A return value of fStatus=0 indicates the method fails because of irrelevant data, and the error message is shown through ShowError(). Another function called OnReset() refreshes the edit and static boxes for a new data entry.

There are six major steps in the development of Code5. This project differs slightly from the skeleton program in Code3A project as the main window is a dialog window derived from MFC's CDialog. A dialog window is a container class that can host several objects from other classes.

Step 1. Create a new Win32 project.
From the menu, choose *File*, followed by *New* and *Project*. Name the project Code5, and press the *OK* button. A new class called CCode5 is created. Declare the project as an empty Windows application project through the check boxes.

From the Solution Explorer, highlight CCode5 and right-click *Properties*. Choose *Use MFC in a Static Library* from the item *Use of MFC* in the combo box. These options allow the application to use the resources in the MFC library.

Step 2. Add the resource file Code5.rc.
This step creates all the friendly tools in the modal window that are the resources in Windows. Right-click *Resource Files* in the Solution Explorer, as shown in Figure 5.9. Choose *Add* from the menu and then *Add Resource*.

Step 3. Add a dialog window.
A dialog window is host to several friendly tools such as buttons, edit boxes, combo boxes, check boxes, and static boxes. To create a dialog window, choose *Dialog* from the *Add Resource* window, as shown in Figure 5.10.

Step 5. Create edit boxes, static boxes, and buttons.
A dialog window as shown in Figure 5.11 appears. This window is automatically assigned with the id IDD_DIALOG1. From the Toolbox, choose *Edit Control*, and draw an edit box, as shown in the figure. Click *Properties Windows*, and assign this edit box with the id IDC_A11 for representing the element a_{11} in the coefficient matrix A.

Continue with the rest of the elements in matrix A (IDC_A12 up to IDC_A55) and vector b (IDC_b1 to IDC_b5) to complete the input interface. Next, click *Toolbox* and choose *Static Text*. Create static boxes for housing the vector x with ids from IDC_x1 to IDC_x5. Figure 5.12 shows the complete interface for this problem with ids on the items in the Windows.

Step 6. Create the files Code5.cpp and Code5.h, and insert their codes.
The main file is Code5.cpp, which drives the application by calling both Code5.h and Code5.rc.

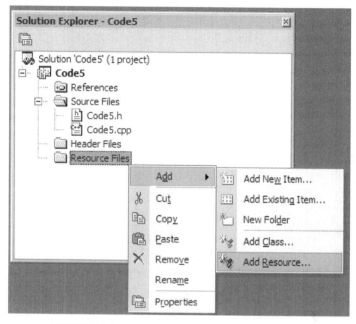

FIGURE 5.9. Adding a resource file called Code5.rc.

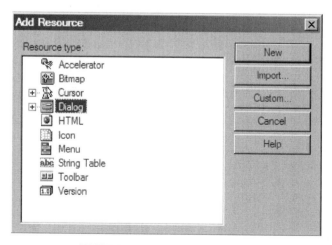

FIGURE 5.10. *Add Resource* items.

The contents of the header file Code5.h differ from the skeleton program in Code3A, as its host window is a dialog window. It is necessary to add #include "resource.h" in the preprocessing area of Code5.h as this file contains all the necessary declarations for the resources created in Code5.rc.

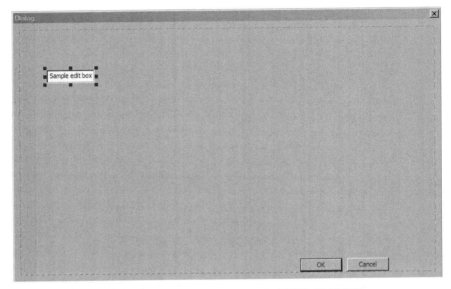

FIGURE 5.11. Dialog window with the id IDD_DIALOG1.

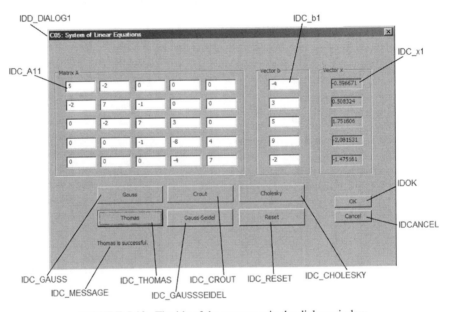

FIGURE 5.12. The ids of the resources in the dialog window.

```
#include <afxdisp.h>
#include <math.h>
#include "resource.h"
#define N 5
```

```
class CCode5 : public CDialog
{
public:
        CCode5(CWnd* pParent = NULL);
        enum { IDD = IDD_DIALOG1 };
        virtual ~CCode5();
        virtual void OnOK();
        void ReadInput(),DisplayResults(),DisplayError(CString);
        afx_msg void OnGauss();
        afx_msg void OnCrout();
        afx_msg void OnCholesky();
        afx_msg void OnThomas();
        afx_msg void OnGaussSeidel();
        afx_msg void OnReset();
        int n;
        double **A,**L,**U,*x,*b;
        CString **sA,*sx,*sb,method;
        DECLARE_MESSAGE_MAP()
};

class CMyWinApp : public CWinApp
{
public:
        virtual BOOL InitInstance();
};
```

The dialog window is created in Code5.cpp. This window is called the *modal window*, as it does not allow background editing if the window is active. Its opposite is the *nonmodal window*, which allows background editing when the window is active. The following code segment in Code5.cpp creates the dialog window:

```
BOOL CMyWinApp::InitInstance()
{
   AfxEnableControlContainer();
   CCode5 dlg;
   m_pMainWnd = &dlg;
   dlg.DoModal();
   return FALSE;
}
```

A method for solving the system of linear equations based on the input data becomes active when one of the push buttons that correspond to the method is left-clicked. Each push button has an id, and it calls the corresponding function through an event mapping given by

TABLE 5.4. Summary of functions, their ids, and their purposes in CCode5

Function	Control Id	Purpose
CCode5()		Constructor of the class CCode5.
~CCode5()		Destructor of the class CCode5.
ReadInput()		Reads input data in matrix A and vector b.
DisplayResults()		Displays the output.
DisplayError()		Displays the errors in the input.
Reset()		Clears all input and output information.
OnOK()	IDOK	Closes the dialog window.
OnGauss()	IDC_GAUSS	Activates the Gaussian elimination method.
OnCrout()	IDC_CROUT	Activates the Crout method.
OnCholesky()	IDC_CHOLESKY	Activates the Cholesky method.
OnThomas()	IDC_THOMAS	Activates the Thomas algorithm.
OnGaussSeidel()	IDC_GAUSSSEIDEL	Activates the Gauss–Seidel method.

```
BEGIN_MESSAGE_MAP(CCode5,CDialog)
    ON_BN_CLICKED(IDC_GAUSS, OnGauss)
    ON_BN_CLICKED(IDC_CROUT, OnCrout)
    ON_BN_CLICKED(IDC_CHOLESKY, OnCholesky)
    ON_BN_CLICKED(IDC_THOMAS, OnThomas)
    ON_BN_CLICKED(IDC_GAUSSSEIDEL, OnGaussSeidel)
    ON_BN_CLICKED(IDC_RESET, OnReset)
END_MESSAGE_MAP()
```

It is obvious in the message map that a click on the *Gauss* button, whose id is IDC_GAUSS, triggers a call on the function OnGauss(). This function reads the data provided by the user in the edit boxes and then solves the problem using the Gaussian elimination method.

Table 5.4 shows the member functions of the class CCode5. Included are eight main member functions representing the buttons in the class. Each function in the class is named according to its duty; for example, OnGauss() responds to the *Gauss* button click, and this function solves the system of linear equations based on the data input from the user. The ids of these buttons are shown in Figure 5.12.

The edit boxes provide input spaces for matrix A and vector b. A system of up to 5×5 linear equations is supported, and this is indicated as the macro N in the header file Code5.h. The program has been designed flexible so as to support other sizes simply filling in the values of A starting from the top left-hand corner. The actual size of the matrix is represented by a variable called n. The size is determined from the filled elements in the diagonal. For example, to have a 3×3 system, or n=3, entries must be made in the first three rows and three columns of the edit boxes beginning from the top left-hand corner.

Beside the button functions, CCode5 has four other application functions, ReadInput(), DisplayResults(), DisplayError(), and OnOK(). These functions are called from the selected method for reading the input data, displaying the output, displaying error messages, and closing the application, respectively.

The constructor in the class is CCode5(). This function activates the dialog box with the id IDD_DIALOG1, which is the one shown in Figure 5.11. This window is derived from the MFC class CDialog. Besides creating the class, CCode5() also initializes several arrays and allocates memory dynamically. This is shown as follows:

```
CCode5::CCode5(CWnd* pParentWnd)
: CDialog (IDD_DIALOG1, pParentWnd)
{
        x=new double [N+1];
        b=new double [N+1];
        sx=new CString [N+1];
        sb=new CString [N+1];
        A=new double *[N+1];
        L=new double *[N+1];
        U=new double *[N+1];
        sA=new CString *[N+1];
        for (int i=0;i<=N;i++)
        {
                A[i]=new double [N+1];
                L[i]=new double [N+1];
                U[i]=new double [N+1];
                sA[i]=new CString [N+1];
        }
}
```

The destructor in the program is ~CCode5(). This function destroys the arrays allocated in the constructor and the class to mark the end of the application runtime. The code is shown as follows:

```
CCode5::~CCode5()
{
        for (int i=0;i<=N;i++)
                delete A[i],L[i],U[i],sA[i];
        delete x,b,sx,sb,A,L,U,sA;
}
```

ReadInput() is a function for reading all input from the user in the form of edit boxes, as shown in Figure 5.11. Input from the user is read through the function GetDlgItemText(). Each edit box has an idd that reflects its matrix element. For example, IDC_A24 represents the edit box for a_{24} in matrix A, whereas IDC_b3 represents b_3 in b. However, the representation is not straightforward. An edit box takes only string (text) input, and this string needs to be converted to a double value before a computation can be performed. Hence, the value input by the user inside the box with the id IDC_A24 is stored as the string sA[2][4], whereas that of IDC_b3 is stored as sb[3]. These values are converted to the double variables called

A[2][4] and b[3], respectively, using the C function atof(). The code is shown, as follows:

```
void CCode5::ReadInput()
{
        GetDlgItemText(IDC_A11,sA[1][1]);
        GetDlgItemText(IDC_A12,sA[1][2]);
        GetDlgItemText(IDC_A13,sA[1][3]);
        GetDlgItemText(IDC_A14,sA[1][4]);
        GetDlgItemText(IDC_A15,sA[1][5]);
        GetDlgItemText(IDC_A21,sA[2][1]);
        GetDlgItemText(IDC_A22,sA[2][2]);
        GetDlgItemText(IDC_A23,sA[2][3]);
        GetDlgItemText(IDC_A24,sA[2][4]);
        GetDlgItemText(IDC_A25,sA[2][5]);
        GetDlgItemText(IDC_A31,sA[3][1]);
        GetDlgItemText(IDC_A32,sA[3][2]);
        GetDlgItemText(IDC_A33,sA[3][3]);
        GetDlgItemText(IDC_A34,sA[3][4]);
        GetDlgItemText(IDC_A35,sA[3][5]);
        GetDlgItemText(IDC_A41,sA[4][1]);
        GetDlgItemText(IDC_A42,sA[4][2]);
        GetDlgItemText(IDC_A43,sA[4][3]);
        GetDlgItemText(IDC_A44,sA[4][4]);
        GetDlgItemText(IDC_A45,sA[4][5]);
        GetDlgItemText(IDC_A51,sA[5][1]);
        GetDlgItemText(IDC_A52,sA[5][2]);
        GetDlgItemText(IDC_A53,sA[5][3]);
        GetDlgItemText(IDC_A54,sA[5][4]);
        GetDlgItemText(IDC_A55,sA[5][5]);
        GetDlgItemText(IDC_b1,sb[1]);
        GetDlgItemText(IDC_b2,sb[2]);
        GetDlgItemText(IDC_b3,sb[3]);
        GetDlgItemText(IDC_b4,sb[4]);
        GetDlgItemText(IDC_b5,sb[5]);

        for (int i=1;i<=N;i++)
                if (sA[i][i]=="")
                {
                        n=i-1; break;
                }
        // convert the string input to double
        for (i=1;i<=n;i++)
        {
                for (int j=1;j<=n;j++)
```

```
            A[i][j]=atof(sA[i][j]);
        b[i]=atof(sb[i]);
    }
}
```

The output in the program is $x = [x_1 \quad x_2 \quad x_3 \quad x_4 \quad x_5]^T$, which is displayed by the function `DisplayResults()`. The vector is represented by the double array x, and it is displayed on the static boxes using `SetDlgItemText()` through the string sx. Conversion from x to the string sx is achieved through the function `Format()`. The static boxes are represented by ids that reflect their variables; for example, `IDC_x1` represents the string sx[1]. The last statement in this function displays the successful message to acknowledge that the given method is successful. The following code shows the contents of `DisplayResults()`:

```
void CCode5::DisplayResults()
{
    for (int i=1;i<=N;i++)
        sx[i].Format("%lf",x[i]);
    SetDlgItemText(IDC_x1,((1<=n)?sx[1]:""));
    SetDlgItemText(IDC_x2,((2<=n)?sx[2]:""));
    SetDlgItemText(IDC_x3,((3<=n)?sx[3]:""));
    SetDlgItemText(IDC_x4,((4<=n)?sx[4]:""));
    SetDlgItemText(IDC_x5,((5<=n)?sx[5]:""));
    SetDlgItemText(IDC_MESSAGE,method+" is successful.");
}
```

`OnGauss()` is the function that responds to the *Gauss* button click identified through the id `IDC_GAUSS`. This function is based on `Code4A.cpp`. A Boolean variable called `fStatus` is introduced to check the value of the pivot element `A[k][k]`, where 0 (FALSE) indicates the element is 0. With this value, the Gaussian elimination method is considered a failure, and an error message is displayed through the function `DisplayError()`. Through this process, a division by zero in `m=A[i][k]/A[k][k]` during the row operations with respect to row k can be avoided. The function is given by

```
void CCode5::OnGauss()
{
    int i,j,k;
    double Sum,p;
    bool fStatus=1;
    CString condition="invertible";
    method="Gauss";
    ReadInput();
```

```
// perform row operations
for (k=1;k<=n-1;k++)
      for (i=k+1;i<=n;i++)
      {
            if (A[k][k]==0)
            {
                  fStatus=0;
                  break;
            }
            else
                  p=A[i][k]/A[k][k];
            for (j=1;j<=n;j++)
                  A[i][j] -= p*A[k][j];
            b[i] -= p*b[k];
      }

// perform backward substitutions
for (i=n;i>=1;i--)
{
      Sum=0;
      x[i]=0;
      for (j=i;j<=n;j++)
      Sum += A[i][j]*x[j];
      x[i]=(b[i]-Sum)/A[i][i];
}
if (fStatus)
      DisplayResults();
else
      DisplayError(condition);
}
```

A click on the *Crout* button whose id is IDC_CROUT activates the function OnCrout(), which computes the problem using the Crout() method. This function also has a Boolean variable called fStatus to check for the zero values in the diagonal elements of matrix L. Successful runtime is achieved through fStatus=1, and the output is displayed through the function DisplayResults(). The code in this function is given, as follows:

```
void CCode5::OnCrout()
{
      int i,j,k,r;
      double Sum,*w;
      bool fStatus=1;
      CString condition="invertible";
      method="Crout";
```

```
w=new double [N+1];
ReadInput();

// initialize L and U
for (i=1;i<=n;i++)
     for (j=1;j<=n;j++)
     {
             L[i][j]=U[i][j]=0;
             if (i==j)
             U[i][j]=1;
     }

// compute L and U
for (i=1;i<=n;i++)
     L[i][1]=A[i][1];
for (j=2;j<=n;j++)
     if (L[1][1]==0)
     {
             fStatus=0;
             break;
     }
     else
             U[1][j]=A[1][j]/L[1][1];
for (j=2; j<=n-1;j++)
{
     for (i=j; i<=n; i++)
     {
             Sum=0;
             for (k=1;k<=j-1;k++)
                   Sum += L[i][k]*U[k][j];
             L[i][j]=A[i][j]- Sum;
     }
     for (k=j+1;k<=n;k++)
     {
     Sum=0;
     for (r=1;r<=j-1;r++)
           Sum += L[j][r]*U[r][k];
     if (L[j][j]==0)
     {
             fStatus=0;
             break;
     }
     else
             U[j][k]=(A[j][k]-Sum)/L[j][j];
     }
}
```

```
Sum=0;
for (k=1;k<=n-1;k++)
        Sum += L[n][k]*U[k][n];
L[n][n]=A[n][n]-Sum;

// forward substitutions for finding w
w[1]=b[1]/L[1][1];
for (j=1;j<=n;j++)
{
        Sum=0;
        for (k=1; k<=j-1;k++)
                Sum += L[j][k]*w[k];
        if (L[j][j]==0)
        {
                fStatus=0;
                break;
        }
        else
                w[j] =(b[j]-Sum)/L[j][j];
}

// backward substitutions for finding x
for (i=n;i>=1;i--)
{
        Sum=0;
        for (k=i+1; k<=n; k++)
                Sum += U[i][k]*x[k];
        x[i]=w[i]-Sum;
}
delete w;
if (fStatus)
        DisplayResults();
else
        DisplayError(condition);
}
```

OnCholesky() responds to the *Cholesky* button click. This function first performs a test on the positive-definiteness of the coefficient matrix, with fStatus=1 indicating positive and fStatus=0 negative. The code is given, as follows:

```
void CCode5::OnCholesky()
{
        int i,j,k;
        double Sum,*w;
        bool fStatus=1;
```

```
CString condition="positive-definite";
method="Cholesky";
w=new double [N+1];
ReadInput();

// check for positive-definite
for (i=1;i<=n;i++)
     for (j=i;j<=n;j++)
          if ((A[i][j]!=A[j][i])
               || A[i][i]<=0
               || (i!=j && A[i][i]<=fabs(A[i][j])))
          {
                    fStatus=0;
                    break;
          }

// initialize L and U
for (i=1;i<=n;i++)
     for (j=1;j<=n;j++)
          L[i][j]=U[i][j]=0;


// compute L and U
for (k=1;k<=n;k++)
{
     Sum=0;
     for (j=1;j<=k-1;j++)
          Sum += pow(L[k][j],2);
     L[k][k]=sqrt(A[k][k]-Sum);
     for (i=1;i<=k-1;i++)
     {
          Sum=0;
          for (j=1;j<=i-1;j++)
               Sum += L[i][j]*L[k][j];
          L[k][i]=(A[k][i]-Sum)/L[i][i];
     }
}

// forward substitutions for finding w
for (j=1;j<=n;j++)
{
     Sum=0;
     for (k=1; k<=j-1;k++)
          Sum += L[j][k]*w[k];
     w[j] =(b[j]-Sum)/L[j][j];
}
```

```
// backward substitutions for finding x
for (i=n;i>=1;i--)
{
        Sum=0;
        for (k=i+1; k<=n; k++)
                Sum += U[i][k]*x[k];
        x[i]=w[i]-Sum;
}
delete w;
if (fStatus)
        DisplayResults();
else
        DisplayError(condition);
}
```

The case of the coefficient matrix in the system is tridiagonal, which leads to the deployment of the Thomas algorithm through the function OnThomas(). Just like other methods, this function uses the Boolean variable fStatus; a value of 1 indicates the matrix is tridiagonal, whereas 0 means the matrix is not tridiagonal. With fStatus=1, the Thomas algorithm is applied to the system, and the output is displayed in DisplayResults(). The code is shown below:

```
void CCode5::OnThomas()
{
        int i,j;
        bool fStatus=1;
        double *e,*f,*g;
        CString condition="tridiagonal";
        method="Thomas";
        e=new double [N+1];
        f=new double [N+1];
        g=new double [N+1];
        ReadInput();

        // check for tridiagonality
        for (i=1;i<=n;i++)
        {
                if (i==1)
                        for (j=3;j<=n;j++)
                                if (A[i][j]!=0)
                                        fStatus=0;
                if (i==n)
                        for (j=1;j<=n-2;j++)
                                if (A[i][j]!=0)
                                        fStatus=0;
                if (i>1 && i<n)
                        if (A[i][i+2]!=0 && A[i][i-2]!=0)
```

```
                              fStatus=0;
      }

      // compute e,f,g
      for (i=1;i<=n;i++)
      {
            f[i]=A[i][i];
            if (i<=n-1)
                  g[i]=A[i][i+1];
            if (i>=2)
                  e[i]=A[i][i-1];
      }
      for (i=2;i<=n;i++)
      {
            e[i]/=f[i-1];
            f[i] -= e[i]*g[i-1];
      }
      for (i=2;i<=n;i++)
            b[i] -= e[i]*b[i-1];
      x[n]=b[n]/f[n];
      for (i=n-1;i>=1;i--)
            x[i]=(b[i]-g[i]*x[i+1])/f[i];
      delete e,f,g;
      if (fStatus)
            DisplayResults();
      else
            DisplayError(condition);
}
```

The iterative technique in the program using Gauss–Seidel is illustrated through the OnGaussSeidel() function. This function checks to make sure the coefficient matrix is diagonally dominant through the Boolean variable fStatus, just like in the previous methods. The code is shown, as follows:

```
void CCode5::OnGaussSeidel()
{
      const int MAX=10;
      double Sum,error;
      double *xOld;
      int i,j,k;
      bool fStatus=1;
      CString condition="diagonally-dominant";
      method="Gauss-Seidel";
      xOld=new double [N+1];
      ReadInput();

      for (i=1;i<=n;i++)
```

```
        {
                Sum=0;
                for (j=1;j<=n;j++)
                        if (i!=j)
                                Sum += fabs(A[i][j]);
                if (fabs(A[i][i])<=Sum)
                        fStatus=0;
        }

        for (i=1;i<=n;i++)
        x[i]=0;

        // perform iterations & update x
        for (k=0;k<=MAX;k++)
        {
                error=0;
                for (i=1;i<=n;i++)
                {
                        Sum=0;
                        for (j=1;j<=n;j++)
                                if (i!=j)
                                        Sum += A[i][j]*x[j];
                        xOld[i]=x[i];
                        x[i]=(b[i]-Sum)/A[i][i];
                        error=((error>fabs(x[i]-xOld[i]))?error:
                                        fabs(x[i]-xOld[i]));
                }
        }
        delete xOld;
        if (fStatus)
                DisplayResults();
        else
                DisplayError(condition);
}
```

The last application function is OnReset(), which responds to a mouse click on the *Reset* button whose id is IDC_RESET. The function resets the data in the edit boxes and prepares for a fresh input. This result is achieved by setting the string arrays sA, sb, and sx to null (""). The code is shown below:

```
void CCode5::OnReset()
{
        int i,j;
        for (i=1;i<=N;i++)
        {
                for (j=1;j<=N;j++)
                        sA[i][j]="";
```

```
            sb[i]="";
            sx[i]="";
        }
        SetDlgItemText(IDC_A11,sA[1][1]);
        SetDlgItemText(IDC_A12,sA[1][2]);
        SetDlgItemText(IDC_A13,sA[1][3]);
        SetDlgItemText(IDC_A14,sA[1][4]);
        SetDlgItemText(IDC_A15,sA[1][5]);
        SetDlgItemText(IDC_A21,sA[2][1]);
        SetDlgItemText(IDC_A22,sA[2][2]);
        SetDlgItemText(IDC_A23,sA[2][3]);
        SetDlgItemText(IDC_A24,sA[2][4]);
        SetDlgItemText(IDC_A25,sA[2][5]);
        SetDlgItemText(IDC_A31,sA[3][1]);
        SetDlgItemText(IDC_A32,sA[3][2]);
        SetDlgItemText(IDC_A33,sA[3][3]);
        SetDlgItemText(IDC_A34,sA[3][4]);
        SetDlgItemText(IDC_A35,sA[3][5]);
        SetDlgItemText(IDC_A41,sA[4][1]);
        SetDlgItemText(IDC_A42,sA[4][2]);
        SetDlgItemText(IDC_A43,sA[4][3]);
        SetDlgItemText(IDC_A44,sA[4][4]);
        SetDlgItemText(IDC_A45,sA[4][5]);
        SetDlgItemText(IDC_A51,sA[5][1]);
        SetDlgItemText(IDC_A52,sA[5][2]);
        SetDlgItemText(IDC_A53,sA[5][3]);
        SetDlgItemText(IDC_A54,sA[5][4]);
        SetDlgItemText(IDC_A55,sA[5][5]);
        SetDlgItemText(IDC_b1,sb[1]);
        SetDlgItemText(IDC_b2,sb[2]);
        SetDlgItemText(IDC_b3,sb[3]);
        SetDlgItemText(IDC_b4,sb[4]);
        SetDlgItemText(IDC_b5,sb[5]);
        SetDlgItemText(IDC_x1,sx[1]);
        SetDlgItemText(IDC_x2,sx[2]);
        SetDlgItemText(IDC_x3,sx[3]);
        SetDlgItemText(IDC_x4,sx[4]);
        SetDlgItemText(IDC_x5,sx[5]);
}
```

5.7 SUMMARY

We have discussed various systems of linear equations, their representation, methods for solving them, and program designs in C++ as their solution. We also developed a

Windows-based version of the program using the Microsoft Foundation Class library, which provides a user-friendly interface for solving the problem.

In solving a system of linear equations, it is important to consider a user-friendly interface that allows interaction between the user and the program. On the problem side, the user must understand the factors affecting the existence and uniqueness of the solution. The solution to a mathematical problem can be drawn based on the solid understanding of the concepts that govern the problem.

Providing a friendly solution to the problem can be summarized to fall into three steps. First, the conceptual and analytic solution based on the mathematical theories must be fully understood. Second, when the first step has been cleared, the next step is to derive the manual solution to the problem using a calculator as a tool. This second approach should be applicable to a small problem, like a system of 3×3 linear equations. Once this step is completed, the third and last step involving programming can now be embarked on comfortably. This step is a realization to the problem where all the tedious work in step one and two is simplified with the click of some buttons. A computer is the ultimate tool that helps to digest the massive calculations involved in solving a problem involving systems of linear equations.

All three steps have been discussed in this chapter, with a special attention on step three. The methods for solving the systems of linear equations in this chapter are the fundamental solution to many problems in engineering and the sciences, as will be discussed later in the book.

NUMERICAL EXERCISES

1. Determine whether each matrix below possesses each of the properties: tridiagonal, positive definite, or diagonally dominant.

a. $\begin{bmatrix} 3 & -1 \\ 2 & 5 \end{bmatrix}$

b. $\begin{bmatrix} 6 & -1 & 0 \\ 3 & 7 & -1 \\ 0 & 2 & 4 \end{bmatrix}$

c. $\begin{bmatrix} 5 & -1 & 0 & 3 \\ -1 & 7 & 2 & 0 \\ 0 & 2 & 8 & 1 \\ 3 & 0 & 1 & 6 \end{bmatrix}$

d. $\begin{bmatrix} 2 & -2 & 1 & 1 \\ 3 & -3 & 2 & 2 \\ 4 & -4 & 3 & 3 \\ 5 & -5 & 4 & 4 \end{bmatrix}$

e. $\begin{bmatrix} 7 & 3 & 0 & 0 & 0 \\ 3 & -5 & 0 & 0 & 0 \\ 0 & 0 & 4 & -2 & 0 \\ 0 & 0 & -2 & 6 & 1 \\ 0 & 0 & 0 & 1 & -3 \end{bmatrix}$

f. $\begin{bmatrix} 5 & -1 & 0 & 0 \\ -1 & 5 & -1 & 0 \\ 0 & -1 & 5 & -1 \\ 0 & 0 & -1 & 5 \end{bmatrix}$

2. Determine whether a system of linear equations whose coefficient matrix is each of the matrices in Problem 1 can be solved using the the Gauss, Gauss with partial pivoting, Gauss–Jordan, Crout, Doolittle, Cholesky, Thomas, Jacobi, and Gauss–Seidel methods.

3. Solve the following systems of linear equations using the indicated method(s):

   a. Gauss, Gauss–Jordan, and Crout.

$$3x - 4y = -1$$
$$2x + 7y = 1$$

   b. Gauss, Gauss with partial pivoting, and Doolittle.

$$x_1 - 2x_2 + x_3 = -1$$
$$3x_1 - x_2 + 2x_3 = 1$$
$$5x_1 - 3x_2 + x_3 = 4$$

   c. Cholesky and Gauss–Seidel.

$$4x_1 - x_2 + x_3 = 3$$
$$-x_1 + 4x_2 + x_3 = 4$$
$$x_1 + x_2 + 4x_3 = 4$$

   d. Cholesky and Gauss–Seidel.

$$5x_1 - x_2 + x_3 + x_4 = -3$$
$$-x_1 + 5x_2 + x_3 + x_4 = -1$$
$$x_1 + x_2 + 5x_3 - x_4 = 2$$
$$x_1 - x_2 - x_3 + 5x_4 = 1$$

   e. Gauss, Thomas algorithm, and Gauss–Seidel.

$$3x_1 - x_2 = -1$$
$$x_1 + 4x_2 + x_3 = -4$$
$$x_2 - 7x_3 + 2x_4 = 2$$
$$2x_3 - 6x_4 - x_5 = 5$$
$$x_4 + 5x_5 = -4$$

Check the results by running `Code5`.

PROGRAMMING CHALLENGES

1. Design standard C++ programs for solving a system of linear equations using the following methods:
   a. Gaussian elimination method with partial pivoting.
   b. Doolittle method.

2. Design a standard C++ program to solve the system of linear equations in Example 5.11 using the Jacobi method.

3. Improve on the Code5 project by adding a few user-friendly features, as follows:
   a. Add the flexibility and scalability of the system. In the Code5 project, only up to a 5×5 system is supported. A flexible system allows scalability where any size of linear systems up to $n \times n$ can be solved. Design an interface to allow any choice of system up to 10×10 to be solved.
   b. Show the staggered results from the calculations in each method applied. For example, it is necessary to display the results from each iteration in the Gauss–Seidel method instead of just the final results. Design a new interface with the parent window from the MFC class CFrameWnd to achieve this objective.

4. Add new buttons and their corresponding functions to the Code5 project using the Gauss–Jordan, Gauss with partial pivoting, Doolittle, and Jacobi methods. In each function, add features to check the runtime errors that may result from properties such as diagonally dominance and singularity of the coefficient matrix.

5. Improve on Code5 by including the file read and open options, specifically to read and store the matrix and vector values in the problem.

Nonlinear Equations

6.1 INTRODUCTION

As described in the previous chapter, a nonlinear equation is an equation that has one or more nonlinear terms in its expression, or an equation that is not in the form of Equation (5.1). In its general form, a nonlinear equation involving n variables, $x_1, x_2, x_3, \ldots, x_n$, can be expressed as

$$f(x_1, x_2, \ldots, x_n) = 0. \tag{6.1}$$

Some examples of nonlinear equations are given below:

$3 - 3x^2 + x^3 = -2,$ nonlinear because of the presence of nonlinear terms x^2 and x^3.

$x + \sin y = -1,$ nonlinear because of the presence of nonlinear term $\sin y$.

$xy = -1,$ nonlinear because the sum of indices in the variable term is 2.

$\frac{1}{1+x},$ nonlinear because variable x is in the denominator.

Finding the root(s) of a nonlinear equation is one of the most fundamental problems involving nonlinear equations. The problem is stated as follows:

Given a function $f(x)$ where $f(x) = 0$ exists, find the value of x.

Alternatively, the problem is also called *finding the zeros of a function*. Graphically, this problem can be described as finding one or more points along the x-axis where the curve $f(x)$ crosses, provided the value(s) exist. Very often, $f(x)$ does not cross the x-axis. In this case, $f(x)$ is said to have no real root.

6.2 EXISTENCE OF SOLUTIONS

Several theorems have been documented to help in finding the zeros of a function.

Theorem 6.1. Intermediate-Value Theorem. Given a function $f(x)$ continuous in $[a, b]$ and a number C, where $f(a) \leq C \leq f(b)$, then a number $p \in (a, b)$ exists such that $f(c) = C$.

Theorem 6.2. Mean-Value Theorem. A function $f(x)$ that is continuous in $[a, b]$ is said to have at least one root in this interval if $f(a)f(b) < 0$.

The intermediate-value theorem and the mean-value theorem guarantee at least one root exists in the interval $[a, b]$ for $f(x)$ if $f(a)f(b) < 0$. However, it does not assume the uniqueness of the solution. $f(a)f(b) < 0$ suggests one of the terms in $f(a)$ or $f(b)$ is positive and the other is negative. This suggestion implies the positive term lies above the x-axis, whereas the negative term is located below the axis. Therefore, the curve must cross the x-axis at least once along the path from $(0, f(a))$ to $(0, f(b))$. This is illustrated in Figure 6.1.

In this chapter, we will concentrate on the case of finding a unique root for $f(x)$ given the solution exists. We will discuss five methods based on iterations, namely, the bisection, false position, Newton–Raphson, secant, and fixed-point methods. In each method, we will emphasize on the construction of C++ code in solving the problem and on the development of its Windows interface. All these methods are summarized in Table 6.1 for their brief description.

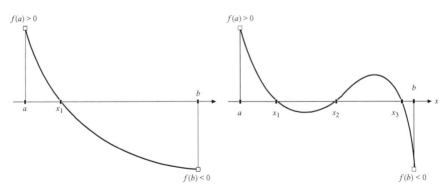

FIGURE 6.1. Mean-value theorem illustration of curves with one root (left) and three roots (right).

TABLE 6.1. Summary of the iterative methods

Method	Type	Summary of Method
Bisection	Closed interval	Update based on bisecting the middle point in the interval.
False position	Closed interval	Update based on the secant line between the end-points of the intervals.
Newton–Raphson	Open interval	Update based on the tangent line intersection with the x-axis.
Secant	Open interval	Update based on the secant line intersection with the x-axis.
Fixed-point iteration	Open interval	Update based on the formulation $x = g(x)$ from $f(x) = 0$.

An iterative technique for finding the zeros of an equation can be classified as either the closed or the open interval approach. A closed interval means the final solution is confined to within the given initial interval, whereas an open interval does not limit the solution to be within this range.

In the closed interval approach, an interval $x \in [a_0, b_0]$ that contains at least one root is identified using the mean-value theorem. The two initial values, a_0 and b_0, serve as the left and right points in the interval. The size of this initial interval is reduced gradually through successive iterations to $[a_1, b_1]$, $[a_2, b_2]$, and so on until the solution converges to an acceptable value x^* based on a stopping condition. This final value is then the final solution to the problem. The beauty of the closed interval approach is convergence to x^* is guaranteed through the mean-value theorem. Some well-known methods in the closed interval approach are the bisection and false position methods.

In the open interval approach, an initial value x_0 is selected as the starting value for x in the iteration. This value is a guess value that should not be far from the solution. A good guess for this value is either a_0 or b_0 in the interval $x \in [a_0, b_0]$ using the mean-value theorem. The interval for the convergence is said to be open in this approach as the generated values are not bound to be within a specified range. Successive iterations are then performed on the initial value, which eventually leads to convergence to its solution after a stopping condition has been reached. The Newton-Raphson, secant and fixed-point iteration methods are some of the most common methods in the open interval approach.

6.3 BISECTION METHOD

The *bisection method* is based on a closed interval that produces a solution by reducing the interval size successively into a tolerable value through a series of iterations. The iterations start at interval $x \in [a_0, b_0]$ whose end points $x = a_0$ and $x = b_0$ are any values that comply with the mean-value theorem requirement, or $f(a_0)f(b_0) < 0$.

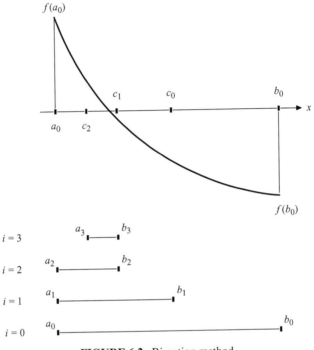

FIGURE 6.2. Bisection method.

Figure 6.2 illustrates the bisection method showing the shrinking intervals after four successive iterations. At each iteration i, the size of the interval $x \in [a_i, b_i]$ from the end points a_i and b_i is reduced into half by updating one of the two end points with the value of the interval midpoint, given by

$$c_i = \frac{a_i + b_i}{2}. \tag{6.2}$$

The update is performed by looking at the sign of $f(a_i)f(c_i)$ to quickly determine the rough location of the root. A value of $f(a_i)f(c_i) < 0$ suggests the root lies in $x \in [a_i, c_i]$, definitely not in $[c_i, b_i]$. Therefore, the end point b_{i+1} is updated with the value of c_i, whereas a_{i+1} takes on a_i, or $b_{i+1} = c_i$ and $a_{i+1} = a_i$, respectively. In a similar note, $f(a_i)f(c_i) > 0$ suggests the root lies in $x \in [c_i, b_i]$, not in $x \in [a_i, c_i]$. In this case, $a_{i+1} = c_i$ and $b_{i+1} = b_i$. In a unique case where $f(a_i)f(c_i) = 0$, the solution has been reached with $x^* = c_i$ as the root of $f(x)$.

The update on the end points of the intervals has the effect of narrowing down the search area to the stage where the difference between the end points is very small. An error given by $e = |c_i - c_{i-1}|$ for $i > 0$ is computed at each iteration. This error value is compared with a preset small value ε to determine whether the iterations are continued or stopped. The iterations are continued if $e > \varepsilon$, which suggests convergence to the solution is not reached yet. Convergence is said to have been reached if $e \leq \varepsilon$, and the whole operation is stopped immediately. The final value of c_i obtained from the last iteration then becomes the desired solution.

Algorithm 6.1 summarizes the computational steps in the bisection method. It is important to set the maximum number of iterations *max* to be a reasonable number such as 20. Depending on the value ε, convergence to the solution is normally reached before this number.

Algorithm 6.1. Bisection Method.
  Given $f(x) = 0$, ε and the initial end points $[a_0, b_0]$, where $f(a_0)f(b_0) < 0$;
  Given *max*=maximum number of iterations;
  For $i = 0$ to *max*
      Compute $c_i = \frac{a_i + b_i}{2}$;
      If $f(a_i)f(c_i) < 0$
          Update $b_{i+1} = c_i$ and $a_{i+1} = a_i$;
      If $f(a_i)f(c_i) > 0$
          Update $a_{i+1} = c_i$ and $b_{i+1} = b_i$;
      If $|c_i - c_{i-1}| < \varepsilon$
          Solution= c_i;
          Stop the iterations;
  Endfor

Figure 6.3 shows a schematic flowchart of the bisection method. The iterations are performed as long as the computed error at each iteration, $e = |c_i - c_{i-1}|$, is less than ε. There are also other ways to determine the error, including $e = |b_i - a_i|$, or $e = |f(c_i)|$.

Example 6.1. Find the root of $f(x) = x^3 - x^2 - 2$ using the bisection method, given the initial values of $a_0 = 1$ and $b_0 = 2$ with iterations until $|c_i - c_{i-1}| < \varepsilon$, where $\varepsilon = 0.005$.

Solution. A quick check confirms $f(a_0)f(b_0) = f(1)f(2) < 0$. Therefore, at least one root lies inside $x \in [1, 2]$. At $i = 0$, $a_0 = 1$ and $b_0 = 2$, and this produces $c_0 = \frac{1+2}{2} = 0.5$. It follows that $f(a_0) = f(1) = -2$ and $f(c_0) = f(1.5) = -0.875$. This gives $f(a_0)f(c_0) > 0$. The update follows with $a_1 = c_0 = 1.5$ and $b_1 = b_0 = 2$.

At $i = 1$, $a_1 = 1.5$ and $b_1 = 2$. This gives $c_1 = \frac{1.5+2}{2} = 1.75$, $f(a_1) = f(1.5) = -0.875$, and $f(c_1) = f(1.75) = 0.296875$. It follows that $f(a_1)f(c_1) < 0$ to produce an update with $b_2 = c_1 = 1.75$ and $a_2 = a_1 = 1.5$. The error is $|c_1 - c_0| = 0.25 > \varepsilon$. Therefore, the iterations continue with $i = 2$.

Table 6.2 shows the results after eight iterations, which eventually leads to convergence when the error is smaller than ε. The final solution is $x^* = c_7 = 1.699219$, which is obtained at $i = 7$ when $|c_7 - c_6| = 0.003906 < \varepsilon$.

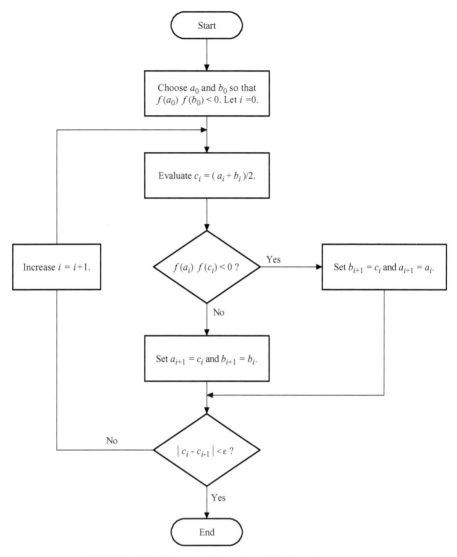

FIGURE 6.3. Schematic flowchart of the bisection method.

6.4 FALSE POSITION METHOD

The *false position method* is based on a linear interpolation of the two end points of the interval for approximating the root. The method starts with the end points a_0 and b_0 in the interval $[a_0, b_0]$, where $f(a_0)f(b_0) < 0$. The gradient of the secant line from $(a_0, f(a_0))$ to $(c_0, 0)$ is given by

$$m_1 = \frac{0 - f(a_0)}{c_0 - a_0},$$

TABLE 6.2. Numerical results from Example 6.1

i	a_i	b_i	c_i	$f(a_i)f(c_i)$	$|c_i - c_{i-1}|$
0	1.000000	2.000000	1.500000	1.750000	
1	1.500000	2.000000	1.750000	−0.259766	0.250000
2	1.500000	1.750000	1.625000	0.305908	0.125000
3	1.625000	1.750000	1.687500	0.014766	0.062500
4	1.687500	1.750000	1.718750	−0.005206	0.031250
5	1.687500	1.718750	1.703125	−0.001669	0.015625
6	1.687500	1.703125	1.695313	0.000068	0.007813
7	1.695313	1.703125	1.699219	−0.000030	0.003906

whereas the gradient from the straight line from $(c_0, 0)$ to $(b_0, f(b_0))$ is

$$m_2 = \frac{f(b_0) - 0}{b_0 - 0}.$$

Since the two lines above are colinear, or $m_1 = m_2$,

$$\frac{f(b_0) - 0}{b_0 - c_0} = \frac{f(b_0) - 0}{b_0 - c_0}.$$

Simplify the above equation, we get

$$c_0 = \frac{a_0 f(b_0) - b_0 f(a_0)}{f(b_0) - f(a_0)},$$

and this equation generalizes to

$$c_i = \frac{a_i f(b_i) - b_i f(a_i)}{f(b_i) - f(a_i)}. \tag{6.3}$$

Figure 6.4 illustrates the false position method. a_0 and b_0 are two guess values that make up the initial interval $[a_0, b_0]$ that is derived from the condition in the mean-value theorem. A secant line is drawn from $(a_0, f(a_0))$ to $(b_0, f(b_0))$. Clearly, this line intersects the x-axis at $(c_0, 0)$. After finding c_0 using Equation (6.3), the interval $[a_0, b_0]$ is divided into $[a_0, c_0]$ and $[c_0, b_0]$. If $f(a_0)f(c_0) < 0$, the root is in $[a_0, c_0]$. An update is performed with $b_1 = c_0$ and $a_1 = a_0$. If $f(a_0)f(c_0) > 0$, then the update involves $a_1 = c_0$ and $b_1 = b_0$ since the root lies in $[c_0, b_0]$. The iterations continue with $i = 1$, and so on, until the stopping criterion is met. The same rule for the stopping criterion as in the bisection method applies, and the iterations stop once the error is smaller than ε. In the case of $f(a_i)f(c_i) = 0$, c_i becomes the solution and the iteration is stopped immediately.

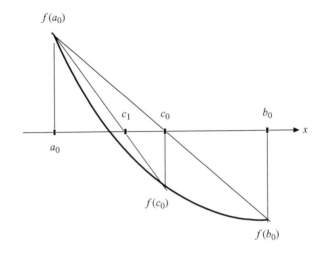

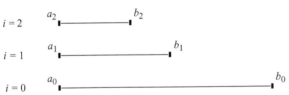

FIGURE 6.4. False position method.

Algorithm 6.2 summarizes the computational steps in the false position method. Obviously, the steps are similar to those in Algorithm 6.1 with the difference in the update formula for c_i only.

Algorithm 6.2. False Position Method.

   Given $f(x) = 0$, ε and the initial end points $[a_0, b_0]$ where $f(a_0)f(b_0) < 0$;
   Given max=maximum number of iterations;
   For $i = 0$ to max
      Compute $c_i = \frac{a_i f(b_i) - b_i f(a_i)}{f(b_i) - f(a_i)}$;
      If $f(a_i)f(c_i) < 0$
         Update $b_{i+1} = c_i$ and $a_{i+1} = a_i$;
      Endif
      If $f(a_i)f(c_i) > 0$
         Update $a_{i+1} = c_i$ and $b_{i+1} = b_i$;
      Endif
      If $|c_i - c_{i-1}| < \varepsilon$
         Solution=c_i;
         Stop the iterations;
      Endif
   Endfor

TABLE 6.3. Numerical results of Example 6.2

i	a_i	b_i	$f(a_i)$	$f(b_i)$	c_i	$f(a_i)f(c_i)$	$\|c_i - c_{i-1}\|$
0	1.000000	2.000000	-2.000000	2.000000	1.500000	1.750000	
1	1.500000	2.000000	-0.875000	2.000000	1.652174	0.192303	0.152174
2	1.652174	2.000000	-0.219775	2.000000	1.686611	0.010291	0.034437
3	1.686611	2.000000	-0.046826	2.000000	1.693781	0.000450	0.007169
4	1.693781	2.000000	-0.009617	2.000000	1.695246	0.000019	0.001465

Example 6.2. Find the root of $f(x) = x^3 - x^2 - 2$ using the false position method, given the initial values of $a_0 = 1$ and $b_0 = 2$, and the error $|c_i - c_{i-1}| < \varepsilon$, where $\varepsilon = 0.005$.

Solution. At $i = 0$, $a_0 = 1$ and $b_0 = 2$. We obtain $f(a_0) = f(1) = -2$ and $f(b_0) = f(2) = 2$. This gives $c_0 = \frac{a_0 f(b_0) - b_0 f(a_0)}{f(b_0) - f(a_0)} = \frac{(1)(2) - (2)(-2)}{(2) - (-2)} = 1.5$ and $f(c_0) = f(1.5) = -0.875$. Therefore, $f(a_0)f(c_0) > 0$, and this gives the first update with $a_1 = c_0 = 1.5$ and $b_1 = b_0 = 2$.

At $i = 1$, $a_1 = 1.5$ and $b_1 = 2$. This gives $f(a_1) = f(1.5) = -0.875$ and $f(b_1) = f(2) = 2$. We get $c_1 = \frac{a_1 f(b_1) - b_1 f(a_1)}{f(b_1) - f(a_1)} = \frac{(1.5)(2) - (2)(-0.875)}{(2) - (-0.875)} = 1.652174$. Subsequently, $f(c_1) = f(1.652) = -0.220$ and $f(a_1)f(c_1) > 0$. An update results in $a_2 = c_1 = 1.652174$ and $b_2 = b_1 = 2$. The error is $|x_1 - x_0| = |c_1 - c_0| = 0.152174 > \varepsilon$.

The iterations continue until $i = 4$, where $|c_4 - c_3| = 0.001465 < \varepsilon$. This produces the final solution with $x* = c_4 = 1.695246$. The full results are shown in Table 6.3.

6.5 NEWTON–RAPHSON METHOD

The Newton–Raphson method is an open interval method that requires one initial value. This iterative method is based on a tangent line at the approximated point that provides an update at the point where this line intersects the x-axis. An initial point is any guess value that is not far from the solution. A good guess value will be $x_0 = a$ or $x_0 = b$ from the mean-value theorem condition, $f(a_0)f(b_0) < 0$. This theorem is not required in the Newton–Raphson method, but it provides a good starting point for the iterations, which contributes in faster convergence to the solution.

Figure 6.5 shows a situation for finding a root of the function $f(x)$ with $x = x_0$ as the initial value. A tangent line is drawn from the point $(x_0, f(x_0))$, which intersects the x-axis at $(x_1, 0)$. The value $x = x_1$ is the new value at iteration $i = 0$. At $i = 1$, another tangent line is drawn from $(x_1, f(x_1))$, which intersects the x-axis at $(x_2, 0)$ to produce an improved value of $x = x_2$. The same step is repeated for $i = 2, 3, \ldots$ until the error given by $e = |x_{i+1} - x_i|$ is smaller than a preset value ε. The final value of x is obtained at the last iteration where $e < \varepsilon$ becomes the solution to the problem.

The formula for the Newton–Raphson method is derived by looking at the tangent line to the point $(x_0, f(x_0))$. The gradient of this tangent line is $f'(x_0)$, which can also

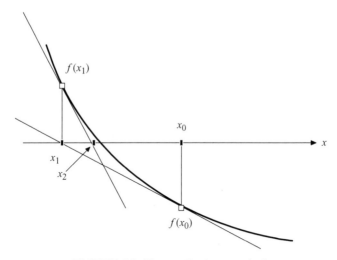

FIGURE 6.5. Newton–Raphson method.

be expressed as

$$f'(x_0) = \frac{f(x_0) - 0}{x_0 - x_1}.$$

The above equation is simplified further to produce

$$x_1 = x_0 - \frac{f(x_0)}{f'(x_0)}.$$

This equation is generalized to produce the Newton–Raphson update formula, as follows:

$$x_{i+1} = x_i - \frac{f(x_i)}{f'(x_i)}. \tag{6.4}$$

The computational steps in the Newton–Raphson method are summarized in Algorithm 6.3. The algorithm requires both $f(x)$ and $f'(x)$.

Algorithm 6.3. The Newton–Raphson Method.
   Given $f(x) = 0$, ε and the initial point x_0;
   Given *max*=maximum number of iterations;
   Find $f'(x)$;
   For $i = 0$ to *max*
      Compute $x_{i+1} = x_i - \frac{f(x_i)}{f'(x_i)}$;
      If $|x_{i+1} - x_i| < \varepsilon$
         Solution=x_{i+1};
         Stop the iterations;
      Endif
   Endfor

TABLE 6.4. Results from Example 6.3

i	x_i	$f(x_i)$	$f'(x_i)$	$|x_{i+1} - x_i|$
0	1.000000	−2.000000	1.000000	2.000000
1	3.000000	16.000000	21.000000	0.761905
2	2.238095	4.201706	10.551020	0.398227
3	1.839868	0.843048	6.475605	0.130188
4	1.709680	0.074396	5.349653	0.013907
5	1.695773	0.000796	5.235391	0.000152
6	1.695621			

Example 6.3. Use the Newton–Raphson method to find the root of $f(x) = x^3 - x^2 - 2$, given the initial value of $x_0 = 1$ with iterations until $|x_i - x_{i-1}| < \varepsilon$, where $\varepsilon = 0.005$.

Solution. With $f(x) = x^3 - x^2 - 2$ we have $f'(x) = 3x^2 - 2x$. At $i = 0$, $x_0 = 1$ and we have $x_1 = x_0 - \frac{f(x_0)}{f'(x_0)} = 3$. This produces an error of $|x_1 - x_0| = |3 - 1| = 2 > \varepsilon$. At $i = 1$, we get $x_2 = x_1 - \frac{f(x_1)}{f'(x_1)} = 2.238095$, and this produces an error of $|x_2 - x_1| = |2.238 - 3| = 0.761905 > \varepsilon$.

The iterations stop at $i = 5$, where $|x_6 - x_5| = 0.000 < \varepsilon$, and this produces the final solution, $x^* = x_6 = 1.695621$. Table 6.4 shows the full results from the iterations.

6.6 SECANT METHOD

One big difficulty with the Newton–Raphson method is in finding the derivative of $f(x)$ at the iterated points. This requirement adds to the overhead on the computer, and it may also add to the computational cost in solving the problem. An alternative approach is provided in the form of the *secant method*, which eliminates the derivative by replacing it with a secant line.

We refer to Figure 6.6 to illustrate the method. The secant method starts with two initial guess points x_0 and x_1, where $f(x_0) \neq f(x_1)$. These two points can be any points that are reliably close to the solution, and they do not have to be on the opposite sides of the x-axis. A good choice for the two initial points will be $x_0 = a$ and $x_1 = b$, which are the end points in the interval $[a, b]$ that abides the mean-value theorem requirement, $f(a)f(b) < 0$.

The first iteration with $i = 0$ produces an approximation at $x = x_2$, which is the x-intercept of the chord that passes through the points $(x_0, f(x_0))$ and $(x_1, f(x_1))$. Similarly, the next approximated value $x = x_3$ is obtained from the x-intercept of the chord that passes through the points $(x_1, f(x_1))$ and $(x_2, f(x_2))$. The same step is repeated until convergence to the solution is reached through the stopping condition $e = |x_{i+2} - x_{i+1}| < \varepsilon$.

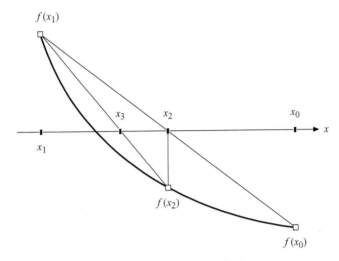

FIGURE 6.6. Secant method.

The iterative formula is derived by referring to Figure 6.6. The gradient of the line from $(x_0, f(x_0))$ to $(x_2, 0)$ is given by

$$m_1 = \frac{f(x_0) - 0}{x_0 - x_2},$$

which equals the gradient of the line from $(x_1, f(x_1))$ to $(x_2, 0)$, given by

$$m_2 = \frac{f(x_0) - 0}{x_0 - x_2} = \frac{f(x_1) - 0}{x_1 - x_2}.$$

The above equation is further simplified, as follows:

$$f(x_0)(x_1 - x_2) = f(x_1)(x_0 - x_2),$$
$$x_2 f(x_0) - x_1 f(x_0) = x_0 f(x_1) - x_2 f(x_1),$$
$$x_2(f(x_1) - f(x_0)) = x_0 f(x_1) - x_1 f(x_0).$$

This produces

$$x_2 = \frac{x_0 f(x_1) - x_1 f(x_0)}{f(x_1) - f(x_0)}.$$

Replacing x_0 with x_i, x_1 with x_{i+1} and x_2 with x_{i+2} in the above equation, we obtain the iterative formula for the secant method:

$$x_{i+2} = \frac{x_i f(x_{i+1}) - x_{i+1} f(x_i)}{f(x_{i+1}) - f(x_i)}. \tag{6.5}$$

The computational steps in the secant method are summarized in Algorithm 6.4.

Algorithm 6.4. Secant Method.
  Given $f(x) = 0$, ε and two initial points, x_0 and x_1;
  Given max=maximum number of iterations;
  For $i = 0$ to max
    Compute $x_{i+2} = \frac{x_i f(x_{i+1}) - x_{i+1} f(x_i)}{f(x_{i+1}) - f(x_i)}$;
    If $|x_{i+2} - x_{i+1}| < \varepsilon$
      Solution=x_{i+2};
      Stop the iterations;
    Endif
  Endfor

Example 6.4. Find the root of $f(x) = x^3 - x^2 - 2$ using the secant method with the initial values of $x_0 = 1$ and $x_1 = 2$. Perform the iterations until $|x_{i+2} - x_{i+1}| < \varepsilon$, where $\varepsilon = 0.005$.

Solution. At $i = 0$, $x_0 = 1$ and $x_1 = 2$. We get $x_2 = \frac{x_0 f(x_1) - x_1 f(x_0)}{f(x_1) - f(x_0)} = \frac{(1)(2) - (2)(-2)}{(2) - (-2)} = $ 1.500. The error is $|x_2 - x_1| = |1.5 - 2| = 0.5 > \varepsilon$. At $i = 1$, $x_3 = \frac{x_1 f(x_2) - x_2 f(x_1)}{f(x_2) - f(x_1)} = $ 1.652174. The error is $|x_3 - x_2| = |1.652174 - 1.500| = 0.152174 > \varepsilon$. The iterations stop at $i = 4$, where $|x_6 - x_5| = 0.000 < \varepsilon$, and this produces the solution, $x^* = x_6 = 1.695619$. The full results from the iterations are shown in Table 6.5.

TABLE 6.5. Numerical results from Example 6.4

i	x_i	$f(x_i)$	$|x_{i+2} - x_{i+1}|$
0	1.000000	−2.000000	0.500000
1	2.000000	2.000000	0.152174
2	1.500000	−0.875000	0.051042
3	1.652174	−0.219775	0.007858
4	1.703216	0.039990	0.000261
5	1.695358		
6	1.695619		

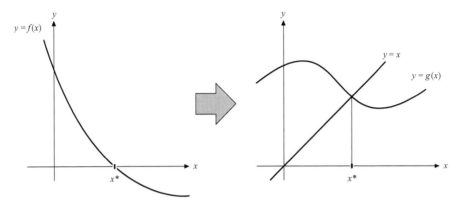

FIGURE 6.7. Reassigning the problem from $f(x) = 0$ to $y = g(x)$.

6.7 FIXED-POINT ITERATION METHOD

Through simplification, the equation $f(x) = 0$ maybe rewritten as $x = g(x)$. This relationship can be obtained by making x the subject of the new equation from $f(x) = 0$. This representation can be viewed as reassigning the problem into finding the intersection point between the straight line $y = x$ and the curve $y = g(x)$ provided the solution exists. Figure 6.7 shows this connection where $g(x)$ is totally a different curve from $f(x)$.

The fixed-point iteration method requires an initial point $x = x_0$, which is a guess value relatively close to the root of $f(x)$. Choosing $x_0 = a$ or $x_0 = b$ from the interval $x \in [a, b]$, where $f(a)f(b) < 0$ is an ideal choice for this value as it is close to the solution. The reassignment from $f(x) = 0$ to $x = g(x)$ gives rise to an iterative relationship given by

$$x_{i+1} = g(x_i) \tag{6.6}$$

for $i = 0, 1, 2, \ldots$. Convergence to the solution is guaranteed if the following condition is met:

$$|g'(x_i)| \leq 1. \tag{6.7}$$

Equation (6.7) suggests $-1 \leq g'(x) \leq 1$, and that the choice for the new equation $x = g(x)$ from $f(x) = 0$ must strictly obey this condition. There is no guarantee for convergence to the solution if this condition is not fulfilled. Figure 6.8 illustrates a case with convergence to the solution according to $x_0 \rightarrow x_1 \rightarrow x_2 \rightarrow x_3$ when the above condition is fulfilled.

The fixed-point iteration method is outlined in Algorithm 6.5. The algorithm is illustrated in Example 6.5.

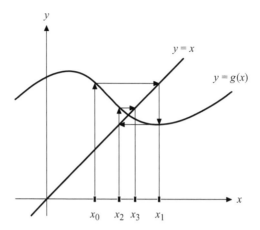

FIGURE 6.8. Convergence in the fixed-point iteration method.

Algorithm 6.5. Fixed-Point Iteration Method.
  Given $f(x) = 0$, ε and an initial points x_0;
  Given *max*=maximum number of iterations;
  Find a suitable equation $x = g(x)$ from $f(x) = 0$ where $-1 \leq g'(x) \leq 1$;
  For $i = 0$ to *max*
      Compute $x_{i+1} = g(x_i)$;
      If $|x_{i+1} - x_i| < \varepsilon$
          Solution=x_{i+1};
          Stop the iterations;
      Endif
  Endfor

Example 6.5. Find the root of $f(x) = x^3 - x^2 - 2$ using the fixed-point iteration method, given the initial value of $x_0 = 1$ with iterations until $|x_{i+1} - x_i| < \varepsilon$, where $\varepsilon = 0.005$.

Solution. Setting $f(x) = 0$ suggests several representations in the form of $x = g(x)$, including $x = \pm\sqrt{x^3 - 2}$, $x = \frac{2}{x^2 - x}$ and $x = (x^2 + 2)^{1/3}$. The first two functions are not suitable as they are not defined at $x_0 = 1$. Obviously, $g(x) = (x^2 + 2)^{1/3}$ is a good candidate as $g'(x) = \frac{2x}{3(x^2+2)^{2/3}}$ is a monotonically increasing function with values between 0 and 1. Hence, $x_{i+1} = g(x_i) = (x_i^2 + 2)^{1/3}$ is a suitable choice here.

  Starting the iterations at $i = 0$ with the initial value of $x_0 = 1$ produces $x_1 = (x_0^2 + 2)^{1/3} = (1^2 + 2)^{1/3} = 1.442250$, with $g'(x_0) = 0.320500$. The error is $|x_1 - x_0| = |1.442250 - 1.000| = 0.442250 > \varepsilon$.

  At $i = 1$, $x_2 = (x_1^2 + 2)^{1/3} = (1.442250^2 + 2)^{1/3} = 1.597925$ and $g'(x_1) = 0.376562$. The error is $|x_2 - x_1| = |1.597925 - 1.442250| = 0.155675 > \varepsilon$.

TABLE 6.6. Numerical results from Example 6.5

i	x_i	$g(x_i)$	$|g'(x_i)|$	$|x_{i+1} - x_i|$
0	1.000000	1.442250	0.320500	0.442250
1	1.442250	1.597925	0.376562	0.155675
2	1.597925	1.657464	0.387772	0.059539
3	1.657464	1.680656	0.391197	0.023192
4	1.680656	1.689743	0.392416	0.009087
5	1.689743	1.693311	0.392877	0.003568
6	1.693311			

Therefore, the process continues with the next iteration. The iterations stop at $i = 5$, where $|x_6 - x_5| = 0.003568 < \varepsilon$. The final solution is $x^* = x_6 = 1.693311$. Table 6.6 shows the complete results from this method.

6.8 VISUAL SOLUTION: CODE6

Code6. User Manual.

1. Select a method from the menu.
2. Enter the input values for the selected method.
3. Click the *Compute* push button to view the results.

Development files: Code6.cpp, Code6.h and MyParser.obj.

MFC provides a rich environment for developing a user-friendly interface for numeric-intensive calculations. In this chapter, we illustrate and discuss a Windows interface for finding the zeros of a function using four methods, namely, the bisection, false position, Newton–Raphson, and secant methods. The other method discussed in the last section, the fixed-point iteration method, is left as an exercise for the reader.

The Windows project for the nonlinear equations is codenamed Code6. Figure 6.9 shows a screen snapshot from the project that consists of an output from the bi-section method. The input in this problem consists of the function $f(x) = 3\sin 2x - 5\cos(4x - 1)$ in $0 \le x \le 1$, and the results are displayed in the list view table and plot-ted as a graph. The solution to $f(x) = 0$ in this problem is displayed as $x = 0.508761$.

The displayed output is made up of four regions. The first region consists of the menu items that are located in the top left corner of the window. The second region consists of the input boxes that are located to the right of the menu items. The results from the calculations that are displayed in a list view table located below the menu items. The last region is located at the bottom, and it displays the solution curve.

The menu region consists of shaded rectangles for the bisection, secant, Newton–Raphson and secant methods. An item in the menu becomes activated from the left button click of the mouse. A click at one of these items prompts the creation of the

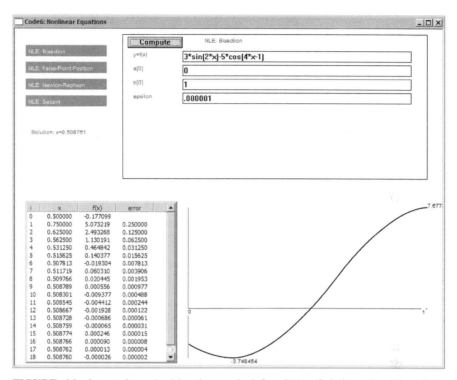

FIGURE 6.9. Output from the bisection method for $f(x) = 3\sin 2x - 5\cos(4x - 1)$ in Code6.

input region. Edit boxes for input according to the selected method appear inside the input region. A push button called *Compute* is also displayed in the input region. A click at the push button causes the input to be read and processed. A complete input for the selected method will produce the desired results in the list view table and its corresponding graph in the curve region.

Figure 6.10 shows the development stages of Code6. A Boolean flag called fStatus monitors the progress of the execution whose initial value of 0 indicates the execution is not yet complete. Another variable called fMenu (not shown in the diagram) stores a value for the selected method: with fMenu=1 for bisection, fMenu=2 for false position, fMenu=3 for Newton–Raphson, and fMenu=4 for secant. A value for fMenu is assigned inside the function OnLButtonDown(). The stage at fStatus=1 indicates the input for the selected method has been completed, and executed. The results are displayed both in the list view table and visualized as a curve with this status.

Code6 consists of two source files, Code6.cpp and Code6.h, and the object file, MyParser.obj. A single class called CCode6 is used in this application. This class is derived from MFC's CFrameWnd, which displays a single window.

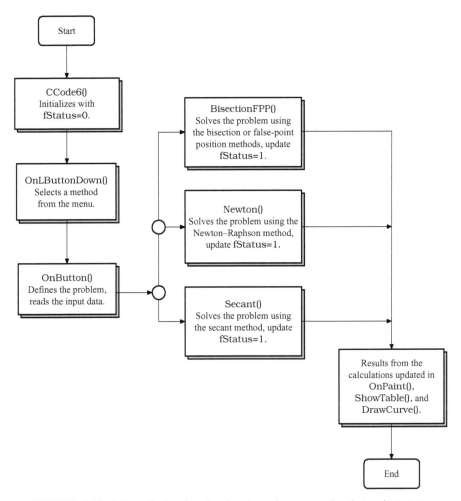

FIGURE 6.10. Schematic drawing showing the main computational steps in `Code6`.

The main variables in this project are represented as four structures: PT, INPUT, MENU, and CURVE. PT defines the points based on the real coordinates, as described in Table 6.7. An array called pt defines the points in the real coordinates in PT.

```
typedef struct
{
      double x,y,yd1,a,b,c,error;
} PT;
PT *pt,max,min,left,right;
```

TABLE 6.7. The elements of PT

Variable	Type	Description
pt[i].y, pt[i].y	double	The point (x_i, y_i).
pt[i].a, pt[i].b	double	The left and right points in the interval $a_i \leq x \leq b_i$.
pt[i].error	double	Error at iteration i.
pt[i].c	double	The point c_i between a_i and b_i in the bisection and false position methods.
pt[i].yd	double	The derivative $y_i' = f'(x_i)$ in the Newton–Raphson method.
max, min	CPoint	The maximum and minimum points in the curve $y = f(x)$.
left, right	CPoint	The left and right points in the curve $y = f(x)$.

TABLE 6.8. Input elements in INPUT

Variable	Type	Description
input[i].item	CString	Value in string input by the user.
input[i].label	CString	Label or title for the corresponding edit box.
input[i].ed	CEdit	ith edit box for collecting the input string.
input[i].hm	CPoint	Home coordinates of the ith edit box.
input[i].rc	CRect	Rectangular object for the ith edit box.
input[i].display	CRect	Rectangular area for displaying the problem.

The number of input items varies with the methods. For example, the bisection method requires four inputs, whereas the secant method requires five. A macro called maxInput is declared to store the maximum number of input, whereas nInputItems is the actual number of inputs for the selected method. Input is made using the edit boxes, which are located in the input region. A structure called INPUT defines the edit boxes and their components, and this is described in Table 6.8.

```
typedef struct
{
     CString label,item;
     CPoint hm;
     CEdit ed;
     CRect rc,display;
} INPUT;
INPUT input[maxInput+1];
```

The menu items are organized into a structure called MENU. A macro called nMenuItems stores the number of menu items whose value is four in this project. The elements of this structure are the menu titles, their rectangular objects, and their home coordinates. The elements are described in Table 6.9.

TABLE 6.9. Menu elements in MENU

Variable	Type	Description
menu[i].item	CString	Item i in the menu.
menu[i].hm	CPoint	Home coordinates of the ith item in the menu.
menu[i].rc	CRect	The rectangle for the ith item in the menu.

```
typedef struct
{
      CString item;
      CPoint hm;
      CRect rc;
} MENU;
MENU menu[nItems+1];
```

A structure called CURVE organizes the generated curve so that it is confined to a rectangular region called rc. The structure has its starting point and end point represented by hm and end, respectively.

```
typedef struct
{
      CPoint hm,end;
      CRect rc;
} CURVE;
CURVE curve;
```

Table 6.10 describes other main variables and objects in Code6.h. The execution of the program is monitored through the value of fStatus, where fStatus=0 is the pre-evaluation stage and fStatus=1 indicates a method has been applied to the problem. Therefore, this variable is suitable to be declared as a Boolean. fMenu is an integer variable for denoting the selected method. Stop is a variable that stores the last iteration number before convergence for all methods, whereas the last value of x from this iteration is stored as Solution, which becomes the solution to the problem.

CCode6 has seven member functions that are described in Table 6.11. BisectionFPP() is a function shared by the bisection and false position methods. It is not necessary to create separate functions for these two methods as the only difference between them is the formula for c_i. The Newton–Raphson and secant methods are represented by Newton() and Secant(), respectively. Other functions in Code6 are more or less similar to functions of similar names as in the previous chapters. They are described briefly in Table 6.11.

There are three events in Code6, and each one of them is handled by its corresponding function. The code is given as

TABLE 6.10. Other variables and objects in Code6

Variable/object	Type	Description
fStatus	bool	A flag whose values are fStatus=0 and fStatus=1, indicating incomplete and complete inputs, respectively.
fMenu	int	Flag for the menu whose value indicates the selected method, fMenu=1 for bisection, fMenu=2 for false position, fMenu=3 for Newton-Raphson, and fMenu=4 for secant.
nInputItems	int	Number of input items in the selected method.
Stop	int	The last iteration whose value is smaller or equals the stopping value, ε.
Solution	double	The last x value in the iterations before convergence.
idc	int	Id for the control resources.
btn	CButton	Push button object called *Compute*.
table	CListCtrl	List view table for displaying the results from iterations.

TABLE 6.11. Member functions in CCode6

Function	Description
CCode6()	Constructor.
~CCode6	Destructor.
BisectionFPP()	Solves the problem using the bisection and false position methods.
Newton()	Solves the problem using the Newton–Raphson method.
Secant()	Solves the problem using the secant method.
DrawCurve()	Draws the curve $y = f(x)$ in the given interval.
ShowTable()	Creates a list view table to display the results.
OnLButtonDown()	Responds to ON_WM_LBUTTONDOWN, which the assigns the method in the menu.
OnButton()	Responds to ON_BN_CLICKED, which reads the input from the user and calls the respective method to produce its solution.
OnPaint()	Displays and updates the output in the main window.

```
BEGIN_MESSAGE_MAP(CCode6,CFrameWnd)
      ON_WM_PAINT()
      ON_WM_LBUTTONDOWN()
      ON_BN_CLICKED(IDC_BUTTON,OnButton)
END_MESSAGE_MAP()
```

The process starts at the constructor. Basically, the constructor allocates memory for the class. It is also in the constructor that several variables are assigned with their initial values. The initial values include the location of the objects, the size of each object, and its title. It is also important to set the initial value for fStatus=0 to indicate no input has been read yet at this level of execution. At this initial stage, the menu has not been displayed, and its flag is set to fMenu=0. The code is given as

```
CCode6::CCode6()
{
        Create(NULL,"Code6: Nonlinear Equations",
                WS_OVERLAPPEDWINDOW,CRect(0,0,800,635),NULL);
        Arial80.CreatePointFont(80,"Arial");
        pt=new PT [m+1];
        fMenu=0; fStatus=0; idc=301;
        menu[1].item="NLE: Bisection";
        menu[2].item="NLE: False-Point Position";
        menu[3].item="NLE: Newton-Raphson";
        menu[4].item="NLE: Secant";
        for (int i=1;i<=nMenuItems;i++)
        {
                menu[i].hm=CPoint(20,30+(i-1)*30);
                menu[i].rc=CRect(menu[i].hm.x,menu[i].hm.y,
                        menu[i].hm.x+150,menu[i].hm.y+20);
        }
        curve.hm=CPoint(320,320); curve.end=CPoint(760,590);
        curve.rc=CRect(curve.hm.x-10,curve.hm.y-10,
                curve.end.x+10,curve.end.y+10);
        input[0].hm=CPoint(200,10);
        input[0].rc=CRect(input[0].hm.x,input[0].hm.y,
                input[0].hm.x+560,input[0].hm.y+260);
        input[0].display=CRect(30,180,170,250);
        for (i=1;i<=maxInput;i++)
                input[i].hm=CPoint(input[0].hm.x+10,input[0].
                        hm.y+30+(i-1)*25);
}
```

OnPaint() displays and updates the output regularly. The initial display in the main window consists of the menu items that are located in shaded rectangular boxes in the menu region. An update is made immediately whenever InvalidateRect() is invoked. The update is made possible through the test

```
if (fMenu!=0)
```

This conditional test separates the initial screen display from the time when a method has been selected. Another test written as

```
if (fStatus)
```

indicates the method has been successfully applied and produces its solution. The message updates the window by drawing the corresponding curve through

DrawCurve() and displaying the solution. If the method fails, for example in the case of the absence of a root in the given interval, then the message *No solution* is displayed. The complete code for OnPaint() is shown below:

```
void CCode6::OnPaint()
{
        CPaintDC dc(this);
        CString str;
        dc.SelectObject(Arial80);
        dc.SetBkColor(RGB(150,150,150));
        dc.SetTextColor(RGB(255,255,255));
        for (int i=1;i<=nMenuItems;i++)
        {
                dc.FillSolidRect(&menu[i].rc,RGB(150,150,150));
                dc.TextOut(menu[i].hm.x+5,menu[i].hm.y+5,menu
                        [i].item);
        }
        dc.SetBkColor(RGB(255,255,255));
        dc.SetTextColor(RGB(100,100,100));
        dc.Rectangle(input[0].rc);
        dc.TextOut(input[0].hm.x+150,
                input[0].hm.y+5,menu[fMenu].item);
        if (fMenu!=0)
        {
                for (i=1;i<=nInputItems;i++)
                        dc.TextOut(input[i].hm.x+10,
                        input[i].hm.y,input[i].label);
                if (fStatus)
                {
                        DrawCurve();
                        str.Format("Solution: x=%lf",Solution);
                        dc.TextOut(input[0].display.left,input
                                [0].display.top,str);
                        fStatus=0;
                }
                else
                        dc.TextOut(input[0].display.left,input
                                [0].display.top,
                                "No solution");
        }
}
```

OnLButtonDown() reponds to ON_WM_LBUTTONDOWN, which corresponds to a click at one of the items in the menu. With the menu displayed, the user has the choice of four methods from which to choose. OnLButtonDown() handles this event and returns true through the test

```
if (menu[k].rc.PtInRect(pt))
```

The above conditional test makes sure that only a left-click at the menu items is valid. The click at one of the menu items causes the assignment of a number to fMenu, with fMenu=1 for bisection, fMenu=2 for false position, fMenu=3 for Newton–Raphson, and fMenu=4 for secant. With this selection, edit boxes and the *Compute* push button appear in the input region. The complete code for OnLButtonDown() is shown below:

```
void CCode6::OnLButtonDown(UINT nFlags,CPoint pt)
{
    int i,k;
    for (k=1;k<=nMenuItems;k++)          // menu items
        if (menu[k].rc.PtInRect(pt))
        {
            fMenu=k;
            InvalidateRect(input[0].rc);
            switch(k)
            {
                case 1:
                case 2:
                    nInputItems=4;
                    input[1].label="y=f(x)";
                    input[2].label="a[0]";
                    input[3].label="b[0]";
                    input[4].label="epsilon";
                    break;
                case 3:
                    nInputItems=5;
                    input[1].label="y=f(x)";
                    input[2].label="y=f'(x)";
                    input[3].label="x[0]";
                    input[4].label="x[m]";
                    input[5].label="epsilon";
                    break;
                case 4:
                    nInputItems=5;
                    input[1].label="y=f(x)";
                    input[2].label="x[0]";
                    input[3].label="x[1]";
                    input[4].label="x[m]";
```

```
                              input[5].label="epsilon";
                              break;
                      }
                      for (int i=1;i<=maxInput-1;i++)
                              input[i].ed.DestroyWindow();
                      btn.DestroyWindow();
                      btn.Create("Compute",WS_CHILD|WS_VISIBLE|
                              BS_DEFPUSHBUTTON,
                              CRect(CPoint(input[0].hm.x+10,
                              input[0].hm.y+5),
                              CSize(100,20)),this,IDC_BUTTON);
                      for (i=1;i<=nInputItems;i++)
                              input[i].ed.Create(WS_CHILD|
                                      WS_VISIBLE|  WS_BORDER,
                                      CRect(input[i].hm.x+100,
                                      input[i].hm.y,input[i].hm.x+520,
                                      input[i].
                                      hm.y+20),this,idc++);
              }
      }
```

The user now completes the input by filling in the values in the edit boxes. Once this is done, a click at *Compute* confirms the entries, and the process proceeds into the next stage. This event is detected as ON_BN_CLICKED, and its message handler is OnButton(). This function reads the input values, and then calls the corresponding function according to its fMenu value.

```
void CCode6::OnButton()
{
      CString str;
      if (fMenu!=0)
      {
              for (int i=1;i<=nInputItems;i++)
                      input[i].ed.GetWindowText(input[i].item);
              if (fMenu==1 || fMenu==2)
                      BisectionFPP();
              if (fMenu==3)
                      Newton();
              if (fMenu==4)
                      Secant();
              if (fStatus)
              {
                      ShowTable();
                      InvalidateRect(curve.rc);
```

```
                    InvalidateRect(input[0].display);
            }
            else
                    InvalidateRect(input[0].display);
        }
}
```

In the above code segment, a conditional test is performed to determine whether the problem has been solved successfully using the selected method. This is done through

```
    if (fStatus)
```

A successful method is one that converges to its solution. This is indicated through the assignment of fStatus=1. A message is also displayed in the main window if the method fails, which is indicated through fStatus=0.

The value of the status flag fStatus is updated inside one of the functions from BisectionFPP(), Newton(), and Secant(). An update of fStatus=1 indicates the input data have been successfully executed in the selected method. A value of 0 means the method fails mostly from wrong end points in locating the root. With fStatus=1, the process now proceeds to the next stage, which is a display of the results in the table through ShowTable(). The graph of $y = f(x)$ is also drawn through an update in OnPaint() using InvalidateRect(). The final solution in the form of an approximated value of the root to $y = f(x)$ is displayed through InvalidateRect(input[0].display).

The bisection and the false position methods are handled by BisectionFPP(). In terms of concepts, the two methods are similar as they are based on a fixed starting interval given by $a_0 \le x \le b_0$. The only difference between them is the way an update is made on c_i. The code segment for BisectionFPP() is given by

```
void CCode6::BisectionFPP()
{
        int i,psi[2];
        double fa,fb,fc,psv[2],epsilon;
        pt[0].a=atof(input[2].item);
        pt[0].b=atof(input[3].item);
        epsilon=atof(input[4].item);
        psi[1]=23; psv[1]=pt[0].a;
        fa=parse(input[1].item,1,psv,psi);
        psv[1]=pt[0].b;
        fb=parse(input[1].item,1,psv,psi);
        if (fa*fb<0)
        {
                for (i=0;i<=maxIter;i++)
                {
```

```
if (fMenu==1)
        pt[i].c=(pt[i].a+pt[i].b)/2;
if (fMenu==2)
        pt[i].c=(pt[i].a*fb-pt[i].b*fa)/(fb-fa);
psv[1]=pt[i].a;
fa=parse(input[1].item,1,psv,psi);
psv[1]=pt[i].b;
fb=parse(input[1].item,1,psv,psi);
psv[1]=pt[i].c; pt[i].x=pt[i].c;
fc=parse(input[1].item,1,psv,psi);pt[i].y=fc;
if (fa*fc>0)
{
        pt[i+1].a=pt[i].c; pt[i+1].b=pt[i].b;
}
else
{
        pt[i+1].b=pt[i].c; pt[i+1].a=pt[i].a;
}
if (i>0)
{
        pt[i].error=fabs(pt[i].c-pt[i-1].c);
        if (pt[i].error<epsilon)
        {
                Stop=i;
                Solution=pt[Stop].c;
                fStatus=1;
                break;
        }
}
        }
    }
}
```

The bisection and false position methods will only produce the desired results if the starting interval complies with the mean-value theorem. In the above code segment, a conditional test is performed for this possibility, as follows:

```
if (fa*fb<0)
```

In the above test, fa and fb are $f(a_0)$ and $f(b_0)$, respectively. This test performs a check on the validity of the starting points in the interval through the mean-value theorem.

The iterations in the bisection and false position methods are governed by a comparison on the error, $|c_i - c_{i-1}| < \varepsilon$ or pt[i].error<epsilon. A check is made

at every iteration to determine whether the stopping criteria have been achieved, through

```
pt[i].error=fabs(pt[i].c-pt[i-1].c);
if (pt[i].error<epsilon)
{
    Stop=i;
    Solution=pt[Stop].c;
    fStatus=1;
    break;
}
```

Several assignments are made if the stopping mark for the iterations has been achieved. First, the last iteration number is stored as Stop. The last value of c_i is then the final solution, and this value is assigned to Solution. Finally, the status flag is assigned with a value of 1 to indicate the method is successful.

The Newton–Raphson method is handled by Newton(). This method is easier to implement than the bisection and false position methods as the method does not involve an update at the end points of the intervals. The code is shown as follows:

```
void CCode6::Newton()
{
    int i,psi[2];
    double epsilon,psv[2];
    pt[0].x=atof(input[3].item);
    pt[m].x=atof(input[4].item);
    epsilon=atof(input[5].item);
    psi[1]=23;
    for (i=0;i<=maxIter;i++)
    {
        psv[1]=pt[i].x;
        pt[i].y=parse(input[1].item,1,psv,psi);
        pt[i].yd1=parse(input[2].item,1,psv,psi);
        if (i<m)
        {
            pt[i+1].x=pt[i].x-pt[i].y/pt[i].yd1;
            pt[i].error=fabs(pt[i+1].x-pt[i].x);
        }
        if (pt[i].error<epsilon)
        {
            Stop=i;
            Solution=pt[Stop].x;
            fStatus=1;
            break;
        }
    }
}
```

A difficulty with the Newton–Raphson method is the requirement of the first derivative in its update formula. Since our application does not support symbolic computing for determining the derivative, it is necessary for the user to enter the input string. In the above code segment, $f'(x)$ is denoted as pt[i].yd1, and its string value is read as input[2].item in the edit box. This string will be processed through parse() to produce the numerical value of $f'(x)$ at the given point.

The error in the Newton–Raphson method is $|x_{i+1} - x_i|$, and this is written as pt[i].error=fabs(pt[i+1].x-pt[i].x). The error is again compared with epsilon at each iteration to determine whether convergence has been achieved.

The secant method is handled by Secant(). The method works on the same concept as in the Newton–Raphson method with the derivative replaced by an approximation using a secant line. The code segment for Secant() is given by

```
void CCode6::Secant()
{
        int i,psi[2];
        double epsilon,psv[2];
        pt[0].x=atof(input[2].item);
        pt[1].x=atof(input[3].item);
        pt[m].x=atof(input[4].item);
        epsilon=atof(input[5].item);
        psi[1]=23; psv[1]=pt[0].x;
        pt[0].y=parse(input[1].item,1,psv,psi);
        for (i=0;i<=maxIter;i++)
        {
                if (i<m-2)
                {
                        psv[1]=pt[i+1].x;
                        pt[i+1].y=parse(input[1].item,1,psv,psi);
                        pt[i+2].x=(pt[i].x*pt[i+1].y
                                -pt[i+1].x*pt[i].y)/(pt[i+1].y-pt[i].y);
                        pt[i].error=fabs(pt[i+2].x-pt[i+1].x);
                }
                if (pt[i].error<epsilon)
                {
                        Stop=i;
                        Solution=pt[Stop].x;
                        fStatus=1;
                        break;
                }
        }
}
```

The two-ahead update in the value of x for the secant method is shown in the above code segment through

```
if (i<m-2)
{
     psv[1]=pt[i+1].x;
     pt[i+1].y=parse(input[1].item,1,psv,psi);
     pt[i+2].x=(pt[i].x*pt[i+1].y-pt[i+1].
         x*pt[i].y)/(pt[i+1].y-pt[i].y);
     pt[i].error=fabs(pt[i+2].x-pt[i+1].x);
}
```

The error in the secant formula is pt[i].error=fabs(pt[i+2].x-pt[i+1].x) or $|x_{i+2} - x_{i+1}|$. As in the other three methods, the error is compared with epsilon at each iteration to determine whether a stopping mark has been reached.

The results from the iterations in the selected method are tabulated in the list view table through ShowTable(). The items displayed are the iteration number i, x_i, $f(x_i)$ and the corresponding error values. The code segment for ShowTable() is given by

```
void CCode6::ShowTable()
{
     CString str;
     CPoint hTable=CPoint(20,310);
     CRect rcTable=CRect(hTable.x,hTable.y,hTable.x+280,
     hTable.y+290);
     table.DestroyWindow();
     table.Create(WS_VISIBLE | WS_CHILD | WS_DLGFRAME |
             LVS_REPORT| LVS_NOSORTHEADER,rcTable,this,idc++);
     table.InsertColumn(0,"i",LVCFMT_CENTER,25);
     table.InsertColumn(1,"x",LVCFMT_CENTER,70);
     table.InsertColumn(2,"f(x)",LVCFMT_CENTER,70);
     table.InsertColumn(3,"error",LVCFMT_CENTER,70);
     for (int i=0;i<=Stop;i++)
     {
             str.Format("%d",i); table.InsertItem(i,str,0);
             str.Format("%lf",pt[i].x);
             table.SetItemText(i,1,str);
             str.Format("%lf",pt[i].y);
             table.SetItemText(i,2,str);
             if (fMenu>2 || ((fMenu<=2) && i>0))
             {
                     str.Format("%lf",pt[i].error);
                     table.SetItemText(i,3,str);
             }
     }
}
```

DrawCurve() is basically the same function as discussed earlier in Chapter 4. There are some modifications to cater the needs in the methods discussed. The graph is not drawn if the selected method fails. The full code segment for DrawCurve() is given as follows:

```
void CCode6::DrawCurve()
{
        CClientDC dc(this);
        int i,psi[2];
        double h,m1,m2,c1,c2,psv[2];
        CString str;
        CPoint px;
        psi[1]=23;
        if (fMenu==1 || fMenu==2)
        {
                left.x=pt[0].a;
                right.x=pt[0].b;
        }
        if (fMenu==3 || fMenu==4)
        {
                left.x=Solution-2;
                right.x=Solution+2;
        }
        h=(right.x-left.x)/(double)m;
        pt[0].x=left.x;
        psv[1]=pt[0].x;
        pt[0].y=parse(input[1].item,1,psv,psi);
        max.y=pt[0].y; min.y=pt[0].y;
        for (i=1;i<=m;i++)
        {
                pt[i].x=pt[i-1].x+h;
                psv[1]=pt[i].x;
                pt[i].y=parse(input[1].item,1,psv,psi);
                if (max.y<pt[i].y)
                        max.y=pt[i].y;
                if (min.y>pt[i].y)
                        min.y=pt[i].y;
        }

        // Cartesian-Windows conversion coordinates
        m1=(double)(curve.end.x-curve.hm.x)/(right.x-left.x);
        c1=(double)curve.hm.x-left.x*m1;
        m2=(double)(curve.hm.y-curve.end.y)/(max.y-min.y);
        c2=(double)curve.end.y-min.y*m2;
```

```
// Draw & label the x,y axis
CPen pGray(PS_SOLID,1,RGB(100,100,100));
dc.SelectObject(pGray);
dc.SelectObject(Arial80);
px=CPoint(m1*0+c1,m2*min.y+c2); dc.MoveTo(px);
px=CPoint(m1*0+c1,m2*max.y+c2); dc.LineTo(px);

px=CPoint(m1*left.x+c1,m2*0+c2); dc.MoveTo(px);
str.Format("%.0lf",left.x); dc.TextOut(px.x,px.y,str);
px=CPoint(m1*right.x+c1,m2*0+c2); dc.LineTo(px);
str.Format("%.0lf",right.x); dc.TextOut(px.x-10,px.y,str);

// draw the curve
CPen pDark(PS_SOLID,2,RGB(50,50,50));
dc.SelectObject(pDark);
for (i=0;i<=m;i++)
{
        px=CPoint((int)(m1*pt[i].x+c1),
                    (int)(m2*pt[i].y+c2));
        if (i==0)
                dc.MoveTo(px);
        else
                dc.LineTo(px);
        if (pt[i].y==max.y)
        {
                str.Format("%.6lf",max.y);
                dc.TextOut(px.x,px.y-10,str);
        }
        if (pt[i].y==min.y)
        {
                str.Format("%.6lf",min.y);
                dc.TextOut(px.x,px.y,str);
        }
    }
}
```

The interval of x for the curve has been set differently the methods. In the case of the bisection and false position methods, DrawCurve() draws a graph based on the input values in the left and right intervals, or $a_0 \le x \le b_0$. This is given by

```
if (fMenu==1 || fMenu==2)
{
      left.x=pt[0].a;
      right.x=pt[0].b;
}
```

6.9 SUMMARY

We discussed five methods for finding the zeros of a function $f(x)$, namely, the bisection, false position, Newton–Raphson, secant, and fixed-point iteration methods. Four methods are illustrated in their visual interfaces using Visual C++. Each method is represented as an item in the menu. The solution for each method is shown in a list view table and is depicted in its solution curve. This interface provides the fundamental requirement for visualization on the nonlinear equation problems.

The nonlinear problem arises in many science and engineering applications. In many cases, the problem appears as a small component from the overall problem. Solving the problem as a component definitely contributes to solving the overall problem. The Windows-friendly interface for handling the numerical computations using C++ helps much in contributing to the solution.

NUMERICAL EXERCISES

1. Referring to the mean-value theorem, determine whether at least one root exists in the given intervals of the following problems:
   a. $f(x) = 1 - 3x, x \in [-1, 2]$.
   b. $f(x) = 1 - 3x + 5x^2, x \in [-1, 2]$.
   c. $f(x) = 1 - 3x + 5x^2 - 6x^4, x \in [-1, 2]$.
   d. $f(x) = 3 \sin x - 5 \cos x, x \in [-1, 2]$.
   e. $f(x) = 3 \sin(2x - 1) - 5 \cos(1 - 2 \sin x)), x \in [-1, 2]$.
   Check the results through their corresponding graphs by running Code6.

2. Solve each problem in Question 1 using the following methods:
   a. Bisection method.
   b. False position method.
   c. Newton–Raphson method.
   d. Secant method.
   e. Fixed-point iteration method.
   Check the results by running Code6.

3. The graph from a given function provides a rough estimate on the location of one or more roots in the function. This is obvious in the case of the occurrence of multiple roots in the function. From the graph, a single root in a refined interval can be obtained. Run Code6 to locate the single roots of the following functions by refining their intervals using this approach:
   a. $f(x) = 1 - x^2 + 2x^3 - 5x^4 + 2x^7, x \in [-2, 2]$.
   b. $f(x) = 3x^4 - 5x^2 + 1, x \in [-1, 3]$.
   c. $f(x) = -3 \cos^2 x + 3 \sin x, x \in [-6, 2]$.
   d. $f(x) = 3 \sin(2x - 1) - 5 \cos(1 - 2 \sin x)), x \in [-1, 2]$.

PROGRAMMING CHALLENGES

1. Improve on Code6 by adding the fixed-point iteration method as an item in the menu. Add a mechanism to check for the convergence of the method by checking for $|g'(x)| < 1$ at every iteration, where $x = g(x)$ is derived from $f(x) = 0$.

2. Code6 generates a graph and its solution only when the product of the left and right end points is negative. Modify the project to display the graph even if the product is positive.

3. An intelligent root finder is one that can determine a small subinterval from a given function where its root is located automatically without the necessity of guessing two end points beforehand. This task can be added to Code6 by adding a routine to check for the occurrence of a root at every small subinterval from a given interval. Modify Code6 to add this option.

CHAPTER 7

Interpolation and Approximation

7.1 CURVE FITTING

The body of a car has been designed in such a way it possesses good aerodynamic features. This is important in order for the car to be comfortable, energy-efficient, cost-effective, and attractive. To achieve these objectives, the body surface of the car is made to be smooth. The normal techniques for designing the body of a car involve computer-aided design tools on the computer. The body is constructed by fitting and blending a set of patches from the B-spline or Bezier surfaces by approximating a set of points. B-spline and Bezier are some two-dimensional curves that are widely used in curve and surface fittings.

In general, curve and surface fitting is useful in many applications, notably in the design of body surfaces such as cars, aircrafts, ships, glasses, pipes, and vases. A patch in the surface is the three-dimensional extension of the B-spline curve which is obtained from a curve fitting technique.

Curve fitting is a generic term for constructing a curve from a given set of points. This objective can be achieved in two ways, through interpolation or approximation. *Interpolation* refers to a curve that passes through all the given points, whereas *approximation* is the case when the curve does not pass through one or more of the given points. The curve obtained from interpolation or approximation is one that best represents all points.

227

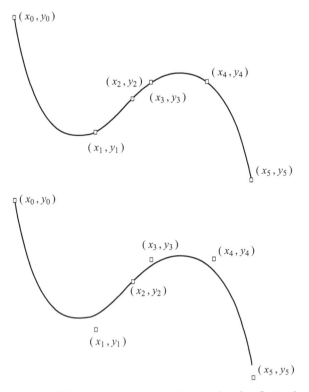

FIGURE 7.1. Interpolation (top) and approximation (bottom).

Figure 7.1 shows two different curves that can be produced from a set of points (x_i, y_i) for $i = 0, 1, \ldots, 5$. The curve at the top is generated through interpolation as it passes through all points. The bottom curve is an approximation as it misses several points.

In this chapter, we will discuss several common interpolation and approximation methods. We will concentrate on the two-dimensional aspect of these methods that provides a strong foundation for three-dimensional or higher problems. The topics include the Lagrange, Newton, and cubic spline methods in interpolation, and the least-squares method in approximation. In discussing the interpolation and approximation methods, the interpolating points are given as (x_i, y_i), for $i = 0, 1, \ldots, n$. There are $n + 1$ points given. The exact values at x_i are $y_i = f(x_i)$, whereas their interpolated values are denoted as $P(x_i)$.

7.2 LAGRANGE INTERPOLATION

An interpolation on two points, (x_0, y_0) and (x_1, y_1), results in a linear equation or a straight line. The standard form of a linear equation is given by

$$y = mx + c, \tag{7.1}$$

where m is the gradient of the line and c is the y-intercept. In the above equation,

$$m = \frac{y_1 - y_0}{x_1 - x_0} \quad \text{and} \quad c = y_0 - mx_0,$$

which results in

$$y = \frac{y_1 - y_0}{x_1 - x_0}x + \frac{x_1 y_0 - x_0 y_1}{x_1 - x_0}.$$

French mathematician, Joseph Louis Lagrange, proposed to rewrite the linear equation so that the two interpolated points, (x_0, y_0) and (x_1, y_1), are directly represented. With this in mind, the linear equation is rewritten as

$$P_1(x) = a_0(x - x_1) + a_1(x - x_0),$$

where a_0 and a_1 are constants. The points x_0 and x_1 in the factors of the above equation are called the *centers*. Applying the equation at (x_0, y_0), we obtain $y_0 = a_0(x_0 - x_1) + a_1(x_0 - x_0)$, or $a_0 = \frac{y_0}{x_0 - x_1}$. At (x_1, y_1), we get $y_1 = a_0(x_1 - x_1) + a_1(x_1 - x_0)$, or $a_1 = \frac{y_1}{x_1 - x_0}$. Therefore, the linear equation becomes

$$P_1(x) = y_0 \frac{(x - x_1)}{x_0 - x_1} + y_1 \frac{(x - x_0)}{x_1 - x_0}. \tag{7.2}$$

The quadratic form of the Lagrange polynomial interpolates three points, (x_0, y_0), (x_1, y_1), and (x_2, y_2). The polynomial has the form of

$$P_2(x) = a_0(x - x_1)(x - x_2) + a_1(x - x_0)(x - x_2) + a_2(x - x_0)(x - x_1),$$

with centers at x_0, x_1, and x_2. At (x_0, y_0),

$$y_0 = a_0(x_0 - x_1)(x_0 - x_2) + a_1(x_0 - x_0)(x_0 - x_2) + a_2(x_0 - x_0)(x_0 - x_1),$$

or

$$a_0 = \frac{y_0}{(x_0 - x_1)(x_0 - x_2)}.$$

Similarly, applying the equation at (x_1, y_1) and (x_2, y_2) yields

$$a_1 = \frac{y_1}{(x_1 - x_0)(x_1 - x_2)},$$

$$a_2 = \frac{y_2}{(x_2 - x_0)(x_2 - x_1)}.$$

This produces a quadratic Lagrange polynomial, given by

$$P_2(x) = y_0 \frac{(x - x_1)(x - x_2)}{(x_0 - x_1)(x_0 - x_2)} + y_1 \frac{(x - x_0)(x - x_2)}{(x_1 - x_0)(x_1 - x_2)} + y_2 \frac{(x - x_0)(x - x_1)}{(x_2 - x_0)(x_2 - x_1)}.$$

$$(7.3)$$

Definition 7.1. The Langrange operator $L_i(x)$ for (x_i, y_i), where $i = 0, 1, \ldots, n$ is defined as

$$L_i(x) = \prod_{\substack{k=1 \\ k \neq i}}^{n} \frac{(x - x_k)}{(x_i - x_k)} = \frac{(x - x_0)(x - x_1) \cdots (x - x_{i-1})(x - x_{i+1}) \cdots (x - x_n)}{(x_i - x_0)(x_i - x_1) \cdots (x_i - x_{i-1})(x_i - x_{i+1}) \cdots (x_i - x_n)}.$$

$$(7.4)$$

In general, the *Lagrange polynomial* of degree n is a polynomial that is produced from an interpolation over a set of points, (x_i, y_i) for $i = 0, 1, \ldots, n$, as follows:

$$P_n(x) = y_0 L_0(x) + y_1 L_1(x) + \cdots + y_n L_n(x). \tag{7.5}$$

There are n factors in both the numerator and the denominator of Equation (7.5). The inequality $k \neq i$ denies zero value in the denominator, which may cause a fatal error in division. It is obvious that $n = 1$ produces a linear curve or a straight line, whereas $n = 2$ produces a quadratic curve or a parabola.

Algorithm 7.1. Lagrange Method.
  Given the interpolating points (x_i, y_i) for $i = 0, 1, \ldots, n$;
  for $i = 0$ to n
      Evaluate $L_i(x) = \prod_{\substack{k=1 \\ k \neq i}}^{n} \frac{(x - x_k)}{(x_i - x_k)}$;
  endfor
  Evaluate $P_n(x) = y_0 L_0(x) + y_1 L_1(x) + \cdots + y_n L_n(x)$;

Example 7.1. Find a polynomial $P(x)$ that interpolates the points $\{(-1, 2), (0, 3), (2, -1), (5, 1)\}$ using the Lagrange method. Hence, find $P(2.5)$.

Solution. There are four given points, and this will produce a polynomial of degree $n = 3$, given as

$$P_3(x) = y_0 L_0(x) + y_1 L_1(x) + y_2 L_2(x) + y_3 L_3(x),$$

$$L_0(x) = \frac{(x - x_1)(x - x_2)(x - x_3)}{(x_0 - x_1)(x_0 - x_2)(x_0 - x_3)} = \frac{(x - 0)(x - 2)(x - 5)}{(-1 - 0)(-1 - 2)(-1 - 5)}$$

$$= -0.055556x(x - 2)(x - 5),$$

$$L_1(x) = \frac{(x - x_0)(x - x_2)(x - x_3)}{(x_1 - x_0)(x_1 - x_2)(x_1 - x_3)} = \frac{(x + 1)(x - 2)(x - 5)}{(0 + 1)(0 - 2)(0 - 5)}$$

$$= -0.1(x + 1)(x - 2)(x - 5),$$

$$L_2(x) = \frac{(x - x_0)(x - x_1)(x - x_3)}{(x_2 - x_0)(x_2 - x_1)(x_2 - x_3)} = \frac{(x + 1)(x - 0)(x - 5)}{(2 + 1)(2 - 0)(2 - 5)}$$

$$= -0.055556x(x + 1)(x - 5),$$

$$L_3(x) = \frac{(x - x_0)(x - x_1)(x - x_2)}{(x_3 - x_0)(x_3 - x_1)(x_3 - x_2)} = \frac{(x + 1)(x - 0)(x - 2)}{(5 + 1)(5 - 0)(5 - 2)}$$

$$= 0.011111x(x + 1)(x - 2).$$

The polynomial as given by Equation (7.1) is

$$P_3(x) = y_0 L_0(x) + y_1 L_1(x) + y_2 L_2(x) + y_3 L_3(x)$$

$$= -0.111112x(x - 2)(x - 5) - 0.3(x + 1)(x - 2)(x - 5)$$

$$-0.055556x(x + 1)(x - 5) + 0.011111x(x + 1)(x - 2).$$

Therefore, $P(2.5) = P_3(2.5) = -2.131944$.

7.3 NEWTON INTERPOLATIONS

The Lagrange method has a drawback. The amount of calculations depends very much on the given number of interpolated points. A large number of points require very tedious calculations on the Lagrange operators, as each of these functions has an equal degree as the interpolating polynomial.

A slightly simpler approach to the Lagrange method is the Newton method which applies to polynomials in the form of Newton polynomials. A Newton polynomial has the following general form:

$$P_n(x) = a_0 + a_1(x - x_0) + a_2(x - x_0)(x - x_1) + \cdots$$

$$+ a_n(x - x_0)(x - x_1) \cdots (x - x_{n-1}). \tag{7.6}$$

In the above equation, a_i for $i = 0, 1, \ldots, n$ are constants whose values are determined by applying the equation at the given interpolated points.

There are several different methods for evaluating the Newton polynomials. They include the divided-difference, forward-difference, backward-difference, and central-difference. We discuss each of these methods in this chapter.

Divided-Difference Method

The divided-difference method is a method for determining the coefficients a_i for $i = 0, 1, \ldots, n$ in Equation (7.6) using the divided-difference constants, which are defined as follows:

Definition 7.2. The divided-difference constant $d_{k,i}$ is defined as ith divided-difference of the function $y = f(x)$ at x_i, where

$$d_{k,i} = \frac{d_{k-1,i+1} - d_{k-1,i}}{x_k - x_i}.$$

The initial values are $d_{0,i} = y_i$ for $i = 0, 1, \ldots, n$ and $k = 1, 2, \ldots, n - 1$.

The general form of the linear equation in Equation (7.1), which interpolates (x_0, y_0) and (x_1, y_1), can also be expressed in the form of divided-difference constants,

$$P(x) = d_{0,0} + d_{1,0}(x - x_0),$$

where x_0 is the center and $d_{0,0}$ and $d_{1,0}$ are special constants called the zeroth and first divided-difference at x_0, respectively. Applying the linear equation at the two points, we obtain

$$d_{0,0} = y_0 \text{ and } d_{1,0} = \frac{y_1 - y_0}{x_1 - x_0}.$$

This gives $P_1(x) = d_{0,0} + d_{1,0}(x - x_0) = y_0 + \frac{y_1 - y_0}{x_1 - x_0}(x - x_0)$.
At the same time, the quadratic form of the Newton polynomial, which interpolates the points (x_0, y_0), (x_1, y_1), and (x_2, y_2) can now be written as

$$P(x) = d_{0,0} + d_{1,0}(x - x_0) + d_{2,0}(x - x_0)(x - x_1).$$

In the above equation, x_0 and x_1 are the centers. Applying the quadratic equation to the three points, we obtain

$$d_{0,0} = y_0,$$

$$d_{1,0} = \frac{d_{0,1} - d_{0,0}}{x_1 - x_0} = \frac{y_1 - y_0}{x_1 - x_0},$$

$$d_{1,1} = \frac{d_{0,2} - d_{0,1}}{x_2 - x_1} = \frac{y_2 - y_1}{x_2 - x_1},$$

$$d_{2,0} = \frac{d_{1,1} - d_{1,0}}{x_2 - x_0} = \frac{\frac{y_2 - y_1}{x_2 - x_1} - \frac{y_1 - y_0}{x_1 - x_0}}{x_2 - x_0}.$$

In general, the divided-difference method for interpolating (x_i, y_i) for $i = 0, 1, \ldots, n$ produces a Newton polynomial of degree n, given by

$$P_n(x) = d_{0,0} + d_{1,0}(x - x_0) + d_{2,0}(x - x_0)(x - x_1) + \cdots$$

$$+ d_{n,0}(x - x_0)(x - x_1)\cdots(x - x_{n-1}). \tag{7.7}$$

Algorithm 7.2 summarizes the divided-difference approach. An example using this algorithm is illustrated in Example 7.2.

Algorithm 7.2. Newton's Divided-Difference Method.
    Given the interpolating points (x_i, y_i) for $i = 0, 1, \ldots, n$;
    Set $d_{0,i} = y_i$, for $i = 0, 1, \ldots, n$;
    Evaluate the divided-difference constants:
        for $i = 0$ to n
            for $k = 1$ to $n - 1$
                Compute $d_{k,i} = \frac{d_{k-1,i+1} - d_{k-1,i}}{x_k - x_i}$;
            endfor
        endfor
    Get $P_n(x)$ using Equation (7.7);

Example 7.2. Find the polynomial from the interpolating points in Example 7.1 using the Newton's divided-difference method.

Solution. From the set $\{(-1, 2), (0, 3), (2, -1), (5, 1)\}$, we obtain

$$x_0 = -1, \quad x_1 = 0, \quad x_2 = 2, \text{ and } x_3 = 5.$$

$$d_{0,0} = y_0 = 2, \quad d_{0,1} = y_1 = 3, d_{0,2} = y_2 = -1, \text{ and } d_{0,3} = y_3 = 1.$$

It follows that

$$d_{1,0} = \frac{d_{0,1} - d_{0,0}}{x_1 - x_0} = -1, \quad d_{1,1} = \frac{d_{0,2} - d_{0,1}}{x_2 - x_1} = -2,$$

$$d_{1,2} = \frac{d_{0,3} - d_{0,2}}{x_3 - x_2} = 0.666667,$$

$$d_{2,0} = \frac{d_{1,1} - d_{1,0}}{x_2 - x_0} = 0.333333, \quad d_{2,1} = \frac{d_{1,2} - d_{1,1}}{x_3 - x_1} = 0.533333,$$

$$d_{3,0} = \frac{d_{2,1} - d_{2,0}}{x_3 - x_0} = 0.033333.$$

Therefore,

$$P_3(x) = d_{0,0} + d_{1,0}(x - x_0) + d_{2,0}(x - x_0)(x - x_1) + d_{3,0}(x - x_0)(x - x_1)(x - x_2)$$

$$= 2 - (x + 1) + 0.333333x(x + 1) + 0.033333x(x + 1)(x - 2).$$

We obtain $P(2.5) = P_3(2.5) = -2.131944$.

Forward-Difference Method

Both the Lagrange and the Newton divided-difference methods can be applied to cases where the x subintervals are uniform or nonuniform. On a special case where the x subintervals are uniform, it may not be necessary to apply the two methods as other methods may prove to be easier. In this case, the divided-difference method with uniform x subintervals can be reduced into a method called *forward-difference*. This method involves an operator called the *forward-difference operator*.

Definition 7.3. The *forward-difference operator* $\Delta_{k,i}$ is defined as the kth forward difference at x_i, or $\Delta_{k,i} = \Delta_{k-1,i+1} - \Delta_{k-1,i}$. The initial values are given by $\Delta_{0,i} = y_i$ for $i = 0, 1, \ldots, n$.

In deriving the Newton polynomial using the forward-difference method, let h be the width of the x subintervals. Uniform subintervals suggest all subintervals in x have equal width given as h. In other words, $h = x_{i+1} - x_i$ for $i = 0, 1, \ldots, n - 1$.

The forward-difference formula is derived from the divided-difference equation. Consider a cubic form of the divided-difference equation from Equation (7.6) given as

$$P_3(x) = d_{0,0} + d_{1,0}(x - x_0) + d_{2,0}(x - x_0)(x - x_1) + d_{3,0}(x - x_0)(x - x_1)(x - x_2).$$

This is simplified into

$$P_3(x) = y_0 + \frac{\Delta_{1,0}}{x_1 - x_0}(x - x_0) + \frac{\Delta_{2,0}}{x_2 - x_0}(x - x_0)(x - x_1)$$

$$+ \frac{\Delta_{3,0}}{x_3 - x_0}(x - x_0)(x - x_1)(x - x_2)$$

$$= y_0 + \frac{\Delta_{1,0}}{h}(x - x_0) + \frac{\Delta_{2,0}}{2h}(x - x_0)(x - x_1)$$

$$+ \frac{\Delta_{3,0}}{3h}(x - x_0)(x - x_1)(x - x_2).$$

It can be proven that the general form of the forward-difference method for interpolating the points (x_i, y_i) for $i = 0, 1, \ldots, n$ can be extended from the cubic case above. The solution is given by

$$P_n(x) = y_0 + \frac{\Delta_{1,0}}{h}(x - x_0) + \frac{\Delta_{2,0}}{2h}(x - x_0)(x - x_1) + \cdots$$

$$+ \frac{\Delta_{n,0}}{nh}(x - x_0)(x - x_1) \cdots (x - x_{n-1}). \tag{7.8}$$

Algorithm 7.3 summarizes the forward-difference approach. This is followed by an illustration using Example 7.3.

Algorithm 7.3. Newton's Forward-Difference Method.
    Given the interpolating points (x_i, y_i) for $i = 0, 1, \ldots, n$;
    Set $\Delta_{0,i} = y_i$, for $i = 0, 1, \ldots, n$;
    Evaluate the forward-difference constants:
        for $i = 0$ to n
            for $k = 1$ to $n - 1$
                Compute $\Delta_{k,i} = \Delta_{k-1,i+1} - \Delta_{k-1,i}$;
            endfor
        endfor
    Get $P_n(x)$ using Equation (7.8);

Example 7.3. Find the Newton polynomial $P(x)$ from the points given by $\{(-2, 2), (0, 3), (2, -1), (4, 1)\}$ using the forward-difference method. Hence, find $P(2.5)$.

Solution. From the set $\{(-2, 2), (0, 3), (2, -1), (4, 1)\}$, we obtain

i	x_i	$\Delta_{0,i} = f(x_i)$	$\Delta_{1,i}$	$\Delta_{2,i}$	$\Delta_{3,i}$
0	-2	2	1	-5	11
1	0	3	-4	6	
2	2	-1	2		
3	4	1			

From the above table, we get the forward-difference values,

$$\Delta_{0,0} = 2, \quad \Delta_{1,0} = 1, \quad \Delta_{2,0} = -5 \text{ and } \Delta_{3,0} = 11.$$

We obtain

$$P_3(x) = \Delta_{0,0} + \frac{\Delta_{1,0}}{1!}(x - x_0) + \frac{\Delta_{2,0}}{2!}(x - x_0)(x - x_1)$$

$$+ \frac{\Delta_{3,0}}{3!}(x - x_0)(x - x_1)(x - x_2)$$

$$= 2 + \frac{1}{1!}(x + 1) - \frac{5}{2!}(x + 1)(x - 0) + \frac{11}{3!}(x + 1)(x - 0)(x - 2)$$

$$= 2 + (x + 1) - 2.5x(x + 1) + 1.833333x(x + 1)(x - 2).$$

It follows that

$$P_3(2.5) = 2 + \frac{1}{1!}(2.5 + 2) + \frac{-5}{2!}(2.5 + 2)(2.5 - 0)$$

$$+ \frac{11}{3}(2.5 + 2)(2.5 - 0)(2.5 - 2) = -11.312500.$$

Backward-Difference Method

The operator $\Delta_{k,i}$ is based on forward difference. It is also possible to do the opposite, that is, backward difference.

Definition 7.4. The *backward difference operator* $\nabla_{k,i}$ is defined as the kth backward operator at x_i or $\nabla_{k,i} = \nabla_{k-1,i} - \nabla_{k-1,i-1}$. The initial values are $\nabla_{0,i} = y_i$ for $i = 0, 1, \ldots, n$.

The backward-difference method is also derived from the Newton divided-difference method. The method requires all x subintervals to have uniform width, given as h. We discuss the case of a quadratic polynomial from the divided-difference method in deriving the formula for the backward-difference method. From Equation (7.7),

$$P_n(x) = f_n + r\nabla f_n + \frac{r(r + 1)}{2!}\nabla^2 f_n + \cdots + \frac{\nabla^n f_n}{n!}\prod_{i=0}^{n-1}(r + i), \tag{7.9}$$

where $\nabla^k f_i = \nabla^{k-1} f_i - \nabla^{k-1} f_{i-1}$ is the backward-difference operator and $r = \frac{x - x_n}{h}$.

Algorithm 7.4 summarizes the steps in the backward-difference method for generating the Newton polynomial. The algorithm is illustrated through Example 7.4.

Algorithm 7.4. Newton's Backward-Difference Method.
  Given the interpolating points (x_i, y_i) for $i = 0, 1, \ldots, n$;
  Set $\Delta_{0,i} = y_i$, for $i = 0, 1, \ldots, n$;
  Evaluate the backward-difference constants:
    for $i = 0$ to n
      for $k = 1$ to $n - 1$
        Compute $\nabla_{k,i} = \nabla_{k-1,i+1} - \nabla_{k-1,i}$;
  Get $P_n(x)$ using Equation (7.9);

Example 7.4. Find the Newton polynomial $P(x)$ from the interpolating points given by $\{(-2, 2), (0, 3), (2, -1), (4, 1)\}$ using the backward-difference method. Hence, find $P(2.5)$ using the backward-difference method.

Solution. From the set $\{(-1, 2), (0, 3), (2, -1), (5, 1)\}$, we obtain

i	x_i	$\nabla_{0,i} = f(x_i)$	$\nabla_{1,i}$	$\nabla_{2,i}$	$\nabla_{3,i}$
0	-2	2			
1	0	3	1		
2	2	-1	-4	-5	
3	4	1	2	6	11

From the above table, we get the forward-difference values,

$$\nabla_{0,3} = 1, \quad \nabla_{1,3} = 2, \quad \nabla_{2,3} = 6, \quad \nabla_{3,3} = 11.$$

This produces

$$P_3(x) = \nabla_{0,3} + \frac{\nabla_{1,3}}{1!}(x - x_3) + \frac{\nabla_{2,3}}{2!}(x - x_3)(x - x_2) + \frac{\nabla_{3,3}}{3!}(x - x_3)(x - x_2)(x - x_1)$$

$$= 1 + \frac{2}{1!}(x - 5) + \frac{6}{2!}(x - 5)(x - 2) + \frac{11}{3!}(x - 5)(x - 2)(x - 0)$$

$$= 1 + 2(x - 5) + 3(x - 5)(x - 2) + 1.833333x(x - 5)(x - 2).$$

Therefore,

$$P(2.5) = 1 + 2(2.5 - 5) + 3(2.5 - 5)(2.5 - 2) + 1.83333(2.5)(2.5 - 5)(2.5 - 2)$$

$$= -13.479163.$$

Stirling's Method

The forward-difference and backward-difference methods may be bias toward the forward and backward directions, respectively, in their construction of Newton polynomials. As a result, the accuracy and precision of the results may differ to some extent. A more practical approach is to consider both the forward and the backward factors within a single formulation. Stirling's method is one such method that considers both factors to produce a more reliable result.

Stirling's method is based on uniform x subintervals, which works best in interpolating points that lie close to the middle of the interval. In interpolating a point x in $x_0 < x < x_n$, the method starts by locating the subinterval $x_i < x < x_{i+1}$, where x_i and x_{i+1} are two given successive points. This determines the index $k = i$, where x_k is the lower-bound value in the subinterval, or $x_k = x_i$. Stirling's solution is expressed in terms of variable r as

$$P(r) = f_k + r\mu_k^1 + \frac{r^2}{2!}\delta_k^2 + \frac{r(r^2 - 1)}{3!}\mu_k^3 + \frac{r^2(r^2 - 1)}{4!}\delta_k^4$$

$$+ \frac{r(r^2 - 1)(r^2 - 2^2)}{5!}\mu_k^5 + \dots, \tag{7.10}$$

where $r = \frac{x - x_k}{h}$ and h is the width of the subintervals. The terms δ_i^j and μ_i^j are called the central-difference constants given by

$$\delta_i^j = \delta_{i+1/2}^{j-1} - \delta_{i-1/2}^{j-1}, \tag{7.11a}$$

$$\mu_i^j = \frac{1}{2}\left[\delta_{i+1/2}^j + \delta_{i-1/2}^j\right], \tag{7.11b}$$

for $i = 0, 1, \dots, n$ and $j = 1, \dots, n$. In Equations (7.11a) and (7.11b), δ_i^j and μ_i^j are the odd and even constants of the Stirling's method, respectively. δ_i^j computes the difference between its left and right point, whereas μ_i^j is the average between the points.

Stirling's method is implemented according to the steps in Algorithm 7.5. The algorithm is illustrated through Example 7.5.

Algorithm 7.5. Stirling's Method.
Given the interpolating points (x_i, y_i) for $i = 0, 1, \dots, n$;
Locate the subinterval $x_i < x < x_{i+1}$ to determine $k = i$;
Evaluate $r = \frac{x - x_k}{h}$ where $x_i < x_k < x_{i+1}$;
Evaluate the central-difference constants δ_i^j for odd j and
$j = 1, 3, \dots, n$;
Evaluate the central-difference constants μ_i^j, for even $j = 2, 4, \dots, n$;
Get $P_n(x)$ using Equation (7.10);

Example 7.5. Find $P(3.7)$ from the following points using the Stirling's method:

i	0	1	2	3	4	5
x_i	3.0	3.2	3.4	3.6	3.8	4.0
$y_i = f(x_i)$	2.5	2.9	2.4	2.0	2.1	2.7

Solution. The width of the subintervals is $h = 0.2$. Since $3.6 < 3.7 < 3.8$, we have $x_k = x_3$ and $k = 3$. Therefore, $y_k = y_3 = 2$ and $r = \frac{x - x_k}{h} = \frac{3.7 - 3.6}{0.2} = 0.5$. We construct a table to determine the values of δ_i^j and μ_i^j based on Equations (7.11a) and (7.11b):

i	x_i	$\delta_i^0 = f(x_i)$	δ_i^1	δ_i^2	δ_i^3	δ_i^4	δ_i^5
0	3	2.5					
			0.400				
1	3.2	2.9		−0.900			
			−0.500		1.000		
2	3.4	2.4		0.100		−0.600	
			−0.400		0.400		0.200
3	3.6	2		0.500		−0.400	
			0.100		0.000		
4	3.8	2.1		0.500			
			0.600				
5	4	2.7					

From Equation (7.10), we obtain

$$P(x = 3.7) = P(r = 0.5) \approx f_3 + r\mu_k^1 + \frac{r^2}{2!}\delta_3^2 + \frac{r(r^2 - 1)}{3!}\mu_3^3 + \frac{r^2(r^2 - 1)}{4!}\delta_3^4$$

$$= 2 + r\left(\frac{-0.400 + 0.100}{2}\right) + \frac{r^2}{2!}(0.500)$$

$$+ \frac{r(r^2 - 1)}{3!}\left(\frac{0.400 + 0.000}{2}\right) + \frac{r^2(r^2 - 1)}{4!}(-0.400)$$

$$= 2 - 0.15r + 0.25r^2 + 0.033r(r^2 - 1) - 0.017r^2(r^2 - 1)$$

$$= 2 - 0.15\,(0.5) + 0.25\,(0.5^2) + 0.033(0.5)\,(0.5^2 - 1)$$

$$- 0.017(0.5^2)\,(0.5^2 - 1) = 1.978.$$

7.4 CUBIC SPLINE

A *spline* is a single curve that is formed from a set of piecewise continuous functions $s_k(x)$ for $k = 1, \ldots, m - 1$ as a result of interpolation over the points (x_k, y_k) for $k = 0, 1, \ldots, m$. A spline made from four pieces of functions, for example, has five interpolating points that appear like a single piece of smooth curve. This is realized as the spline is continuous at all joints. Not only that, the spline is also continuous in terms of derivatives at these points, which contribute to the smoothness.

A spline of degree n is constructed from piecewise polynomials of the same degree. The simplest spline is the *linear spline*, which consists of straight lines connecting the points successively. A *quadratic spline* consists of quadratic polynomials connecting

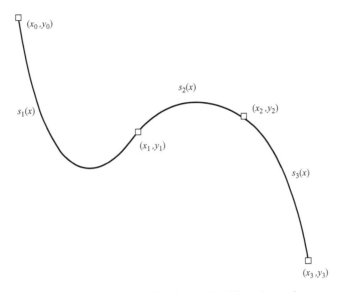

FIGURE 7.2. A cubic spline interpolated from four points.

the points where the first derivatives at the interpolated points exist. As the name suggests, a *cubic spline* connects the points through cubic polynomials where, beside the continuity, the curve has the main property as it is differentiable in its first and second derivatives at each connecting point between any two pieces of function.

Figure 7.2 shows a cubic spline interpolated over four points (x_i, y_i), for $i = 0, 1, 2, 3$, whose pieces are denoted as $s_k(x)$ for $k = 1, 2, 3$. Each piece of the spline is a continuous cubic function in the given subinterval. The spline appears to be very smooth at each interior node as its first and second derivatives exist there.

Cubic spline is a spline of degree three in the interval $x_0 \leq x \leq x_m$, which is made up of a set of piecewise polynomials $s_k(x)$. The general form of a cubic spline is expressed as

$$s_k(x) = a_k + b_k x + c_k x_k^2 + d_k x_k^3. \tag{7.12}$$

In the above equation, a_k, b_k, c_k, and d_k are constants for $k = 1, 2, \ldots, m$ that make up the cubic polynomials. The spline interpolates $m + 1$ points, (x_i, y_i) for $i = 0, 1, \ldots, m$.

A cubic spline has the following properties that satisfy the requirements of continuity and differentiability at the interpolating nodes:

Property 1. The spline is continuous in $x_0 \leq x \leq x_n$, or $s_1(x_0) = y_0$ and $s_k(x_k) = y_k$ for $k = 1, 2, \ldots, m$. This property states that the spline passes through all $m + 1$ given points.

Property 2. The points at the end of each interior node must be equal or $s_k(x_{k+1}) = s_{k+1}(x_{k+1})$ for $k = 1, 2, \ldots, m - 1$. This property states that the value of the spline from the left of the interior node must be the same as the value from its right.

Property 3. The first derivative at the end of each interior knot is continuous or $s_k'(x_{k+1}) = s_{k+1}'(x_{k+1})$ for $k = 1, 2, \ldots, m - 1$. This property is derived from fundamental calculus where a derivative at a point is said to exist if its value from the left equals that from the right.

Property 4. The second derivative is continuous at each interior knot or $s_k''(x_{k+1}) = s_{k+1}''(x_{k+1})$ for $k = 1, 2, \ldots, m - 1$. The same property from calculus applies where the second derivative at a point is said to exist if its derivative from the left equals that from the right.

Property 5. The end knots are *free boundaries*, which means the second derivative at the end knots are zero or $s_1''(x_0) = s_n''(x_n) = 0$. This property suggests the two end nodes must be tied to zero.

Property 5 above suggests the condition of $s_1''(x_0) = s_n''(x_n) = 0$ referred to as free boundaries at the end nodes. The spline with this condition is referred to as a *natural cubic spline*. Alternatively, another type of spline called the *clamped cubic spline* is produced if the condition at the end nodes is changed to $s_1''(x_0) = s_n''(x_n) \neq 0$.

The first derivative of the Equation (7.11) is a quadratic function whereas its second derivative is linear. Let $w_k = s''(x_k)$ for $0 \leq k \leq m$. From the condition $s''(x_0) = s''(x_m) = 0$, we have $w_0 = w_m = 0$. The second derivative can be written as

$$s_k''(x) = w_{k-1} + q_{k-1}(x - x_{k-1}),$$

where $q_{k-1} = \frac{w_k - w_{k-1}}{x_k - x_{k-1}}$ is the slope of $s_k''(x)$. This equation simplifies to

$$s_k''(x) = \frac{w_{k-1}}{x_k - x_{k-1}}(x_k - x) + \frac{w_k}{x_k - x_{k-1}}(x - x_{k-1}).$$

Integrating the above term twice gives an alternative form of Equation (7.12),

$$s_k(x) = A_k(x_k - x)^3 + B_k(x - x_{k-1})^3 + C_k(x_k - x) + D_k(x - x_{k-1}). \tag{7.13}$$

for $k = 1, 2, \ldots, m - 1$. The constants A_k, B_k, C_k, and D_k are found to be

$$A_k = \frac{w_{k-1}}{6(x_k - x_{k-1})}, \tag{7.14a}$$

$$B_k = \frac{w_k}{6(x_k - x_{k-1})}, \tag{7.14b}$$

$$C_k = \frac{y_{k-1}}{x_k - x_{k-1}} - \frac{w_{k-1}}{6(x_k - x_{k-1})}, \tag{7.14c}$$

$$D_k = \frac{y_k}{x_k - x_{k-1}} - \frac{w_k}{6(x_k - x_{k-1})}. \tag{7.14d}$$

The value of w_k can be found from the condition $s'_k(x_k) = s'_{k+1}(x_k)$. From Equation (7.13), we obtain the relationship given by

$$(x_{k+1} - x_k)w_{k-1} + 2(x_{k+1} - x_{k-1})w_k + (x_{k+1} - x_k)w_{k+1}$$

$$= 6\left[\frac{y_{k+1} - y_k}{x_{k+1} - x_k} + \frac{y_{k-1} - y_k}{x_k - x_{k-1}}\right], \tag{7.15}$$

for $k = 1, 2, \ldots, m - 1$. This produces a tridiagonal system of linear equations, given by

$$\begin{bmatrix} f_1 & g_1 & 0 & \cdots & 0 & 0 & 0 \\ e_2 & f_2 & g_2 & \cdots & 0 & 0 & 0 \\ 0 & e_3 & f_3 & \cdots & 0 & 0 & 0 \\ \cdots & \cdots & \cdots & \cdots & \cdots & \cdots & \cdots \\ 0 & 0 & 0 & \cdots & f_{m-3} & g_{m-3} & 0 \\ 0 & 0 & 0 & \cdots & e_{m-2} & f_{m-2} & g_{m-2} \\ 0 & 0 & 0 & \cdots & 0 & e_{m-1} & f_{m-1} \end{bmatrix} \begin{bmatrix} w_1 \\ w_2 \\ w_3 \\ \cdots \\ w_{m-3} \\ w_{m-2} \\ w_{m-1} \end{bmatrix} = \begin{bmatrix} r_1 \\ r_2 \\ r_3 \\ \cdots \\ r_{m-3} \\ r_{m-2} \\ r_{m-1} \end{bmatrix}, \tag{7.16}$$

whose three-band diagonal elements are

$$e_{k+1} = x_{k+1} - x_k, \tag{7.17a}$$

$$f_k = 2(x_{k+1} - x_{k-1}), \tag{7.17b}$$

$$g_k = x_{k+1} - x_k, \tag{7.17c}$$

$$r_k = 6\left[\frac{y_{k+1} - y_k}{x_{k+1} - x_k} + \frac{y_{k-1} - y_k}{x_k - x_{k-1}}\right]. \tag{7.17d}$$

With w_k determined from Equation (7.16), we obtain the values of A_k, B_k, C_k, and D_k using Equations (7.14a), (7.14b), (7.14c), and (7.14d), respectively.

Algorithm 7.6. Cubic Spline Method.

Given the interpolating points (x_i, y_i) for $i = 0, 1, \ldots, m$;

For $k = 1$ to $m - 1$

Compute $f_k = 2(x_{k+1} - x_{k-1})$;

Compute $r_k = 6\left[\frac{y_{k+1} - y_k}{x_{k+1} - x_k} + \frac{y_{k-1} - y_k}{x_k - x_{k-1}}\right]$;

If $k > 1$

Compute $e_{k+1} = x_{k+1} - x_k$;

Endif
If $k < m - 1$
    Compute $g_k = x_{k+1} - x_k$;
Endif
Endfor
Solve the system of linear equations to find w_k;
Find A_k, B_k, C_k and D_k from Equation (7.14);

The Thomas algorithm is the most practical method for solving the tridiagonal system of linear equations as the method uses a small number of variables and, therefore, consumes a small amount of memory. Example 7.6 shows an example of the cubic spline interpolation where the generated system of linear equations is solved using the Thomas algorithm method.

Example 7.6. Find a cubic spline that interpolates $(x_k, y_k) = \{(-1, 2), (0, 3), (2, -1), (5, 1), (6, 5)\}$.

Solution. In this problem, there are five points, and $m = 4$, where $w_0 = w_4 = 0$. From Algorithm 7.6, we get a system of three linear equations given by

$$
\begin{bmatrix} f_1 & g_1 & 0 \\ e_2 & f_2 & g_2 \\ 0 & e_3 & f_3 \end{bmatrix} \begin{bmatrix} w_1 \\ w_2 \\ w_3 \end{bmatrix} = \begin{bmatrix} r_1 \\ r_2 \\ r_3 \end{bmatrix}.
$$

The values of e_k, f_k, g_k, and r_k for $k = 1$ to $k = m$ are computed using Equations (7.17a), (7.17b), (7.17c), and (7.17d). The results are displayed in the following table:

k	x_k	y_k	e_k	f_k	g_k	r_k
0	-1.0	2.0				
1	0.0	3.0		6.000000	2.000000	-18.000000
2	2.0	-1.0	2.000000	10.000000	3.000000	16.000000
3	5.0	1.0	3.000000	8.000000		20.000000
4	6.0	5.0				

This produces the following system of linear equations:

$$
\begin{bmatrix} 6 & 2 & 0 \\ 2 & 10 & 3 \\ 0 & 3 & 8 \end{bmatrix} \begin{bmatrix} w_1 \\ w_2 \\ w_3 \end{bmatrix} = \begin{bmatrix} -18 \\ 16 \\ 20 \end{bmatrix}.
$$

The above system is solved to produce $w_1 = -3.588832$, $w_2 = 1.766497$, and $w_3 = 1.837563$. We obtain A_k, B_k, C_k, and D_k using Equations (7.14a), (7.14b), (7.14c), and (7.14d), as shown in the following table:

k	x_k	y_k	A_k	B_k	C_k	D_k
0	−1.0	2.0				
1	0.0	3.0	0.000000	−0.598139	2.000000	6.588832
2	2.0	−1.0	−0.299069	0.147208	1.799069	−0.647208
3	5.0	1.0	0.098139	0.102087	−0.431472	0.267706
4	6.0	5.0	0.306261	0.000000	0.693739	5.000000

7.5 LEAST-SQUARES APPROXIMATION

The *least-squares method* is an approximation method for a set of points based on the sum of the square of the errors. The method is popularly applied in many applications, such as in the statistical analysis involving multiple data regression. Multiple regression involves approximation on several variables based on straight lines or linear equations. Its advantage of using low-degree polynomials contributes to providing tools for forecasting, experimental designs, and other forms of statistical modeling.

In the least-squares method, an error at a point is defined as the difference between the true value and the approximated value. The method has the advantage over the Lagrange and Newton methods as the approximation is independent of the number of points. This allows low-degree polynomials for fitting a finite number of points.

The least-squares method generates a low-degree polynomial for approximating the given points by minimizing the sum of the squares of the errors. The solution to the problem is obtained by solving a system of linear equations that is generated from the minimization.

Approximation using the least-squares method can be achieved in either continuous or discrete forms. The difference between these two forms rests on the use of integral in the former and summation in the latter. The continuous least-squares method is appropriate in applications requiring the use of continuous variables and in analog-based applications. On the other hand, the discrete least-squares method is good at handling applications that have finite data. We will limit our discussion only to the discrete least-squares method in this chapter.

The discrete least-squares form of the problem is based on m interpolated points (x_i, y_i) for $i = 0, 1, \ldots, m - 1$. The curve to be fitted is a low-degree polynomial $P(x)$ that best represents all points. The most common function used in the least-squares method is the linear function $P(x) = a_0 + a_1 x$, which is good enough for many applications. Occasionally, some applications also require quadratic or cubic polynomials.

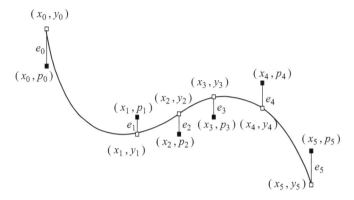

FIGURE 7.3. Approximation using a polynomial in the least-squares method.

In the least-squares method, the error between the interpolated point y_i and the approximated value $p_i = P(x_i)$ at the point $x = x_i$ is given by

$$e_i = y_i - p_i. \tag{7.18}$$

The sum of the squares of the errors e_i at the points $x = x_i$ for $i = 0, 1, \ldots, m - 1$ is expressed as an objective function E, as follows:

$$E = \sum_{i=0}^{m-1} e_i^2 = \sum_{i=0}^{m-1} (y_i - p_i)^2. \tag{7.19}$$

Figure 7.3 shows a case of $m = 6$ points with the interpolated points, (x_i, y_i) in white squares and the approximated points (x_i, p_i) in dark squares. At each point x_i, the error $e_i = y_i - p_i$ is computed. The objective function in Equation (7.18) is obtained by adding the sum of squares of all these errors.

The curves to be fitted in the least-squares approximations are normally low-degree polynomials, such as

$$P_1(x) = c_0 + c_1 x, \text{ for a linear function,}$$

$$P_2(x) = c_0 + c_1 x + c_2 x^2, \text{ for a quadratic polynomial,}$$

$$P_3(x) = c_0 + c_1 x + c_2 x^2 + c_3 x^3, \text{ for a cubic polynomial.}$$

In linear approximation, a straight line equation is used to approximate the points. The objective function becomes

$$E = \sum_{i=0}^{m-1} (y_i - P_1(x_i))^2 = \sum_{i=0}^{m-1} (y_i - (c_0 + c_1 x_1))^2. \tag{7.20}$$

Our objective here is to find the values of c_0 and c_1 by minimizing the sum of the square of the errors. The minimization requires setting $\frac{\partial E}{\partial c_0} = 0$ and $\frac{\partial E}{\partial c_1} = 0$ to produce two linear equations that will be sufficient to solve for c_0 and c_1. The partial derivatives are obtained as follows:

$$\frac{\partial E}{\partial c_0} = -2 \sum_{i=0}^{m-1} (y_i - (c_0 + c_1 x)),$$

$$\frac{\partial E}{\partial c_1} = -2 \sum_{i=0}^{m-1} x_i (y_i - (c_0 + c_1 x)).$$

Setting $\frac{\partial E}{\partial a_0} = 0$, we obtain the first linear equation

$$2 \sum_{i=0}^{m-1} (y_i - (c_0 + c_1 x)) = 0,$$

$$\sum_{i=0}^{m-1} y_i - \sum_{i=0}^{m-1} c_0 - \sum_{i=0}^{m-1} c_1 x_i = 0,$$

$$c_0 \sum_{i=0}^{m-1} 1 + c_1 \sum_{i=0}^{m-1} x_i = \sum_{i=0}^{m-1} y_i.$$

The second linear equation is obtained by setting $\frac{\partial E}{\partial c_1} = 0$:

$$\sum_{i=0}^{m-1} x_i (y_i - (c_0 + c_1 x)) = 0,$$

$$c_0 \sum_{i=0}^{m-1} x_i + c_1 \sum_{i=0}^{m-1} x_i^2 = \sum_{i=0}^{m-1} x_i y_i.$$

The two equations can be written in matrix form as follows:

$$\begin{bmatrix} \sum_{i=0}^{m-1} 1 & \sum_{i=0}^{m-1} x_i \\ \sum_{i=0}^{m-1} x_i & \sum_{i=0}^{m-1} x_i^2 \end{bmatrix} \begin{bmatrix} c_0 \\ c_1 \end{bmatrix} = \begin{bmatrix} \sum_{i=0}^{m-1} y_i \\ \sum_{i=0}^{m-1} y_i x_i \end{bmatrix}. \tag{7.21}$$

A least-squares approximation using the quadratic function $P_2(x_i) = c_0 + c_1 x + c_2 x^2$ produces a system of 3×3 linear equations. The objective function is

$$E = \sum_{i=0}^{m-1} (y - (c_0 + c_1 x + c_2 x^2))^2. \tag{7.22}$$

The approximation is obtained by minimizing the sum of the squares of the errors through $\frac{\partial E}{\partial c_0} = 0$, $\frac{\partial E}{\partial c_1} = 0$, and $\frac{\partial E}{\partial c_2} = 0$. The first equation is obtained through the following steps:

$$\frac{\partial E}{\partial c_0} = -2 \sum_{i=0}^{m-1} \left(y_i - \left(c_0 + c_1 x + c_2 x^2 \right) \right),$$

$$\frac{\partial E}{\partial c_0} = 0 : \quad \sum_{i=0}^{m-1} \left(y_i - \left(c_0 + c_1 x + c_2 x^2 \right) \right) = 0,$$

$$c_0 \sum_{i=0}^{m-1} 1 + c_1 \sum_{i=0}^{m-1} x_i + c_2 \sum_{i=0}^{m-1} x_i^2 = \sum_{i=0}^{m-1} y_i.$$

The second equation is generated in the same manner, as follows:

$$\frac{\partial E}{\partial c_1} = -2 \sum_{i=0}^{m-1} x_i \left(y_i - \left(c_0 + c_1 x + c_2 x^2 \right) \right),$$

$$\frac{\partial E}{\partial c_1} = 0 : \quad \sum_{i=0}^{m-1} x_i \left(y_i - \left(c_0 + c_1 x + c_2 x^2 \right) \right) = 0,$$

$$c_0 \sum_{i=0}^{m-1} x_i + c_1 \sum_{i=0}^{m-1} x_i^2 + c_2 \sum_{i=0}^{m-1} x_i^3 = \sum_{i=0}^{m-1} x_i y_i.$$

We also obtain the third equation from similar steps as above

$$\frac{\partial E}{\partial a_2} = -2 \sum_{i=0}^{m-1} x_i^2 \left(y_i - \left(c_0 + c_1 x + c_2 x^2 \right) \right),$$

$$\frac{\partial E}{\partial a_2} = 0 : \quad \sum_{i=0}^{m-1} x_i^2 \left(y_i - \left(c_0 + c_1 x + c_2 x^2 \right) \right) = 0,$$

$$c_0 \sum_{i=0}^{m-1} x_i^2 + c_1 \sum_{i=0}^{m-1} x_i^3 + c_2 \sum_{i=0}^{m-1} x_i^4 = \sum_{i=0}^{m-1} x_i^2 y_i.$$

The three equations are formulated in matrix form, as follows:

$$\begin{bmatrix} \sum_{i=0}^{m-1} 1 & \sum_{i=0}^{m-1} x_i & \sum_{i=0}^{m-1} x_i^2 \\ \sum_{i=0}^{m-1} x_i & \sum_{i=0}^{m-1} x_i^2 & \sum_{i=0}^{m-1} x_i^3 \\ \sum_{i=0}^{m-1} x_i^2 & \sum_{i=0}^{m-1} x_i^3 & \sum_{i=0}^{m-1} x_i^4 \end{bmatrix} \begin{bmatrix} c_0 \\ c_1 \\ c_2 \end{bmatrix} = \begin{bmatrix} \sum_{i=0}^{m-1} y_i \\ \sum_{i=0}^{m-1} y_i x_i \\ \sum_{i=0}^{m-1} y_i x_i^2 \end{bmatrix}. \tag{7.23}$$

Equations (7.21) and (7.23) can be generalized into approximating a polynomial of degree n. The least-squares method produces the following system of $(n+1) \times (n+1)$ linear equations:

$$
\begin{bmatrix}
\sum_{i=0}^{m-1} 1 & \sum_{i=0}^{m-1} x_i & \cdots & \sum_{i=0}^{m-1} x_i^{n-1} & \sum_{i=0}^{m-1} x_i^n \\[2mm]
\sum_{i=0}^{m-1} x_i & \sum_{i=0}^{m-1} x_i^2 & \cdots & \sum_{i=0}^{m-1} x_i^n & \sum_{i=0}^{m-1} x_i^{n+1} \\[2mm]
\cdots & \cdots & \cdots & \cdots & \cdots \\[2mm]
\sum_{i=0}^{m-1} x_i^n & \sum_{i=0}^{m-1} x_i^{n+1} & \cdots & \sum_{i=0}^{m-1} x_i^{2n-2} & \sum_{i=0}^{m-1} x_i^{2n-1} \\[2mm]
\sum_{i=0}^{m-1} x_i^{n+1} & \sum_{i=0}^{m-1} x_i^{n+2} & \cdots & \sum_{i=0}^{m-1} x_i^{2n-1} & \sum_{i=0}^{m-1} x_i^{2n}
\end{bmatrix}
\begin{bmatrix}
c_0 \\ c_1 \\ \cdots \\ c_{n-1} \\ c_n
\end{bmatrix}
=
\begin{bmatrix}
\sum_{i=0}^{m-1} y_i \\[2mm]
\sum_{i=0}^{m-1} y_i x_i \\[2mm]
\cdots \\[2mm]
\sum_{i=0}^{m-1} y_i x_i^{n-1} \\[2mm]
\sum_{i=0}^{m-1} y_i x_i^n
\end{bmatrix}.
$$

$$(7.24)$$

Equation (7.24) is a generalization for fitting a polynomial of degree n into a set of m points using the least-squares approximation method. Letting $S_i = \sum_{k=0}^{m-1} x_k^j$ and $v_i = \sum_{k=0}^{m-1} y_k x_k^i$ for $i = 0, 1, \ldots n$, and $j = 0, 1, \ldots, 2n$ this equation can be rewritten as

$$
\begin{bmatrix}
S_0 & S_1 & \cdots & S_{n-1} & S_n \\
S_1 & S_2 & \cdots & S_n & S_{n+1} \\
\cdots & \cdots & \cdots & \cdots & \cdots \\
S_{n-1} & S_n & \cdots & S_{2n-2} & S_{2n-1} \\
S_n & S_{n+1} & \cdots & S_{2n-1} & S_{2n}
\end{bmatrix}
\begin{bmatrix}
c_0 \\ c_1 \\ \cdots \\ c_{n-1} \\ c_n
\end{bmatrix}
=
\begin{bmatrix}
v_0 \\ v_1 \\ \cdots \\ v_{n-1} \\ v_n
\end{bmatrix}.
$$

$$(7.25)$$

As a word of caution, since the equation involves a power of zero, a value of $x = 0$ produces $0°$, which should be treated as 1.

Algorithm 7.7. Least-Squares Method.
  Given the points (x_i, y_i) for $i = 0, 1, \ldots, m$;
  Select the polynomial $P(x) = c_0 + c_1 x + \cdots + c_n x^n$;
  Find $S_i = \sum_{k=0}^{m-1} x_k^j$ and $v_i = \sum_{k=0}^{m-1} y_k x_k^i$ from Equation (7.25).
  Solve the system of linear equations to find a_i for $i = 0, 1, \ldots, n$;

Example 7.7. Find a least-squares polynomial of degree 2 that approximates the points $(x_k, y_k) = \{(-1, 2), (0, 3), (2, -1), (5, 1)\}$.

Solution. In this problem, the number of points is $m = 4$, whereas the polynomial degree is $n = 2$. The quadratic polynomial is $P_2(x) = c_0 + c_1 x + c_2 x^2$, and the objective here is to find the values of c_0, c_1, and c_2. From Equation (7.24), a 3×3 system of equations is produced, as follows:

$$
\begin{bmatrix}
\sum_{i=0}^{3} 1 & \sum_{i=0}^{3} x_i & \sum_{i=0}^{3} x_i^2 \\
\sum_{i=0}^{3} x_i & \sum_{i=0}^{3} x_i^2 & \sum_{i=0}^{3} x_i^3 \\
\sum_{i=0}^{3} x_i^2 & \sum_{i=0}^{3} x_i^3 & \sum_{i=0}^{3} x_i^4
\end{bmatrix}
\begin{bmatrix}
c_0 \\ c_1 \\ c_2
\end{bmatrix}
=
\begin{bmatrix}
\sum_{i=0}^{3} y_i \\
\sum_{i=0}^{3} y_i x \\
\sum_{i=0}^{3} y_i x_i^2
\end{bmatrix}.
$$

which becomes

$$
\begin{bmatrix}
4 & 6 & 30 \\
6 & 30 & 134 \\
30 & 134 & 642
\end{bmatrix}
\begin{bmatrix}
c_0 \\ c_1 \\ c_2
\end{bmatrix}
=
\begin{bmatrix}
5 \\ 1 \\ 23
\end{bmatrix}.
$$

Solving the above system of linear equations, we obtain $c_0 = 1.339713$, $c_1 = -1.698565$ and $c_2 = 0.327751$. Therefore, the approximated polynomial is $P_3(x) = 1.339713 - 1.698565x + 0.327751x^2$.

7.6 VISUAL SOLUTION: CODE7

Code7. User Manual.

1. Left-click the points in the graphical area in the order from left to right.

2. Select a method from the menu for displaying the corresponding curve.

Development files: Code7.cpp and Code7.h.

Our discussion on interpolation and approximation methods will not be complete without looking at the visual interface of the problems. The project is called Code7, and it displays curves from the points clicked on the displayed window. Curves can be chosen from the methods of Lagrange, Newton's divided-difference, cubic spline, and least-squares.

Figure 7.4 shows a sample output from Code7 showing the Lagrange polynomial that interpolates five points. The main window in the output is split into three regions. The first region consists of the menu items from the four methods. Second is the input region for the points, which also displays the curve from the problem. Third is a table of values of x and their interpolated/approximated polynomial values $P(x)$.

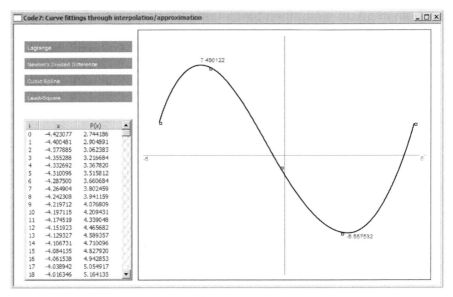

FIGURE 7.4. Output from Code7.

Input for the problem is made by left-clicking the mouse in the input region in the order from left to right. A maximum of 20 points can be clicked in as the input points. A curve corresponding to the selected method is generated by left-clicking an item in the menu once input has been completed.

Figure 7.5 is a schematic drawing illustrating the computational steps in Code7. The execution progress in Code7 is monitored through fStatus with fStatus=0 indicating the pre-curve drawing stage and fStatus=1 when the curve has been generated. The program starts with the initialization of variables and objects in the constructor, CCode7(). The initial display consists of the menu items and the input region. Input is performed by left-clicking anywhere in the curve region in the order from left to right. A click anywhere in the input region is recorded as an input point, and a small rectangle is drawn at the point. The event is handled by OnLButtonDown(). With at least two points drawn, a click at a menu item activates OnLButtonDown() again, and a call to the selected method is made. This changes the fStatus value from 0 to 1, and a curve corresponding to the selected method is drawn. The solution is displayed in the list view table and plotted as a graph in the input region.

Code7 consists of three files, Code7.cpp and Code7.h. The application is based on a single class called CCode7, which is inherited from CFrameWnd. The main variables and objects are organized into four structures, PT, MENU, CURVE, and OUTPUT. Basically, PT has the same set of objects as before. A new addition is the array ipt, which represents the plotted points from the mouse clicks. The array has the size of nPt+1, where nPt is the total number of plotted points. Therefore, ipt is not the same as pt. The latter is an array that represents all points that make up a curve. PT is given as follows:

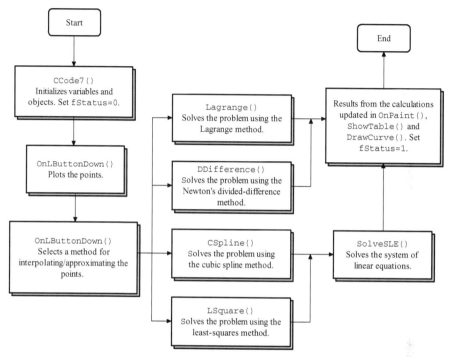

FIGURE 7.5. Schematic drawing of the computational steps in Code7.

```
typedef struct
{
    double x,y;
}PT;
PT *pt,*ipt,max,min,left,right;
```

OUTPUT represents the output items that are accessed through output:

```
typedef struct
{
    CPoint hm,end;
    CRect rc;
}OUTPUT;
OUTPUT output;
```

MENU and CURVE are structures that represent the menu and curve items. The structures are accessed through menu and curve, respectively. The two structures are the same as the ones discussed in Chapter 6. Hence, their contents and components will not be discussed any more here.

TABLE 7.1. Other variables and objects in `CCode7`

Variable/Object	Type	Description
`maxPt`	`int` `(macro)`	Maximum number of plotted points allowed.
`nMenuItems`	`int` `(macro)`	Total number of menu items.
`m`	`int` `(macro)`	Total number of subintervals of x for drawing the curve.
`fStatus`	`bool`	Flag whose values are fStatus=0 and fStatus=1, indicating incomplete and complete inputs, respectively.
`fMenu`	`int`	Flag for the menu whose values indicate the selected method, fMenu=1 for bisection, fMenu=2 for false-point position, fMenu=3 for Newton–Raphson, and fMenu=4 for secant.
`m1,c1,m2,c2`	`double`	Conversion variables from the real coordinates to Windows.
`h`	`double`	Width between the subintervals of x for drawing the curve.
`nPt`	`int`	Actual number of plotted points from the mouse clicks.
`idc`	`int`	Id for the control resources.
`btn`	`CButton`	Push button object called *Compute*.
`table`	`CListCtrl`	List view table for displaying the results from iterations.

Table 7.1 lists other main variables and objects in `CCode7`. The maximum number of points allowed to be plotted is 20, and the number is represented by a macro called `maxPt`. The actual number of plotted points is the integer `nPt` whose initial value is 0. This value is incremented by one each time a point is clicked in the input region. There are five items in the menu, and they are represented by `nMenuItems`. The selected item in the menu is identified through `fMenu`, where `fMenu=1` is the Lagrange method, `fMenu=2` is the Newton's divide-difference method, `fMenu=3` is the cubic spline method, `fMenu=4` is the linear least-squares method, and `fMenu=5` is the cubic least-square method. Conversion variables from the real coordinates to Windows are represented by `m1`, `c1`, `m2`, and `c2`. These variables are declared to be global as they are called by more than one function in this project.

Table 7.2 lists the functions in `CCode7`. Each interpolation and approximation method is represented by a function: `Lagrange()` for the Lagrange method, `DDifference()` for the Newton's divided-difference method, `CSpline()` for the cubic spline method, and `LSquare()` for the linear and cubic least-squares methods.

Mouse's Left-Click Event

In `Code7`, the left button click is an event that represents two separate jobs, selecting an item in the menu and plotting the points in the input region. The two jobs are

TABLE 7.2. Member functions in `CCode7`

Function	Description
`CCode7()`	Constructor.
`~CCode7()`	Destructor.
`Lagrange()`	Interpolates the plotted points using the Lagrange method.
`DDifference()`	Interpolates the plotted points using the Newton's divided-difference method.
`CSpline()`	Interpolates the plotted points using the cubic spline method.
`LSquare()`	Approximates the plotted points using the linear and cubic least-squares methods.
`DrawCurve()`	Draws the curve $y = f(x)$ in the given interval.
`ShowTable()`	Creates a list view table to display the plotted points.
`OnLButtonDown()`	Responds to `ON_WM_LBUTTONDOWN`, which allows points to be clicked in the input region and a menu item to be selected.
`OnPaint()`	Displays and updates the output in the main window.

recognized inside `OnLButtonDown()` through `menu[k].rc.PtInRect(px)` and `curve.rc.PtInRect(px)`, respectively. The first option causes the program to call a method for interpolating or approximating the points and to produce the desired output. The second option invalidates the main window by drawing a small rectangle for each plotted point. The code fragments for this function are given as

```
void CCode7::OnLButtonDown(UINT nFlags,CPoint px)
{
    CRect rc;
    for (int k=1;k<=nMenuItems;k++)
        if (menu[k].rc.PtInRect(px))
            if (nPt>1)
            {
                fMenu=k;
                table.DestroyWindow();
                switch(fMenu)
                {
                    case 1:
                        Lagrange(); break;
                    case 2:
                        DDifference(); break;
                    case 3:
                        CSpline(); break;
                    case 4:
                    case 5:
                        LSquare(); break;
                }
```

```
                    fStatus=1;
                    ShowTable();
                    InvalidateRect(curve.rc);
              }
    if (curve.rc.PtInRect(px))
    {
              ipt[nPt].x=(px.x-c1)/m1;
              ipt[nPt].y=(px.y-c2)/m2;
              rc=CRect(px.x,px.y,px.x+5,px.y+5);
              InvalidateRect(rc);
              nPt++;
    }
}
```

Main Window Update

The main display is constantly updated in OnPaint() whenever a point is clicked or the menu is activated. A point click in the input area causes a small rectangle to be drawn, whereas a click on the menu activates the selected method for solving the problem and displays the curve in the main window. The code fragments are

```
void CCode7::OnPaint()
{
    int i,k;
    CPaintDC dc(this);
    CString str;
    CPoint px;
    CRect rc;
    dc.SetBkColor(RGB(150,150,150));
    dc.SetTextColor(RGB(255,255,255));
    dc.SelectObject(Arial80);
    for (i=1;i<=nMenuItems;i++)
    {
        dc.FillSolidRect(&menu[i].rc,RGB(150,150,150));
        dc.TextOut(menu[i].hm.x+5,menu[i].hm.y+5,menu[i].item);
    }
    dc.SetBkColor(RGB(255,255,255));
    dc.SetTextColor(RGB(100,100,100));
    dc.Rectangle(curve.rc);

    // Draw & label the x,y axes
    CPen pGray(PS_SOLID,1,RGB(100,100,100));
    dc.SelectObject(pGray);
    px=CPoint(m1*0+c1,m2*min.y+c2); dc.MoveTo(px);
    px=CPoint(m1*0+c1,m2*max.y+c2); dc.LineTo(px);
```

```
px=CPoint(m1*left.x+c1,m2*0+c2); dc.MoveTo(px);
px=CPoint(m1*right.x+c1,m2*0+c2); dc.LineTo(px);
dc.SelectObject(Arial80);
str.Format("%.0lf",left.x); dc.TextOut(px.x,px.y,str);
str.Format("%.0lf",right.x); dc.TextOut(px.x-10,px.y,str);
if (fMenu==0)
{
        px.x=m1*ipt[nPt-1].x+c1;
        px.y=m2*ipt[nPt-1].y+c2;
        rc=CRect(px.x,px.y,px.x+5,px.y+5);
        dc.Rectangle(rc);
}
if (fStatus)
{
        for (i=0;i<=nPt-1;i++)
        {
                px.x=m1*ipt[i].x+c1;
                px.y=m2*ipt[i].y+c2;
                dc.Rectangle(px.x,px.y,px.x+5,px.y+5);
        }
        DrawCurve();
}
}
```

Lagrange's Solution

Lagrange's method is handled by Lagrange(). The function performs the computation using Algorithm 7.1. The code segment is given as follows:

```
void CCode7::Lagrange()
{
    int i,j,k;
    double L, h=(ipt[nPt-1].x-ipt[0].x)/m;
    pt[0].x=ipt[0].x;
    pt[0].y=ipt[0].y;
    for (i=0;i<=m;i++)
    {
        pt[i].y=0;
        for (j=0;j<=nPt-1;j++)
        {
            L=1;
            for (k=0;k<=nPt-1;k++)
                    if (j!=k)
                            L*= (pt[i].x-ipt[k].x)/
                                (ipt[j].x-ipt[k].x);
```

```
                    pt[i].y += ipt[j].y*L;
            }
            if (i<m)
                    pt[i+1].x=pt[i].x+h;
        }
}
```

The Lagrange operator in this function is represented by L, and calculations are made based on the product in Equation (7.4). Interpolation is obtained through the following code segment:

```
for (i=0;i<=m;i++)
{
        pt[i].y=0;
        for (j=0;j<=nPt-1;j++)
        {
                L=1;
                for (k=0;k<=nPt-1;k++)
                        if (j!=k)
                                L *= (pt[i].x-ipt[k].x)/
                                    (ipt[j].x-ipt[k].x);
                pt[i].y += ipt[j].y*L;
        }
}
```

Newton's Divided-Difference Solution

Newton's divided-difference method is slightly simpler to execute than the Lagrange method as it does not require the operator $L(x)$ to be determined from the total number of points. The code fragments for this method, as outlined in Algorithm 7.2, are written in DDifference() as

```
void CCode7::DDifference()
{
        int i,j,k;
        double *d,product;
        double h=(ipt[nPt-1].x-ipt[0].x)/(double)m;
        d=new double *[maxPt+1];
        for (i=0;i<=maxPt+1;i++)
                d[i]=new double [maxPt+1];
        pt[0].x=ipt[0].x;
        for (i=0;i<=nPt-1;i++)
                d[0][i]=ipt[i].y;
        for (k=1;k<=nPt-1;k++)
                for (i=0;i<=nPt-1-k;i++)
```

```
                    d[k][i]=(d[k-1][i+1]-d[k-1][i])/
                            (ipt[k+i].x-ipt[i].x);
        for (i=0;i<=m;i++)
        {
              pt[i].y=d[0][0];
              for (j=1;j<=nPt-1;j++)
              {
                    product=1;
                    for (k=0;k<=j-1;k++)
                          product *= (pt[i].x-ipt[k].x);
                    pt[i].y += d[j][0]*product;
              }
              if (i<m)
                    pt[i+1].x=pt[i].x+h;
        }
}
```

The solution in the divide-difference method is provided by Equation (7.7), and the code for this equation is given by

```
for (i=0;i<=m;i++)
{
      pt[i].y=d[0][0];
      for (j=1;j<=nPt-1;j++)
      {
            product=1;
            for (k=0;k<=j-1;k++)
                  product *= (pt[i].x-ipt[k].x);
            pt[i].y += d[j][0]*product;
      }
      if (i<m)
            pt[i+1].x=pt[i].x+h;
}
```

Cubic Spline Solution

The cubic spline solution is provided through CSpline(). The function has been developed based on Algorithm 7.6, and the following code segment lists the full contents of this function:

```
void CCode7::CSpline()
{
   int i,j,k;
   double *e,*f,*g,*r,*w;
   double h;
```

```
e=new double [maxPt+1];
f=new double [maxPt+1];
g=new double [maxPt+1];
r=new double [maxPt+1];
w=new double [maxPt+1];

// form the tridiagonal system
e[0]=0; e[1]=0;
f[0]=0; g[0]=0; r[0]=0;
g[nPt-2]=0;
for (k=1;k<=nPt-2;k++)
{
    e[k+1]=ipt[k+1].x-ipt[k].x;
    f[k]=2*(ipt[k+1].x-ipt[k-1].x);
    g[k]=ipt[k+1].x-ipt[k].x;
    r[k]=6*((ipt[k+1].y-ipt[k].y)/(ipt[k+1].x-ipt[k].x)
        +(ipt[k-1].y-ipt[k].y)/(ipt[k].x-ipt[k-1].x));
}

// Thomas algorithm to solve the system
w[0]=0; w[nPt-1]=0;
for (k=2;k<=nPt-2;k++)
{
    e[k]/=f[k-1];
    f[k]-=e[k]*g[k-1];
}
for (k=2;k<=nPt-2;k++)
    r[k]-=e[k]*r[k-1];
w[nPt-2]=r[nPt-2]/f[nPt-2];
for (k=nPt-3;k>=1;k--)
    w[k]=(r[k]-g[k]*w[k+1])/f[k];
for (k=1;k<=nPt-1;k++)
{
    A[k]=w[k-1]/(6*(ipt[k].x-ipt[k-1].x));
    B[k]=w[k]/(6*(ipt[k].x-ipt[k-1].x));
    C[k]=ipt[k-1].y/(ipt[k].x-ipt[k-1].x)
        -w[k-1]/6*(ipt[k].x-ipt[k-1].x);
    D[k]=ipt[k].y/(ipt[k].x-ipt[k-1].x)
        -w[k]/6*(ipt[k].x-ipt[k-1].x);
}
h=(ipt[nPt-1].x-ipt[0].x)/m;
pt[0].x=ipt[0].x;
pt[0].y=ipt[0].y;
for (i=0;i<=m;i++)
```

```
    {
        for (k=1;k<=nPt-1;k++)
            if (pt[i].x>=ipt[k-1].x && pt[i].x<=ipt[k].x)
                pt[i].y=A[k]*pow(ipt[k].x-pt[i].x,3)
                    +B[k]*pow(pt[i].x-ipt[k-1].x,3)
                    +C[k]*(ipt[k].x-pt[i].x)
                    +D[k]*(pt[i].x-ipt[k-1].x);
        if (i<m)
            pt[i+1].x=pt[i].x+h;
    }
    ShowTable();
    delete e,f,g,r,w;
}
```

The main objective in CSpline() is to solve for the constants A_k, B_k, C_k, and D_k in Equation (7.13), for $k = 1, 2, \ldots, m - 1$. The solution is provided by Equations (7.14), which in turn requires the evaluation of w_k, for $k = 1, 2, \ldots, m - 1$. In finding the values of w_k, two main steps are involved. The first step is the formation of the tridiagonal system of linear equations, as given by Equation (7.16). The second step is the solution to the linear system using any suitable method.

The tridiagonal system of linear equations is formed based on Equations (7.17a), (7.17b), (7.17c), and (7.17d), as follows:

```
// form the tridiagonal system
e[0]=0; e[1]=0;
f[0]=0; g[0]=0; r[0]=0;
g[nPt-2]=0;
for (k=1;k<=nPt-2;k++)
{
    e[k+1]=ipt[k+1].x-ipt[k].x;
    f[k]=2*(ipt[k+1].x-ipt[k-1].x);
    g[k]=ipt[k+1].x-ipt[k].x;
    r[k]=6*((ipt[k+1].y-ipt[k].y)/(ipt[k+1].x-ipt[k].x)
        +(ipt[k-1].y-ipt[k].y)/(ipt[k].x-ipt[k-1].x));
}
```

The tridiagonal system of linear equations is solved using the Thomas algorithm, as discussed in Chapter 5. The code fragments are given by

```
// Thomas algorithm to solve the system
w[0]=0; w[nPt-1]=0;
for (k=2;k<=nPt-2;k++)
{
    e[k]/=f[k-1];
```

```
        f[k]-=e[k]*g[k-1];
}
for (k=2;k<=nPt-2;k++)
        r[k]-=e[k]*r[k-1];
w[nPt-2]=r[nPt-2]/f[nPt-2];
for (k=nPt-3;k>=1;k--)
        w[k]=(r[k]-g[k]*w[k+1])/f[k];
```

The computed values of w_k are applied to determine the values of the constants A_k, B_k, C_k, and D_k in Equations (7.14), as follows:

```
for (k=1;k<=nPt-1;k++)
{
        A[k]=w[k-1]/(6*(ipt[k].x-ipt[k-1].x));
        B[k]=w[k]/(6*(ipt[k].x-ipt[k-1].x));
        C[k]=ipt[k-1].y/(ipt[k].x-ipt[k-1].x)
                -w[k-1]/6*(ipt[k].x-ipt[k-1].x);
        D[k]=ipt[k].y/(ipt[k].x-ipt[k-1].x)
                -w[k]/6*(ipt[k].x-ipt[k-1].x);
}
```

The rest of the code in CSpline() generates (x_i, y_i) for $i = 0, 1, \ldots, m$ for plotting the cubic spline. The code is given by

```
for (i=0;i<=m;i++)
{
    for (k=1;k<=nPt-1;k++)
            if (pt[i].x>=ipt[k-1].x && pt[i].x<=ipt[k].x)
                pt[i].y=A[k]*pow(ipt[k].x-pt[i].x,3)
                        +B[k]*pow(pt[i].x-ipt[k-1].x,3)
                        +C[k]*(ipt[k].x-pt[i].x)+D[k]*(pt[i].x
                        -ipt[k-1].x);
    if (i<m)
            pt[i+1].x=pt[i].x+h;
}
```

Least-Squares Solution

In the least-squares approximation, linear and cubic polynomials are illustrated in items 4 and 5 in the menu. A single function called LSquare() represents the solution to both problems, as the only difference between them is the size of the systems of linear equations generated, two and four, respectively. The solution is provided by Algorithm 7.7, and the code fragments are given by

```
void CCode7::LSquare()
{
    int i,j,k,sa,sv;
    double **a,*b,*S,*v,sum;
    double h=(ipt[nPt-1].x-ipt[0].x)/(double)m;
    pt[0].x=ipt[0].x;
    pt[0].y=ipt[0].y;
    max.y=pt[0].y; min.y=pt[0].y;
    switch (fMenu)
    {
        case 4: sa=2; sv=1; break;
        case 5: sa=6; sv=3; break;
    }
    a=new double *[sa+1];
    S=new double [sa+1];
    b=new double [sv+1];
    v=new double [sv+1];
    for (i=1;i<=sa+1;i++)
        a[i]=new double [sa+1];
    for (k=0;k<=sa;k++)
    {
        S[k]=0;
        if (k<=sv)
            v[k]=0;
    }
    for (k=0;k<=sa;k++)
        for (i=0;i<=nPt-1;i++)
        {
            S[k] += ((ipt[i].x==0&&k==0)?1:pow(ipt[i].x,k));
            if (k<=sv)
                v[k] += ipt[i].y*((ipt[i].
                    x==0&&k==0)?1:pow(ipt[i].x,k));
        }

    //Form the SLE
    for (i=1;i<=sv+1;i++)
    {
        for (j=1;j<=i;j++)
        {
            a[i][j]=S[i+j-2];
            a[j][i]=a[i][j ];
        }
        b[i]=v[i-1];
    }
```

```
// Solve the SLE
double m1,Sum;
for (k=1;k<=sv;k++)
      for (i=k+1;i<=sv+1;i++)
      {
            m1=a[i][k]/a[k][k];
            for (j=1;j<=sv+1;j++)
                  a[i][j] -= m1*a[k][j];
            b[i] -= m1*b[k];
      }
for (i=sv+1;i>=1;i--)
{
      Sum=0;
      c[i-1]=0;
      for (j=i;j<=sv+1;j++)
            Sum += a[i][j]*c[j-1];
      c[i-1]=(b[i]-Sum)/a[i][i];
}
for (i=0;i<=m;i++)
{
      if (fMenu==4)
            pt[i].y=c[0]+c[1]*pt[i].x;
      if (fMenu==5)
            pt[i].y=c[0]+c[1]*pt[i].x+c[2]*pow(pt[i].x,2)
                  +c[3]*pow(pt[i].x,3);
      if (i<m)
            pt[i+1].x=pt[i].x+h;
}
}
```

In the above code, linear polynomial approximation is recognized through the assignment fMenu=4, whereas cubic approximation is recognized through fMenu=5. Two major steps are involved in the least-squares method. The first step is the formation of the system of linear equations, and the second is the solution to this system using any suitable method. Linear approximation is represented by $y = c_0 + c_1 x$. There are two unknowns to be determined from this equation, and this leads to the formation of a 2×2 system of linear equations, given by

$$\begin{bmatrix} S_0 & S_1 \\ S_1 & S_2 \end{bmatrix} \begin{bmatrix} c_0 \\ c_1 \end{bmatrix} = \begin{bmatrix} v_0 \\ v_1 \end{bmatrix}.$$

Similarly, a cubic polynomial approximation in $y = c_0 + c_1 x + c_2 x^2 + c_3 x^3$ requires the formation of a 4×4 system of linear equations, given as

$$\begin{bmatrix} S_0 & S_1 & S_2 & S_3 \\ S_1 & S_2 & S_3 & S_4 \\ S_2 & S_3 & S_4 & S_5 \\ S_3 & S_4 & S_5 & S_6 \end{bmatrix} \begin{bmatrix} c_0 \\ c_1 \\ c_2 \\ c_3 \end{bmatrix} = \begin{bmatrix} v_0 \\ v_1 \\ v_2 \\ v_3 \end{bmatrix}.$$

In LSquare(), the size of the matrix is held by sa, which receives its assignment from fMenu. From this value, the values of S_i and v_k can be determined, and this leads to the formation of the system of linear equations. The code for the above equation is given by

```
for (k=0;k<=sa;k++)
{
    S[k]=0;
    if (k<=sv)
        v[k]=0;
}
for (k=0;k<=sa;k++)
    for (i=0;i<=nPt-1;i++)
    {
        S[k] += ((ipt[i].x==0&&k==0)?1:pow(ipt[i].x,k));
        if (k<=sv)
            v[k] += ipt[i].y*((ipt[i]
                    .x==0&&k==0)?1:pow(ipt[i].x,k));
    }

//Form the SLE
for (i=1;i<=sv+1;i++)
{
    for (j=1;j<=i;j++)
    {
        a[i][j]=S[i+j-2];
        a[j][i]=a[i][j];
    }
    b[i]=v[i-1];
}
```

The system of linear equations is then solved using the Gaussian elimination method to produce the values of c_i, as shown in the following code segment:

```
// Solve the SLE
double m1,Sum;
```

```
for (k=1;k<=sv;k++)
    for (i=k+1;i<=sv+1;i++)
    {
        m1=a[i][k]/a[k][k];
        for (j=1;j<=sv+1;j++)
            a[i][j] -= m1*a[k][j];
        b[i] -= m1*b[k];
    }
for (i=sv+1;i>=1;i--)
{
    Sum=0;
    c[i-1]=0;
    for (j=i;j<=sv+1;j++)
        Sum += a[i][j]*c[j-1];
    c[i-1]=(b[i]-Sum)/a[i][i];
}
```

The final section of LSquare() generates the points (x_i, y_i) for $i = 0, 1, \ldots, m$ for generating the corresponding curves.

```
for (i=0;i<=m;i++)
{
    if (fMenu==4)
        pt[i].y=c[0]+c[1]*pt[i].x;
    if (fMenu==5)
        pt[i].y=c[0]+c[1]*pt[i].x+c[2]*pow(pt[i].x,2)
                +c[3]*pow(pt[i].x,3);
    if (i<m)
        pt[i+1].x=pt[i].x+h;
}
```

7.7 SUMMARY

Interpolation and approximation concepts are widely used in science and engineering applications such as in data analysis, curve and surface fittings, and forecasting. The output from interpolation and approximation is often expressed as a polynomial that becomes the generalization to the problem.

In this chapter, interpolation and approximation have been illustrated as polynomials and display as curves. The techniques discussed are the Lagrange, Newton's divided-difference, cubic spline, and least-squares methods. Each of these methods has its strength and weaknesses in interpolating or approximating the points depending on the objectives. For example, in producing low-degree polynomials, it is wise to deploy the least-squares or spline fitting methods.

Interpolation and approximation techniques are used widely in computer graphics and computer-aided designs (CAD). In computer graphics, interpolation and approximation contribute to producing graphs that provide the relationship between the given points. In computer-aided designs, various bodies and structures are constructed based on interpolative and approximated techniques. In either case, interpolation and approximation are regarded as important tools for modeling and simulation.

NUMERICAL EXERCISES

1. Apply the Lagrange, Newton's divided-difference and cubic spline methods for interpolating the following sets of data:
   a. $\{(0, 3), (1, 5), (4, -1)\}$.
   b. $\{(0, 3), (1, 5), (4, -1), (5, 0)\}$.
   c. $\{(0, 3), (1, 5), (2, -1), (3, 0)\}$.

2. The chapter discusses interpolation methods using polynomials only. It is also possible to use other functions, including nonpolynomials in interpolation. Find the values of a and b in $f(x) = a \sin x + b \cos x$, which interpolates the points $(0, 0.5)$ and $(\pi/2, -0.3)$.

3. Do some research to study the properties of a spline of degree two or quadratic spline. Hence, find the quadratic spline that interpolates the points $\{(0, 3), (1, 5), (4, -1), (5, 0)\}$.

4. Use the Stirling's method to approximate the value of $P(1.5)$ from a data set given by $\{(0, 3), (1, 5), (2, -1), (3, 0)\}$.

PROGRAMMING CHALLENGES

1. Run Code7 to check for the existence of at least one root of the functions in the given intervals in the following problems.

2. Code7 produces curves based on points clicked in the rectangular region provided in the window. This approach provides an easy mode of input for the user, but at the same time, a difficulty arises where it is not possible to click the exact points using a mouse. To overcome this difficulty, it is wise to provide another interface where input for the points is allowed using edit boxes. Improve on Code7 by adding this mode of input to the interface.

3. In Code7, the points for plotting the curves must be plotted in the order from left to right only. Breaking this rule will produce some undesirable results on the generated curve. Improve on the code to allow the points to be plotted from any direction.

4. A quadratic spline, instead of a cubic spline, may prove to be a better choice in interpolating points for producing a low-degree polynomial. Find some suitable references for studying the properties of this method, and produce a program for generating its curve.

5. The Newton's forward-difference and backward-difference methods and Stirling's method are based on equal-width subintervals. Devise a program for these methods using the mouse clicks as their input. The input should start with a single interval whose left and right end points are marked through the mouse clicks. The program then divides the intervals into m equal-width subintervals automatically where m in the input value is supplied by the user.

CHAPTER 8

Differentiation and Integration

8.1 INTRODUCTION

The derivative of a function gives a clear indication of the rate of increase or decrease of the function with respect to its domain. The rate of increase or decrease of a function contributes to the overall understanding of a system that is governed by such parameters. In general, the derivatives of a function contribute to the modeling of a given problem, which describes the properties and dynamics of the elements in the problem. For example, in studying the electrostatic field properties of an area, the solution requires a graphing of the first and second derivatives of the points in the area. We will discuss some useful properties of the derivatives of a function for modeling in the next few chapters.

The *analytical* derivative of a function $f(x)$ with respect to the variable x is denoted by $f'(x)$ or $\frac{df}{dx}$. This derivative returns the exact solution in the form of a function of x. For example, if $f(x) = x^2 - \sin x$, then its analytical derivative is $f'(x) = 2x - \cos x$.

By default, a digital computer does not have the processing capability to produce the analytical derivative of a function. However, this analytical solution can still be produced through a software that stores a list of primitive functions and its derivatives in a numerical database. On the computer, the analytical solution to a derivative is a complex problem that requires several recursive calls to the functions in its numerical database. *Symbolic computing* is one area of study that addresses this problem, which involves the construction of a database of mathematical functions for generating the analytical solution to their derivatives.

The computer gives a good approximation to the derivatives of a function based on some finite points in the given domain. A *numerical* approximation to the derivative of as a function $f(x)$ is a solution expressed as a number, rather than as a function of x. The domain of $f(x)$ is first decomposed into its discrete form, x_i for $i = 1, 2, \ldots, m$, and the derivatives are computed at these points based on their relative points.

An integral of a function does the opposite of derivative. Integral returns the differentiated function to its original value. A numerical approach for finding the integral of a function is needed in cases where the integral is difficult to compute.

In this chapter, we will discuss several numerical approaches to finding the derivatives and integral of a function. The approximated values returned from the methods are reasonably close to the exact values and are acceptable in most cases.

8.2 NUMERICAL DIFFERENTIATION

The starting point in finding a derivative is the nth-order Taylor series expansion of a discrete function $y_i = f(x_i)$ given as follows:

$$y_{i+1} \approx y_i + \frac{h}{1!}y_i' + \frac{h^2}{2!}y_i'' + \frac{h^3}{3!}y_i''' + \cdots + \frac{h^n}{n!}y_i^{(n)}. \tag{8.1}$$

Equation (8.1) gives the one-term forward expansion of $y = f(x)$. In this equation, $h = \Delta x$ is the width of the x subintervals that are assumed to be uniform. The term $y_{i+1} = f(x_{i+1})$ above is equivalent to $y_{i+1} = f(x_{i+1}) = f(x_i + h)$.

Figure 8.1 shows the relative position of the discrete points of $f(x)$ in the interval $x_{i-2} \le x \le x_{i+2}$. Taking only up to the second derivative term, this equation reduces to

$$y_{i+1} \approx y_i + \frac{h}{1!}y_i' + \frac{h^2}{2!}y_i''. \tag{8.2a}$$

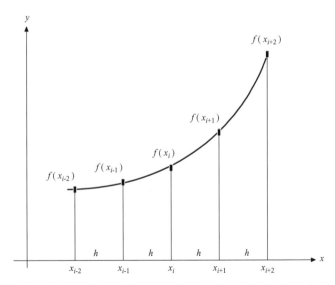

FIGURE 8.1. Relative position of discrete points of $f(x)$ in the Taylor series expansion.

Obviously, the one-term forward expansion in Taylor series involves $x_{i+1} = x_i + h$. In a similar manner, we can define the two-term forward expansion by replacing h with $2h$, or $x_{i+2} = x_i + 2h$, to produce

$$y_{i+2} \approx y_i + \frac{2h}{1!} y_i' + \frac{(2h)^2}{2!} y_i'' = y_i + \frac{2h}{1!} y_i' + \frac{4h^2}{2!} y_i''. \tag{8.2b}$$

It follows that the one-term backward expansion is obtained by replacing h with $-h$, or $x_{i-1} = x_i + (-h) = x_i - h$, to produce

$$y_{i-1} \approx y_i + \frac{(-h)}{1!} y_i' + \frac{(-h)^2}{2!} y_i'' = y_i - \frac{h}{1!} y_i' + \frac{h^2}{2!} y_i''. \tag{8.2c}$$

Subsequently, the two-term backward expansions of $y = f(x)$ involve $x_{i-2} = x_i + (-2h) = x_i - 2h$ and are given by

$$y_{i-2} \approx y_i + \frac{(-2h)}{1!} y_i' + \frac{(-2h)^2}{2!} y_i'' = y_i - \frac{2h}{1!} y_i' + \frac{4h^2}{2!} y_i''. \tag{8.2d}$$

The simplest rule for the first derivative is obtained by taking y_i' as a subject in the first two terms of Equation (8.1). It produces the *forward-difference rule for the first derivative*,

$$y_{i+1} \approx y_i + \frac{h}{1} y_i',$$

$$y_i' \approx \frac{y_{i+1} - y_i}{h}. \tag{8.3a}$$

The *forward-difference rule for the second derivative* is obtained by subtracting twice Equation (8.2c) from Equation (8.2d) and by simplifying the terms to produce

$$y_i'' \approx \frac{y_{i+2} - 2y_{i+1} + y_i}{h^2}. \tag{8.3b}$$

There is also the *backward-difference rule for the first derivative*, obtained by taking y_i' as a subject in the first two terms of Equation (8.2c), to produce

$$y_i' \approx \frac{y_i - y_{i-1}}{h}. \tag{8.4a}$$

Similarly, subtracting twice Equation (8.2c) from Equation (8.2d) gives the *backward-difference rule for the second derivative*:

$$y_i'' \approx \frac{y_i - 2y_{i-1} + y_{i-2}}{h^2}. \tag{8.4b}$$

As the names suggest, the forward-difference and backward-difference rules have the disadvantage in that they are bias toward the forward and backward points, respectively. A more balanced approach is the central-difference rules, which consider both forward and backward points. Subtracting Equation (8.2a) from (8.2c) and simplifying terms, we obtain the *central-difference rule for the first derivative*,

$$y_i' \approx \frac{y_{i+1} - y_{i-1}}{2h}. \tag{8.5a}$$

By adding Equations (8.2a) and (8.2c) and simplifying the terms, we obtain the *central-difference rule for the second derivative*,

$$y_i'' \approx \frac{y_{i+1} - 2y_i + y_{i-1}}{h^2}. \tag{8.5b}$$

Example 8.1. Given $y = x \sin x$, find y' and y'' at the points along $0 \le x \le 2$ whose subintervals have an equal width given by $h = 0.3333$ using the forward, backward, and central-difference methods.

Solution. In this problem, $m = (2 - 0)/.3333 = 6$. The analytical derivatives are given by $y' = \sin x + x \cos x$ and $y'' = 2 \cos x - x \sin x$. Table 8.1 plots the values of x_i and y_i for $i = 0, 1, \dots, 6$. The analytical derivatives at these points are shown in the fourth and fifth columns of the table.

At $i = 0$, y_0' and y_0'' can be evaluated using the forward-difference rules using Equation (8.3a) and (8.3b):

$$y_0' \approx \frac{y_1 - y_0}{h} = \frac{0.1090 - 0}{0.3333} = 0.3270,$$

$$y_0'' \approx \frac{y_2 - 2y_1 + y_0}{h^2} = \frac{0.4122 - 2(0.1090) + 0}{0.3333^2} = 1.7481.$$

It is not possible to evaluate y_0' and y_0'' using the backward and central-difference rules as both methods involve the terms y_{-1} and y_{-2}, which do not exist. The results at other points are shown in Table 8.1. Through comparison with the analytical (exact) method, the central-difference method produces the closest approximations for the derivatives.

TABLE 8.1. The numerical results from Example 8.1

i	x_i	y_i	Analytical Solutions		Forward-difference		Backward-difference		Central-difference	
			y_i'	y_i''	y_i'	y_i''	y_i'	y_i''	y_i'	y_i''
0	0	0	0	2	0.3270	1.7481	void	void	void	void
1	0.3333	0.1090	0.6421	1.7809	0.9097	1.1342	0.3270	void	0.6183	1.7481
2	0.6667	0.4122	1.1423	1.1594	1.2877	0.2277	0.9097	1.7481	1.0987	1.1342
3	1.000	0.8414	1.3818	0.2391	1.3636	−0.8228	1.2877	1.1342	1.3257	0.2277
4	1.3333	1.2959	1.2856	−0.8255	1.0894	−1.8319	1.3636	0.2277	1.2265	−0.8228
5	1.6667	1.6590	0.8358	−1.8515	0.4788	void	1.0894	−0.8228	0.7841	−1.8319
6	2.000	1.8186	0.0770	−2.6509	void	void	0.4788	−1.8319	void	void

8.3 NUMERICAL INTEGRATION

An integral of the form $\int_a^b f(x)\,dx$ in the interval $a \le x \le b$ has many applications, especially in science and engineering. In its fundamental form, the integral evaluates the area enclosed between the function $f(x)$ and the x-axis (see Figure 8.2). Area in this sense is a symbolic represention of many other quantities and problems in engineering such as moment, volume, mass, density, pressure, and temperature. Therefore, a numerical solution to a definite integral contributes as part of the whole solution to a problem.

The exact methods for evaluating definite integrals have been widely discussed in elementary calculus classes. Some popular methods involve techniques such as direct integration, substitutions, integration by parts, partial fractions, and substitutions with the use of trigonometric functions. However, not all definite integral problems can be solved using such methods. This may be the case when the function $f(x)$ is too difficult to integrate, or in the case where $f(x)$ is given in the form of a discrete data set. Furthermore, it may not be possible to implement the exact method on the computer as the computer does not have the analytical skill like a human does. Therefore, approximations using numerical methods may prove to be an alternative approach and practical for implementation here.

Numerical solutions for evaluating definite integrals are obtained using several methods. Some of the most fundamental methods include the trapezium, Simpson, Simpson's 3/8, and Gaussian quadrature methods. We discuss these methods in this section.

Trapezium Method

The trapezium method is a classic technique for approximating the definite integral of a function $\int_a^b f(x)\,dx$ in the interval $a \le x \le b$. As the name suggests, the method

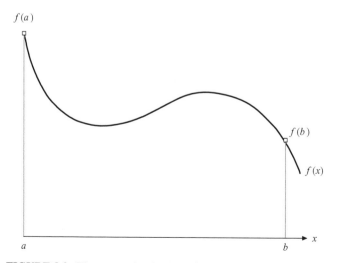

FIGURE 8.2. The area under the curve between $f(x)$ and the x-axis.

is based on an approximation of the area under the curve between the function and the x-axis as several trapeziums taken over some finite subintervals.

In the trapezium method, the x interval in $a \leq x \leq b$ is divided into m equal-width subintervals, each with width h. A straight line is connected from $(x_i, f(x_i))$ to $(x_{i+1}, f(x_{i+1}))$ in the subinterval $[x_i, x_{i+1}]$, and this forms a trapezium with $f(x_i)$ and $f(x_{i+1})$ as the parallel sides whose distance between them is h. The area under the curve in this subinterval is then approximated as the area of the trapezium, given as

$$\int_{x_i}^{x_{i+1}} f(x)\,dx \approx \frac{h}{2}(f(x_i) + f(x_{i+1})).$$

With m subintervals, $\int_a^b f(x)dx$ is approximated as the sum of all areas of the trapeziums, given as

$$\int_a^b f(x)\,dx \approx \sum_{i=0}^{m-1} \frac{h}{2}(f(x_i) + f(x_{i+1}))$$

$$= \frac{h}{2}[f(x_0) + f(x_m) + 2\{(f(x_1) + f(x_2) + \cdots + f(x_{m-1})\}]. \quad (8.6)$$

Equation (8.6) gives the *trapezium method* for finding the definite integral of a function $f(x)$ in the interval from $x = a$ to $x = b$ with m equal subintervals. The method is illustrated in Figure 8.3 using four equal subintervals. From the figure, it is clear that $\int_{x_0}^{x_4} f(x)\,dx$ for $x_0 \leq x \leq x_4$ is a problem in finding the area under the curve in the given interval. The interval in this problem is divided into four subintervals with $m = 4$ and $h = (x_4 - x_0)/m$. The total area is then approximated as the sum of

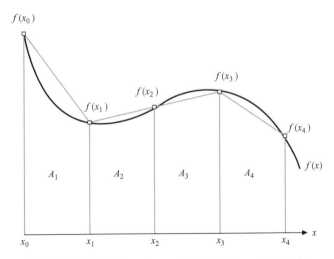

FIGURE 8.3. The trapezium method with four subintervals.

the subareas A_i for $i = 1, 2, 3, 4$, given as

$$\int_a^b f(x)\,dx \approx A_1 + A_2 + A_3 + A_4$$

$$= \frac{h}{2}(f(x_0) + f(x_1)) + \frac{h}{2}(f(x_1) + f(x_2)) + \frac{h}{2}(f(x_2) + f(x_3)) + \frac{h}{2}(f(x_3) + f(x_4))$$

$$= \frac{h}{2}[f(x_0) + f(x_4) + 2(f(x_1) + f(x_2) + f(x_3))].$$

Example 8.2. Find $\int_0^3 x \sin x \, dx$ using the trapezium method with nine subintervals.

Solution. In this problem, $m = 9$, $x_0 = 0$, and $x_m = 3$. The width of each subinterval is $h = \frac{3-0}{9} = 0.333333$. With nine subintervals, there are 10 points on $y = f(x) = x \sin x$, and their values are shown in the following table:

i	0	1	2	3	4	5	6	7	8	9
x_i	0	0.3333	0.6667	1.0000	1.3333	1.6667	2.0000	2.3333	2.6667	3.0000
y_i	0	0.1090	0.4122	0.8414	1.2959	1.6590	1.8186	1.6872	1.2194	0.4234

Therefore, $\int_0^3 x \sin x \, dx \approx \frac{h}{2}[y_0 + y_9 + 2(y_1 + y_2 + y_3 + y_4 + y_5 + y_6 + y_7 + y_8)] = 3.0846$.

Simpson's Method

The trapezium method uses a straight line approximation on two successive $f(x)$ values in the subintervals for evaluating $\int_a^b f(x)\,dx$ for $a \le x \le b$. A straight line approximation may not produce a good solution especially when the number of subintervals is small, because a straight line does not approximate a given curve well. From the large error imposed, the trapezium method requires a large number of subintervals in order to produce a good solution.

The accuracy of the trapezium method can be improved by replacing the straight line with a quadratic function. This idea makes sense as a quadratic curve lies closer to the real curve in the given function than a straight line. Since a quadratic function requires three points for interpolation, the method suggests pairs of two subintervals in the approximation. The approach is called the *Simpson method*.

As the Simpson method requires two subintervals in each pair, the total number of subintervals in the given interval must be an even number. Consider the subintervals $[x_i, x_{i+1}]$ and $[x_{i+1}, x_{i+2}]$, which involve the points, $(x_i, f(x_i))$, $(x_{i+1}, f(x_{i+1}))$, and $(x_{i+2}, f(x_{i+2}))$. Figure 8.4 shows this scenario with the dotted curve as the quadratic approximation to the real curve. This curve has an equation given by

$$y = a_0 + a_1 x + a_2 x^2,$$

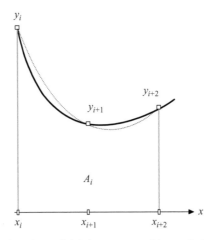

FIGURE 8.4. Quadratic polynomial fitting on two subintervals in the Simpson's method.

where a_0, a_1, and a_2 are the constants that are to be determined. The quadratic curve interpolates (x_i, y_i), (x_{i+1}, y_{i+1}) and (x_{i+2}, y_{i+2}). This produces a system of linear equations, given as

$$y_i = a_0 + a_1 x_i + a_2 x_i^2,$$
$$y_{i+1} = a_0 + a_1 x_{i+1} + a_2 x_{i+1}^2,$$
$$y_{i+2} = a_0 + a_1 x_{i+2} + a_2 x_{i+2}^2.$$

The above system is solved to produce the Simpson's formula, given by

$$\int_{x_i}^{x_{i+2}} f(x)\, dx \approx \frac{h}{3} (y_i + 4y_{i+1} + y_{i+2}). \tag{8.7}$$

Equation (8.7) can be extended into the case of m even subintervals for $x = x_0$ to $x = x_m$. The total area is found as a sum of $m/2$ subareas, where each subarea combines two subintervals. Applying Equation (8.7), we obtain the Simpson's method for m subintervals, as follows:

$$\int_{x_0}^{x_m} f(x)\, dx \approx \frac{h}{3}(y_0 + 4y_1 + y_2) + \frac{h}{3}(y_2 + 4y_3 + y_4) + \cdots + \frac{h}{3}(y_{m-2} + 4y_{m-1} + y_m)$$

$$= \frac{h}{3}[(y_0 + y_m) + 4(y_1 + y_3 + \cdots + y_{m-1}) + 2(y_2 + y_4 + \cdots + y_{m-2})]. \tag{8.8}$$

Example 8.3. Find $\int_0^3 x \sin x\, dx$ using the Simpson's method with nine subintervals.

Solution. In this problem, $m = 9$, $x_0 = 0$, $x_m = 3$, and $h = \frac{3-0}{9} = 0.333333$. The problem refers to the same table as in Example 8.1. Therefore,

$$\int_0^3 f(x)\,dx \approx \frac{h}{3}[(y_0 + y_9) + 4(y_1 + y_3 + y_5 + y_7) + 2(y_2 + y_4 + y_6 + y_8)]$$

$$= 3.0111.$$

An extension to the Simpson's method is the *Simpson's 3/8 method*. In this method, three subintervals are combined to produce one subarea based on a cubic polynomial. This approach requires four points for each subarea, and therefore, it provides a more accurate approximation to the problem. With three subintervals for each subarea, the total number of subintervals in the Simpson's 3/8 method becomes a multiple of three.

Interpolation over four points for each subarea in the Simpson's 3/8 method produces a system of four linear equations. The interpolating curve is a cubic polynomial given by

$$y = a_0 + a_1 x + a_2 x^2 + a_3 x^3,$$

where a_0, a_1, and a_2 are constants. The above curve interpolates (x_i, y_i), (x_{i+1}, y_{i+1}), (x_{i+2}, y_{i+2}), and (x_{i+3}, y_{i+3}) to produce an approximated subarea, given by

$$\int_{x_i}^{x_{i+3}} f(x)\,dx \approx \frac{3h}{8}(y_i + 3y_{i+1} + 3y_{i+2} + y_{i+3}). \tag{8.9}$$

With m subintervals, Equation (8.9) can be extended to produce

$$\int_{x_0}^{x_m} f(x)\,dx \approx \frac{3h}{8}[(y_0 + y_m) + 2(y_3 + y_6 + \cdots + y_{m-3})$$

$$+ 3(y_1 + y_2 + y_4 + y_5 + \cdots + y_{m-2} + y_{m-1})]. \tag{8.10}$$

Example 8.4. Find $\int_0^3 x \sin x\,dx$ using the Simpson's 3/8 method with nine subintervals.

Solution. In this problem, $m = 9$, $x_0 = 0$, $x_m = 3$, and $h = \frac{3-0}{9} = 0.333333$. The problem refers to the same table of values as in Example 8.1. Therefore,

$$\int_0^3 f(x)\,dx \approx \frac{3h}{8}[(y_0 + y_9) + 2(y_3 + y_6) + 3(y_1 + y_2 + y_4 + y_5 + y_6 + y_7)]$$

$$= 3.1112.$$

TABLE 8.2. Legendre polynomials

n	$P_n(x)$
0	1
1	x
2	$(3x^2 - 1)/2$
3	$(5x^3 - 3x)/2$
4	$(35x^4 - 30x^2 + 3)/8$
5	$(63x^5 - 70x^3 + 15x)/8$
6	$(231x^6 - 315x^4 + 105x^2 - 5)/16$

Gaussian Quadrature

The three methods outlined above are applicable to problems whose data are normally given in discrete form. Quite often data are given in the form of a function $f(x)$ whose integral is difficult to find. A method called the *Gaussian quadrature* specializes in tackling this kind of problem. The method is based on an approximation on Legendre polynomials.

A Legendre polynomial of degree n is given in the form of an ordinary differential equation of order n, as follows:

$$P_n(x) = \frac{1}{2^n n} \frac{d^n}{dx^n} \left((x^2 - 1)^n \right). \tag{8.11}$$

Table 8.2 lists some Legendre polynomials of lower degrees. Legendre polynomials have the characteristics of being orthogonal over $x \in (-1, 1)$ and satisfy

$$\int_{-1}^{1} P_m(x) P_n(x) dx = \frac{2}{2n + 1} \delta_{mn},$$

where δ_{mn} is the Kronecker delta, which is 1 if $m = n$, and 0 otherwise.

The Gaussian quadrature method consists of two major steps. First, the integral $\int_a^b f(x) dx$ in the interval $a \leq x \leq b$ is transformed into the form of $\int_{-1}^{1} g(t) dt$. This is achieved through a linear relationship given by

$$x = \frac{(b - a)t + (b + a)}{2}, \tag{8.12}$$

where $g(t)$ is a new function that is continuous in $-1 \leq t \leq 1$. It can be verified that substituting $x = a$ and $x = b$ into the linear equation produces $t = -1$ and $t = 1$, respectively. The transformation preserves the area under the curve in $\int_{-1}^{1} g(t) dt$, as illustrated in Figure 8.5. We have

$$f(x) dx = \left(\frac{b - a}{2} \right) f \left(\frac{(b - a)t + (b + a)}{2} \right) dt.$$

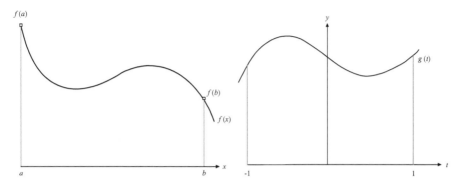

FIGURE 8.5. Transformation from $\int_a^b f(x)\,dx$ to $\int_{-1}^1 g(t)\,dt$ in the Gauss–Legendre method.

The integral becomes

$$\int_a^b f(x)\,dx = \int_a^b \left(\frac{b-a}{2}\right) f\left(\frac{(b-a)t+(b+a)}{2}\right) dt.$$

This gives the relationship between the $f(x)$ and $g(t)$ as

$$g(t) = \left(\frac{b-a}{2}\right) f\left(\frac{(b-a)t+(b+a)}{2}\right). \tag{8.13}$$

The second step in this method consists of an approximation to the integral $\int_{-1}^1 g(t)\,dt$ through a quadrature, given by

$$\int_{-1}^1 g(t)\,dt \approx \sum_{i=1}^n w_i g(t_i). \tag{8.14}$$

The above equation is called the n-point Gauss quadrature, where n is an integer number greater than 1. The constant w_i is called the weight, whereas t_i is the argument in the quadrature. The argument t_i in Equation (8.14) is actually a root of the Legendre polynomial given in Table 8.1.

Equation (8.14) suggests the original problem is approximated as a sum of n quadratures whose weights and arguments are determined from the quadrature properties. The weights and arguments of some of the Gaussian quadrature are tabulated in Table 8.3.

Example 8.5. Find $\int_0^3 x \sin x\,dx$ using the three-point Gaussian quadrature method.

TABLE 8.3. Weights and arguments in the Gaussian quadrature

n	Weight, w_i	Argument, t_i
2	$w_1 = 1$	$t_1 = -\sqrt{1/3} = -0.577350$
	$w_2 = 1$	$t_2 = \sqrt{1/3} = 0.577350$
3	$w_1 = 0.555556$	$t_1 = -\sqrt{3/5} = -0.774597$
	$w_2 = 0.888889$	$t_2 = 0$
	$w_3 = 0.555556$	$t_3 = \sqrt{3/5} = 0.774597$
4	$w_1 = 0.347855$	$t_1 = -0.861136$
	$w_2 = 0.652145$	$t_2 = -0.339981$
	$w_3 = 0.652145$	$t_3 = 0.339981$
	$w_4 = 0.347855$	$t_4 = 0.861136$
5	$w_1 = 0.236927$	$t_1 = -0.90610$
	$w_2 = 0.478629$	$t_2 = -0.538469$
	$w_3 = 0.568889$	$t_3 = 0$
	$w_4 = 0.478629$	$t_4 = 0.538469$
	$w_5 = 0.236927$	$t_5 = 0.90610$

Solution. We have $x = \frac{(3-0)t+(3+0)}{2} = \frac{3t+3}{2}$.

$$\int_0^3 x \sin x \, dx = \int_{-1}^1 \frac{3t+3}{2}\left(\sin\frac{3t+3}{2}\right)\frac{3}{2}dt$$

$$= \int_{-1}^1 \frac{9(t+1)}{4}\left(\sin\frac{3t+3}{2}\right)dt.$$

We get $g(t) = \frac{9(t+1)}{4}(\sin\frac{3t+3}{2})$. From Table 8.3, we obtain

$$g(t_1) = g(-0.7746) = \frac{9(-0.7746+1)}{4}\left(\sin\frac{3(-0.7746)+3}{2}\right) = 0.1682,$$

$$g(t_2) = g(0) = \frac{9(0+1)}{4}\left(\sin\frac{3(0)+3}{2}\right) = 2.2443,$$

$$g(t_3) = g(0.7746) = \frac{9(0.7746+1)}{4}\left(\sin\frac{3(0.7746)+3}{2}\right) = 1.8427.$$

Therefore,

$$\int_0^3 x \sin x \, dx \approx w_1 g(t_1) + w_2 g(t_2) + w_3 g(t_3)$$

$$= 0.5556(0.1682) + 0.8889(2.2443) + 0.5556(1.8427)$$

$$= 3.1123.$$

8.4   VISUAL SOLUTION: CODE8

Code8. User Manual.

1. Select an item in the menu.
2. Fill in the input in the input boxes.
3. Left-click the *Compute* button to see the results.

Development files: Code8.cpp, Code8.h and MyParser.obj.

The programming solutions to the numerical differentiation and integration of a function are discussed here in a project called Code8. The project has two items in the menu, namely, differentiation and integration. The first item focuses on computing the first and second derivatives of the named function using the central-difference rules. The second item computes the definite integral of the named function according to its interval using the trapezium, Simpson, Simpson's 3/8, and Gaussian quadrature methods.

Figure 8.6 is an output from the project that shows the results and graphs of $y = x \sin x$ and its first and second derivatives at x_i for $i = 0, 1, \ldots, m$. The approximation

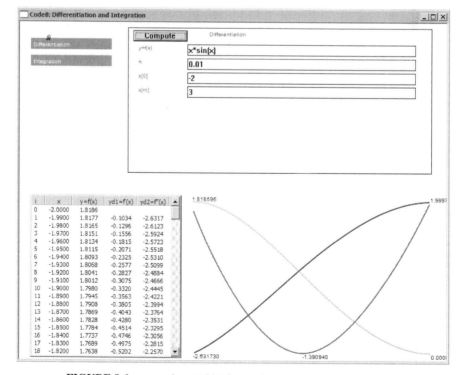

FIGURE 8.6. $y = x \sin x$ and its first and second derivatives in Code8.

of the derivatives using the central-difference rules are valid in the interval from $i = 1$ to $i = m - 1$, as $f'(x_i)$ and $f''(x_i)$ are not defined at $i = 0$ and $i = m$. In addition, each curve has been scaled so that its maximum and minimum points touch the upper and lower portions of the drawing region.

Code8 includes the source files Code8.cpp and Code8.h. A single class called CCode8 is used in this application. The global variables and objects in CCode8 are organized into four structures, namely, PT, INPUT, MENU, and CURVE. Basically, the structures in CCode8 are very similar to the structures of the same name in the previous chapters. PT represents the points in the functions in real coordinates with two additional members, the first and second derivatives of the functions denoted as yd1 and yd2, respectively.

```
typedef struct
{
        double x,y,yd1,yd2;
} PT;
PT *pt,max,min,left,right;
```

The functions in this class are summarized in Table 8.4. Differentiation() and Integration() are two key functions that represent the solutions to the differentiation and integration problems, respectively. These functions are called from OnButton(), which responds to the left-click on the *Compute* push button. The full results are shown in a list value table through ShowTable() and are displayed as graphs through DrawCurve().

Figure 8.7 shows the schematic drawing of the computational stages in Code8. A variable called fStatus monitors the runtime progress, with its initial value of fStatus=0. A value of fStatus=1 indicates the selected method for the problem has been successfully executed. There are only two items in the menu, differentiation

TABLE 8.4. Member functions in CCode8

Function	Description
CCode8()	Constructor.
~CCode8()	Destructor.
Differentiation()	Computes the first and second derivatives of the given function using the central-difference rules.
Integration()	Computes the integral of the given function using the trapezium, Simpson, Simpson's 3/8, and Gaussian quadrature methods.
DrawCurve()	Draws the curve, and its first and second derivatives.
ShowTable()	Shows the results in a list view table.
OnLButtonDown()	Responds to ON_WM_LBUTTONDOWN, which allows points to be clicked in the input region and a menu item to be selected.
OnButton()	Responds to ON_BN_CLICKED by calling the corresponding function for solving the given problem.
OnPaint()	Displays and updates the output in the main window.

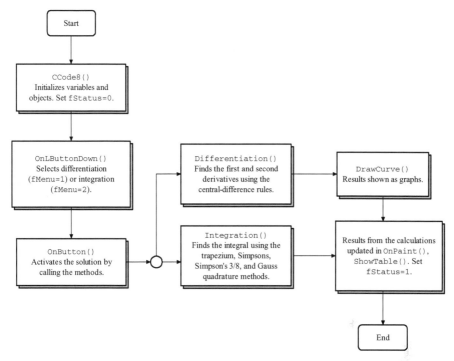

FIGURE 8.7. Schematic drawing showing the computational stages in Code8.

and integration. The selected item is recognized through fMenu=1 for differentiation and fMenu=2 for integration.

The process starts at the constructor function, CCode8(). The main window is created, and all global variables and objects are initialized in this function. A status flag called fStatus is set to zero to indicate its initial state of execution. A menu consisting of two items, *Differentiation* and *Integration*, appears. A click at one of its items activates OnLButtonDown(), which assigns fMenu=1 for *Differentiation* and fMenu=2 for *Integration*. A selected item produces edit boxes for collecting input data in the problem. The process is passed to OnButton() for branching the solution to either Differentiation() or Integration(), according to its fMenu value. Differentiation() causes the graph from $f(x)$ and its first and second derivatives to be drawn in the graphical region. Integration() causes the evaluation of the integral using the trapezium, Simpson, Simpson's 3/8, and Gaussian quadrature methods. The solutions are displayed in ShowTable() and plotted as graphs in DrawCurve(). A successful execution is reflected with fStatus=1 in the final stage of the execution.

Differentiation

The solution to a differentiation problem is handled by Differentiation(). The function evaluates the first and second derivatives of the given function at (x_i, y_i), for

$i = 1, 2, \ldots, m - 1$, using the central-difference rules as given by Equations (8.5a) and (8.5b). The function is given by

```
void CCode8::Differentiation()
{
        int i,psi[6];
        double psv[6];
        CString str;
        fStatus=1;
        pt[0].x=atof(input[3].item);
        h=atof(input[2].item);
        m=(int)(atof(input[4].item)-atof(input[3].item))/h;
        m=((m<M)?m:M);
        pt[m].x=pt[0].x+(double)m*h;
        psi[1]=23; psv[1]=pt[0].x;
        pt[0].y=parse(input[1].item,1,psv,psi);
        for (i=0;i<=m;i++)
                if (i<m)
                {
                        pt[i+1].x=pt[i].x+h;
                        psv[1]=pt[i+1].x;
                        pt[i+1].y=parse(input[1].item,1,psv,psi);
                        if (i>0)
                        {
                                pt[i].yd1=(pt[i+1].y-pt[i-1].y)/(2*h);
                                pt[i].yd2=(pt[i+1].y-2*pt[i].y
                                        +pt[i-1].y)/(h*h);
                        }
                }
}
```

In `Differentiation()`, the input string is read as `input[1].item`. The first derivative is represented by `pt[i].yd1`, whereas the second is `pt[i].yd2`. With the central-difference rules, no derivatives can be evaluated at (x_0, y_0) and (x_m, y_m) as the solution requires one-point forward and one-point backward. `Differentiation()` computes the first and second derivatives of the input function $f(x)$ that is read as `input[1].item`. The derivatives are computed at each point (x_i, y_i), which is represented by `pt[i].x` and `pt[i].y` in the coding.

Integration

A function called `Integration()` handles the solution to the integration problem using the trapezium, Simpson, and Gaussian quadrature methods. The Simpson and Simpson's 3/8 methods produce the results correctly when the number of input subintervals m is a multiple of two and three, respectively. The function is given by

```
void CCode8::Integration()
{
        int i,psi[2];
        double psv[2];
        aTrap=0; aSimp=0; aSimp38=0; aGL=0;
        fStatus=1;
        pt[0].x=atof(input[3].item);
        m=atoi(input[2].item); m=((m<M)?m:M);
        h=(atof(input[4].item)-atof(input[3].item))/(double)m;
        pt[m].x=pt[0].x+(double)m*h;
        psi[1]=23;
        for (i=0;i<=m;i++)
        {
                psv[1]=pt[i].x;
                pt[i].y=parse(input[1].item,1,psv,psi);
                if (i>0 && i<m)
                {
                        aTrap += 2*pt[i].y;
                        if (m%2==0)
                        {
                                if (i%2==1)
                                aSimp += 4*pt[i].y;
                                if (i%2==0)
                                aSimp += 2*pt[i].y;
                        }
                        if (m%3==0)
                        {
                                if (i%3==0)
                                aSimp38 += 2*pt[i].y;
                                else
                                aSimp38 += 3*pt[i].y;
                        }
                }
                if (i<m)
                        pt[i+1].x=pt[i].x+h;
        }
        aTrap += pt[0].y+pt[m].y;
        aTrap *= h/2;
        if (m%2==0)
        {
                aSimp += pt[0].y+pt[m].y;
                aSimp *= h/3;
        }
        if (m%3==0)
```

```
        {
                aSimp38 += pt[0].y+pt[m].y;
                aSimp38 *= 3*h/8;
        }
        double g1,g2,g3,t1,t2,t3,a,b;
        a=pt[0].x; b=pt[m].x;
        t1=-0.774597; psv[1]=((b-a)*t1+(b+a))/2;
        g1=(b-a)/2*parse(input[1].item,1,psv,psi);
        t2=0; psv[1]=((b-a)*t2+(b+a))/2;
        g2=(b-a)/2*parse(input[1].item,1,psv,psi);
        t3=0.774597; psv[1]=((b-a)*t3+(b+a))/2;
        g3=(b-a)/2*parse(input[1].item,1,psv,psi);
        aGL=5*g1/9+8*g2/9+5*g3/9;
}
```

The number of subintervals m is denoted as m in the coding. The maximum number allowed is 200, which is stored as the macro M. A simple statement can be added to the code to control the user's input so that this maximum value is not breached, through

```
m=atoi(input[2].item); m=((m<M)?m:M);
```

The trapezium and Simpson methods are solved by performing iterations from $i = 0$ to $i = m$. Because of their constraints, the points in the Simpson and Simpson's 3/8 methods become relevant through the expressions m%2=0 and m%3=0, which indicate the multiplicity of two and three, respectively. The following code fragments show how the three methods are handled:

```
for (i=0;i<=m;i++)
{
        psv[1]=pt[i].x;
        pt[i].y=parse(input[1].item,1,psv,psi);
        if (i>0 && i<m)
        {
                aTrap += 2*pt[i].y;
                if (m%2==0)
                {
                        if (i%2==1)
                        aSimp += 4*pt[i].y;
                        if (i%2==0)
                        aSimp += 2*pt[i].y;
                }
                if (m%3==0)
                {
                        if (i%3==0)
                        aSimp38 += 2*pt[i].y;
                        else
```

```
                        aSimp38 += 3*pt[i].y;
                }
        }
        if (i<m)
                pt[i+1].x=pt[i].x+h;
}
aTrap += pt[0].y+pt[m].y;
aTrap *= h/2;
if (m%2==0)
{
        aSimp += pt[0].y+pt[m].y;
        aSimp *= h/3;
}
if (m%3==0)
{
        aSimp38 += pt[0].y+pt[m].y;
        aSimp38 *= 3*h/8;
}
```

The Gauss quadrature three-point method involves a transformation of coordinates from x to t through the relationship given by Equation (8.12). The method is handled through the following code fragments:

```
double g1,g2,g3,t1,t2,t3,a,b;
a=pt[0].x; b=pt[m].x;
t1=-0.774597; psv[1]=((b-a)*t1+(b+a))/2;
g1=(b-a)/2*parse(input[1].item,1,psv,psi);
t2=0; psv[1]=((b-a)*t2+(b+a))/2;
g2=(b-a)/2*parse(input[1].item,1,psv,psi);
t3=0.774597; psv[1]=((b-a)*t3+(b+a))/2;
g3=(b-a)/2*parse(input[1].item,1,psv,psi);
aGL=5*g1/9+8*g2/9+5*g3/9;
```

Table and Graph Output

Output in Code8 is expressed as the iterated values in a list view table and is plotted as graphs. The functions for these operations are ShowTable() and DrawCurve(), which are quite similar to the functions of the same names in the previous chapters. In DrawCurve(), two additional curves are plotted on top of the graph of $y = f(x)$. They are the graphs of the first-derivative $y' = f'(x)$ and the second-derivative $y'' = f''(x)$. The three graphs are plotted in the same window to illustrate their relationship at the points.

8.5 SUMMARY

Numerical differentiation and integration are two important problems in numerical analysis. The two problems contribute to providing the approximated solution, which is needed in several application problems. A typical requirement in differentiation is the numerical approximations to the first and second derivatives. Some fundamental methods for these approximations are provided using the forward-difference, backward-difference and central-difference, rules. The central-difference method has the advantage over the other two methods, as the method considers points from both left and right of the approximated point. Therefore, this method produces more accurate approximations to the problem. We illustrate this method in the visual interface.

The chapter also discusses numerical integration through the trapezium, Simpson, and Gauss–Legendre methods. The first two methods are based on the given points along the interval that are uniformly spaced. The last method does not require evenly distributed points. Instead, the method produces a linear transformation that preserves the area under the given curve.

NUMERICAL EXERCISES

1. Find the first and second derivatives at each point in the following table using the Newton forward-, backward-, and central-difference methods:

x	0	0.25	0.5	0.75	1.0	1.25	1.5
y	-1	1	2.5	4	2	1.5	1

2. Evaluate the following integrals on the given data using the trapezium, Simpson, and Simpson's 3/8 methods:

x	0	0.25	0.5	0.75	1.0	1.25	1.5
y	-1	1	2.5	4	2	1.5	1

3. Evaluate $\int_{-1}^{3} 3x^2 \cos 2x \, dx$ using

   a. The trapezium method with six subintervals.
   b. The Simpson method with six subintervals.
   c. The Simpson's 3/8 method with six subintervals.
   d. The Gaussian three-point quadrature method.
   e. The Gaussian four-point quadrature method.

4. Evaluate $\int_{-1}^{3} \frac{1-2x}{3x^2+1} \, dx$ using

   a. The trapezium method with six subintervals.
   b. The Simpson method with six subintervals.

c. The Simpson's 3/8 method with six subintervals.

d. The Gaussian three-point quadrature method.

e. The Gaussian four-point quadrature method.

PROGRAMMING CHALLENGES

1. Modify the graphs of the function and its derivatives in Code8 so that the vertical values follow the scale in $y = f(x)$.

2. The Newton–Raphson method in Chapter 6 requires the first derivative of the function $f(x)$ to be determined before iterations can be performed to find the root of the function. Modify the program in Code6 so that the first derivative of this function is approximated using the central-difference rule.

3. The expansion of a function $y = f(x)$ at $x = x_{i+1} = x_i + h$, where h is a small increment, using Taylor series of order 2 is given by

$$y_{i+1} \approx y_i + \frac{h}{1}y_i' + \frac{h^2}{2}y_i''.$$

Design a program to find the values of y_i for $i = 2, 3, \ldots, 50$, given $x_0 = 0$, $y_0 = 0$, $y_1 = 1$, and $h = \Delta x = 0.1$, using the Newton backward-difference rules.

CHAPTER 9

Eigenvalues and Eigenvectors

9.1 EIGENVALUES AND THEIR SIGNIFICANCE

The eigenvalues of a square matrix are important parameters in determining things like the stability of the structure where the matrix is based. A bridge depends on the strength of its beam across a given length whose stability is determined by its eigenvalues. In image processing, the eigenvalues of a matrix that represent an image hold the key to the quality of the image. With a proper technique, a blur image can be transformed into a crisp one once the eigenvalues of its corresponding matrix are known.

Therefore, finding the eigenvalues of a square matrix and their corresponding eigenvectors has become an important problem in many applications. The eigenvalue problem can be stated as follows:

> Given a matrix A, find the nonzero vectors v such that $Av = \lambda v$, where λ is the eigenvalue and v is its corresponding eigenvector. The real eigenvalues and its eigenvectors of the matrix will not exist if no value of v can satisfy $Av = \lambda v$.

In many applications, it may not be necessary to find all the eigenvalues of a given matrix. In this case, the eigenvalues whose values are extreme become the focus. An eigenvalue whose modulus is the largest is called the *most dominant eigenvalue*, whereas one that is smallest is called the *least dominant eigenvalue*. In most applications, it may not be necessary to find all the eigenvalues of a given matrix. Finding the most dominant eigenvalue or the least dominant eigenvalue may be sufficient in providing the solution to a given problem.

We discuss two iterative methods called the power method and the shifted power method, for finding the most dominant and least dominant eigenvalues of a matrix, respectively.

9.2 EXACT SOLUTION AND ITS EXISTENCE

In a transformation, an *eigenvector* is a nonzero vector v, which transforms the given quantity A into a new vector b that has the same direction as v. This transformation can be written as $T[v] = \lambda v = b$ or

$$Av = \lambda v = b. \tag{9.1}$$

In the above equation, λ is the scale factor for this transformation, and it is called the *eigenvalue* λ of the quantity A. In most cases, the quantity mentioned above can be expressed as a matrix . We denote (λ, v) as an *eigen-pair* of A for this transformation.

In a system of linear equations, A represents a square matrix in $Av = b$. An eigenpair (λ, v) exists, where $Av = \lambda v = b$. In general, a given square matrix A of size $n \times n$ has n eigenpairs given as (λ_i, v_i), for $i = 1, 2, \ldots, n$, where each eigenvalue can be a unique or repetitious real or complex number.

Since λ is a constant in Equation (9.1), an identity matrix I can be inserted into the equation so that $A = \lambda I$. The term $P_n(\lambda) = A - \lambda I$ is then a function called the *characteristic polynomial* of degree n. The exact method for finding the eigenvalues of a matrix then becomes the problem of solving the following equation:

$$|A - \lambda I| = 0, \tag{9.2}$$

as $(A - \lambda I)v = 0$ has unique solutions if $|A - \lambda I| = 0$. Equation (9.2) suggests an exact method for finding the eigenvalues of a given matrix A. The method is illustrated through Example 9.1.

Example 9.1. Find the eigenvalues λ and its corresponding eigenvector v of a matrix A, given as

$$A = \begin{bmatrix} 3 & -1 & 0 \\ -1 & 2 & -1 \\ 0 & -1 & 3 \end{bmatrix}.$$

Solution.

$$|A - \lambda I| = \begin{vmatrix} \begin{pmatrix} 3 & -1 & 0 \\ 0 & 2 & -1 \\ 0 & -1 & 3 \end{pmatrix} - \lambda \begin{pmatrix} 1 & 0 & 0 \\ 0 & 1 & 0 \\ 0 & 0 & 1 \end{pmatrix} \end{vmatrix}$$

$$= \begin{vmatrix} 3 - \lambda & -1 & 0 \\ -1 & 2 - \lambda & -1 \\ 0 & -1 & 3 - \lambda \end{vmatrix}.$$

Setting $(A - \lambda I)v = 0$, we have $-\lambda^3 + 8\lambda^2 - 19\lambda + 12 = -(\lambda - 1)(\lambda - 3)(\lambda - 4) = 0$. This produces $\lambda_1 = 1$, $\lambda_2 = 3$ and $\lambda_3 = 4$. From Equation (9.2), the eigenvector v_1 in the eigenpair (λ_1, v_1) is found by setting $\lambda_1 = 1$, as follows:

$$\begin{pmatrix} 3 - \lambda_1 & -1 & 0 \\ -1 & 2 - \lambda_1 & -1 \\ 0 & -1 & 3 - \lambda_1 \end{pmatrix} \begin{bmatrix} v_{11} \\ v_{12} \\ v_{13} \end{bmatrix} = \begin{pmatrix} 2 & -1 & 0 \\ -1 & 1 & -1 \\ 0 & -1 & 2 \end{pmatrix} \begin{bmatrix} v_{11} \\ v_{12} \\ v_{13} \end{bmatrix} = \begin{bmatrix} 0 \\ 0 \\ 0 \end{bmatrix}.$$

This produces

$$2v_{11} - v_{12} = 0,$$
$$-v_{11} + v_{12} - v_{13} = 0, .$$
$$-v_{12} + 2v_{13} = 0.$$

Solving the above system of linear equations by setting $v_{11} = k_1$, we get

$$\begin{bmatrix} v_{11} \\ v_{12} \\ v_{13} \end{bmatrix} = k_1 \begin{bmatrix} 1 \\ 2 \\ 1 \end{bmatrix}.$$

This gives the eigenvector for $\lambda_1 = 1$ as $v_1 = [\begin{smallmatrix} 1 \\ 2 \\ 1 \end{smallmatrix}]$. For $\lambda_2 = 3$:

$$\begin{pmatrix} 3 - \lambda & -1 & 0 \\ -1 & 2 - \lambda & -1 \\ 0 & -1 & 3 - \lambda \end{pmatrix} \begin{bmatrix} v_{21} \\ v_{22} \\ v_{23} \end{bmatrix} = \begin{pmatrix} 0 & -1 & 0 \\ -1 & -1 & -1 \\ 0 & -1 & 0 \end{pmatrix} \begin{bmatrix} v_{21} \\ v_{22} \\ v_{23} \end{bmatrix} = \begin{bmatrix} 0 \\ 0 \\ 0 \end{bmatrix}.$$

This produces

$$-v_{22} = 0,$$
$$-v_{21} - v_{22} - v_{23} = 0, .$$
$$-v_{22} = 0.$$

Setting $v_{21} = k_2$, we get $[v_{21}\ v_{22}\ v_{23}]^T = k_2[1\ 0\ -1]^T$. This gives the eigenvector for $\lambda_2 = 3$ as $v_2 = [1\ 0\ -1]^T$. Similarly, for $\lambda_3 = 4$,

$$\begin{pmatrix} 3 - \lambda & -1 & 0 \\ -1 & 2 - \lambda & -1 \\ 0 & -1 & 3 - \lambda \end{pmatrix} \begin{bmatrix} v_{31} \\ v_{32} \\ v_{33} \end{bmatrix} = \begin{pmatrix} -1 & -1 & 0 \\ -1 & -2 & -1 \\ 0 & -1 & -1 \end{pmatrix} \begin{bmatrix} v_{31} \\ v_{32} \\ v_{33} \end{bmatrix} = \begin{bmatrix} 0 \\ 0 \\ 0 \end{bmatrix}.$$

This produces the following system of linear equations:

$$-v_{31} - v_{32} = 0,$$
$$-v_{31} - 2v_{32} - v_{33} = 0,$$
$$-v_{32} - v_{33} = 0.$$

Finally, we get $[v_{31}\ v_{32}\ v_{33}]^T = k_3[1\ -1\ -1]^T$ and the eigenvector for $\lambda_3 = 4$ is $v_3 = [1\ -1\ -1]^T$.

9.3 POWER METHOD

A given matrix A of size $n \times n$ has n eigenvalues, denoted as λ_i for $i = 1, 2, \ldots, n$, which may be real or complex, unique, or repeated. λ_A is said to be the *most dominant eigenvalue* of A if its modulus is the largest, $|\lambda_A| \geq |\lambda_i|$, provided it exists. At the same time, λ_a is the *least dominant eigenvalue* of the matrix if $|\lambda_a| \leq |\lambda_i|$, provided it exists.

The most dominant eigenvalue of a matrix and its corresponding eigenvector can be found using an iterative method called the power method. In this method, iterations are performed to update the vector according to

$$v^{(k+1)} = \frac{1}{\lambda_{k+1}} A v^{(k)}. \tag{9.3}$$

For consistency, we denote the most dominant eigenvalue of A as λ_A and its corresponding eigenvector as v_A. The iterated values of these two variables at iteration k are λ_k and v_k. The given values are A, and the error tolerance is a small number close to 0, denoted as ε. The error is defined as $|\lambda_{k+1} - \lambda_k|$, and the iterations will only stop when $|\lambda_{k+1} - \lambda_k| < \varepsilon$.

A search for λ_A and v_A begins by defining the initial values of v. A suitable value is the unit vector, with one element having the value of 1 and the rest with 0. For example, $v_0 = [0\ \ 1\ \ 0]^T$ or $v_0 = [1\ \ 0\ \ 0]^T$ are some of the possible initial values for a 3×3 matrix.

The iterations start at $k = 0$ by evaluating Av_0. The elements in this vector are compared with one whose absolute value is the largest assigned to λ_1. It follows from Equation (9.3) that $v^{(1)} = \frac{1}{\lambda_1} Av^{(0)}$. A test on the error with $|\lambda_1 - \lambda_0| < \varepsilon$ is performed to determine whether convergence has been achieved. The iterations continue with $k = 1$ by repeating the same process if the test is not complied.

The iterations will only stop if the criteria of $|\lambda_{k+1} - \lambda_k| < \varepsilon$ are achieved. The number of iterations depends much on the given value of ε. A given value such as 0.00005 will require a lot more iterations than a bigger number such as 0.05. Convergence is said to have been achieved at iteration k when $|\lambda_{k+1} - \lambda_k| < \varepsilon$. The solutions are obtained as $\lambda_A = \lambda_{k+1}$ and $v_A = v^{(k+1)}$, which are the last values in the iterations.

Algorithm 9.1 outlines the summary of the steps for finding λ_A and v_A. The maximum number of iterations is indicated by K. This algorithm is illustrated in Example 9.2 using a 3×3 matrix.

Algorithm 9.1. Power Method.
Given a square matrix A, an initial vector v_0, and error tolerance
value of ε;
For $k = 0$ to K
  Find $\lambda_{k+1} = Av_r^k$ where $|Av_r^k| \geq |Av_i^k|$ for $i = 1, 2, \ldots, n$;
  Compute $v^{(k+1)} = \frac{1}{\lambda_{k+1}} Av^{(k)}$;
  If $|\lambda_{k+1} - \lambda_k| < \varepsilon$
    $\lambda_A = \lambda_{k+1}$ and $v_A = v^{(k+1)}$;
    Stop the iterations;
  Endif
Endfor

Example 9.2. Given $A = \begin{bmatrix} 2 & 2 & -1 \\ -1 & 3 & -1 \\ 0 & -1 & -2 \end{bmatrix}$, find the most dominant eigenvalue
and its corresponding eigenvector using the power method whose error tolerance is
$|\lambda_{k+1} - \lambda_k| < \varepsilon$, where $\varepsilon = 0.005$ and $v^{(0)} = (0, 1, 0)$.

Solution. Start the iterations at $k = 0$ with $v^{(0)} = (0, 1, 0)$. We get

$$Av^{(0)} = \begin{bmatrix} 2 & -2 & 1 \\ -1 & 3 & -1 \\ 0 & -1 & 2 \end{bmatrix} \begin{bmatrix} 0 \\ 1 \\ 0 \end{bmatrix} = \begin{bmatrix} -2 \\ 3 \\ -1 \end{bmatrix}.$$

Therefore, $\lambda_1 = 2.000$ and $v^{(1)} = \frac{1}{\lambda_1} Av^{(0)} = (-0.666667, 1, -0.333333)$. Repeating
the same steps with $k = 1$, and we get

$$Av^{(1)} = \begin{bmatrix} 2 & -2 & 1 \\ -1 & 3 & -1 \\ 0 & -1 & 2 \end{bmatrix} \begin{bmatrix} .666667 \\ 1 \\ -.333333 \end{bmatrix} = \begin{bmatrix} -3.666667 \\ 4 \\ -1.666667 \end{bmatrix}.$$

This gives $\lambda_2 = 4.000000$ and $v^{(2)} = \frac{1}{\lambda_2} Av^{(1)} = (-0.916667, 1, -0.416667)$. The
error is $|\lambda_2 - \lambda_1| = 1 > \varepsilon$, which indicates the iterations will continue. Table 9.1
summarizes the results obtained until convergence at $k = 5$, where $|\lambda_6 - \lambda_5| = 0.000583 < \varepsilon$. The solutions are $\lambda_A \approx \lambda_6 = 4.415430$ and $v_A = v_6 = (-0.999776, 1, -0.415099)$.

9.4 SHIFTED POWER METHOD

The shifted power method is a slight extension of the power method for finding
the least dominant eigenvalue of a given matrix and its corresponding eigenvector.
In finding the least dominant eigenvalue of the matrix A, or λ_a, two separate sets of

TABLE 9.1. Numerical results from Example 9.2

k	$v^{(k)}$			λ_{k+1}	$|\lambda_{k+1} - \lambda_k|$
0	0	1	0	3.000000	
1	−0.666667	1.000000	−0.333333	4.000000	1.000000
2	−0.916667	1.000000	−0.416667	4.333333	0.333333
3	−0.980769	1.000000	−0.423077	4.403846	0.070513
4	−0.995633	1.000000	−0.419214	4.414847	0.011001
5	−0.999011	1.000000	−0.416419	4.415430	0.000583
6	−0.999776	1.000000	−0.415099		

iterations are performed. First, iterations are performed using the same power method discussed in the last section to find the most dominant eigenvalue of A or λ_A. From this finding, a new matrix called B is created from the relationship given by

$$B = A - \lambda_A I, \tag{9.4}$$

where I is the identity matrix. The second set of iterations using the power method is then applied to find the most dominant eigenvalue of B, or λ_B, and its corresponding eigenvector, v_B. The power method formula in this case is

$$v^{(k+1)} = \frac{1}{\lambda_{k+1}} B v^{(k)}. \tag{9.5}$$

The least dominant eigenvalue of A is obtained by shifting the result as

$$\lambda_a = \lambda_A + \lambda_B. \tag{9.6}$$

Algorithm 9.2 outlines the steps in the shifted power method. The algorithm is further illustrated in Example 9.3.

Algorithm 9.2. Shifted Power Method.
Given a square matrix A, v_0 and ε;
Find λ_A and v_A from Algorithm 9.1;
Let $B = A - \lambda_A I$;
For $k = 0$ to K
    Find $\lambda_{k+1} = B v_r^k$ where $|B v_r^k| \geq |B v_i^k|$ for $i = 1, 2, \ldots, n$;
    Compute $v^{(k+1)} = \frac{1}{\lambda_{k+1}} B v^{(k)}$;
    If $|\lambda_{k+1} - \lambda_k| \leq \varepsilon$
        $\lambda_B = \lambda_{k+1}$ and $v_B = v^{(k+1)}$;
        Stop the iterations;
    Endif
    Get $\lambda_a = \lambda_A + \lambda_B$ and $v_a = v_B$;
Endfor

TABLE 9.2. Numerical results from Example 9.3

k	$v^{(k)}$			λ_{k+1}	$|\lambda_{k+1} - \lambda_k|$
0	0.000000	1.000000	0.000000	−2.000000	
1	1.000000	0.707437	0.500000	−3.329749	1.329749
2	1.000000	0.751088	0.575081	−3.341970	0.012221
3	1.000000	0.789288	0.640292	−3.353158	0.011188
4	1.000000	0.822220	0.696511	−3.362804	0.009646
5	1.000000	0.850436	0.744678	−3.371068	0.008264
6	1.000000	0.874482	0.785727	−3.378111	0.007043
7	1.000000	0.894882	0.820552	−3.384086	0.005975
8	1.000000	0.912121	0.849982	−3.389135	0.005049
9	1.000000	0.926643	0.874772	−3.393389	0.004253
10	1.000000	0.938842	0.895597		

Example 9.3. Given $A = \begin{bmatrix} 2 & 2 & -1 \\ -1 & 3 & -1 \\ 0 & -1 & -2 \end{bmatrix}$, find the least dominant eigenvalue and its corresponding eigenvector using the power method. The error tolerance is $|\lambda_{k+1} - \lambda_k| < \varepsilon$, where $\varepsilon = 0.005$ and $v^{(0)} = (0, 1, 0)$.

Solution. From Example 9.2, we have $\lambda_A \approx \lambda_6 = 4.415430$ and $v_A = v_6 = (-0.999776, 1, -0.415099)$. We obtain

$$B = A - \lambda_A I = \begin{bmatrix} -2.414875 & -2 & 1 \\ -1 & -1.414875 & -1 \\ 0 & -1 & -2.414875 \end{bmatrix}.$$

Iterations are performed to find λ_B and v_B. Table 9.2 shows the results from the iterations.

We obtain $\lambda_B \approx \lambda_{10} = -3.393389$ and $v_B \approx v_{10} = (1, 0.938842, 0.895597)$. Finally, we get the solutions:

$$\lambda_a = \lambda_A + \lambda_B = 4.415430 - 3.393389 = 1.021486,$$

$$v_a = v_B = (1, 0.938842, 0.895597).$$

9.5 QR METHOD

A *symmetric matrix* is a square matrix that has the fundamental property of the entries in the ith column equal those in the ith row, or

$$A = A^T, \text{ or } a_{ij} = a_{ij}. \tag{9.7}$$

A symmetric matrix appears in many problems in science and engineering. Its importance is realized in problems where matrices play an important role. For example, symmetric matrices are commonly encountered in computer graphics where they are used in many algebraic operations.

Another important type of matrix is the *orthogonal matrix*. A square matrix A is said to be orthogonal if

$$A^T = A^{-1}, \text{ or } AA^T = I. \tag{9.8}$$

A symmetric matrix A whose contents are real numbers possesses another important property in that it can be diagonalized by an orthogonal matrix or

$$D = Q^T AQ = Q^{-1}AQ, \tag{9.9}$$

where D is its diagonal form and Q is an orthogonal matrix. Two matrices A and B are said to be *similar* if $A = C^{-1}BC$, where C is any nonsingular matrix.

A real Hermitian matrix consists of a symmetric matrix or $A = A^T$. This type of matrix has the property where all its eigenvalues are real. In addition, the eigenvectors of a symmetric matrix are orthogonal, and the matrix consists of an orthonormal basis of eigenvectors.

A symmetric matrix has an advantage over a nonsymmetric matrix as all its eigenvalues can be determined through some algebraic steps. Two main steps are involved in finding these eigenvalues. First, the symmetric matrix is reduced into a tridiagonal matrix using a series of transformation called the Householder transformation. The second step is applying a technique called the QR method to the tridiagonal matrix to produce the eigenvalues.

Householder Transformation

The Householder transformation converts a symmetric matrix into a similar symmetric tridiagonal matrix. Let $w \in \mathbb{R}^n$ be a vector and $w^T w = 1$; then

$$P = I - 2ww^T. \tag{9.10}$$

The $n \times n$ matrix P produced in the transformation from the above transformation is symmetric and orthogonal, or $P^{-1} = P^T = P$.

The Householder transformation involves several iterative steps In producing a symmetric tridiagonal matrix from a similar symmetric matrix. The steps are outlined in Algorithm 9.3. This algorithm is illustrated through Example 9.4.

Algorithm 9.3. Householder transformation.

Given a square symmetric matrix $A = [a_{ij}]$ for $i, j = 1, 2, \ldots, n$;

Compute $\alpha = -\text{sgn}(a_{k+1,k})(\sum_{j=k+1}^{n} (a_{jk})^2)^{1/2}$;

Compute $r = (\frac{1}{2}\alpha^2 - \frac{1}{2}\alpha a_{k+1,k})^{1/2}$;

Let $w_1 = w_2 = \cdots = w_k = 0$;

For $k = 1$ to $n - 2$

  Compute $w_{k+1} = \frac{a_{k+1,k}-\alpha}{2r}$ and $w_j = \frac{a_{jk}}{2r}$; for $j = k+2, k+3, \ldots, n$;

  Compute $P^{(k)} = I - 2ww^T$;

  Update $A^{(k+1)} = P^{(k)} A^{(k)} P^{(k)}$;

Endfor

Example 9.4. Use the Householder's method to transform the following symmetric matrix into a similar symmetric tridiagonal matrix:

$$A = \begin{bmatrix} 2 & -1 & 1 & 4 \\ -1 & 2 & -1 & 1 \\ 1 & -1 & 2 & -2 \\ 4 & 1 & -2 & 3 \end{bmatrix}.$$

Solution. Let $A^{(1)} = A$. At iteration $k = 1$,

$$\alpha = 4.2426,$$

$$r = 3.3349,$$

$$w = [0 \quad -0.786 \quad 0.1499 \quad 0.5997].$$

$$P^{(1)} = I - 2ww^T$$

$$= \begin{bmatrix} 1 & 0 & 0 & 0 \\ 0 & -0.2357 & 0.2357 & 0.9428 \\ 0 & 0.2357 & 0.9550 & -0.1798 \\ 0 & 0.9428 & -0.1798 & 0.2807 \end{bmatrix}.$$

We obtain

$$A^{(2)} = P^{(1)} A^{(1)} P^{(1)} = \begin{bmatrix} 2 & 4.2426 & 0 & 0 \\ 4.2426 & 1.6667 & -1.4515 & 1.0295 \\ 0 & -1.4515 & 2.1844 & -1.6127 \\ 0 & 1.0295 & -1.6127 & 3.1490 \end{bmatrix}.$$

At iteration $k = 2$:

$$\alpha = 1.7795,$$

$$r = 1.6955,$$

$$w = \begin{bmatrix} 0 & 0 & -0.9528 & 0.3036 \end{bmatrix}.$$

$$P^{(2)} = I - 2ww^T$$

$$= \begin{bmatrix} 1 & 0 & 0 & 0 \\ 0 & 1 & 0 & 0 \\ 0 & 0 & -0.8156 & 0.5785 \\ 0 & 0 & 0.5785 & 0.8156 \end{bmatrix}.$$

$$A^{(3)} = P^{(2)} A^{(2)} P^{(2)} = \begin{bmatrix} 2 & 4.2426 & 0 & 0 \\ 4.2426 & 1.6667 & 1.7795 & 0 \\ 0 & 1.7795 & 4.0292 & 0.9883 \\ 0 & 0 & 0.9883 & 1.3041 \end{bmatrix}.$$

$A^{(3)}$ is a symmetric tridiagonal matrix transformation of A, where

$$A^{(3)} = P^{(2)} P^{(1)} A P^{(1)} P^{(2)} = P^{(2)T} P^{(1)T} A P^{(1)} P^{(2)} = \left(P^{(2)} P^{(1)} \right)^T A P^{(1)} P^{(2)}.$$

Hence, $A^{(3)}$ and A are similar and they share the same eigenvalues.

QR Factorization

QR factorization is a technique for factorization of a symmetric tridiagonal matrix A into the product of an orthogonal matrix Q and an upper triangular matrix R, or

$$A = QR$$

$$= \begin{bmatrix} u_1 & u_2 & \cdots & u_n \end{bmatrix} \begin{bmatrix} r_{11} & r_{12} & \cdots & r_{1n} \\ 0 & r_{22} & \cdots & r_{2n} \\ \vdots & \ddots & \ddots & \vdots \\ 0 & \cdots & 0 & r_{nn} \end{bmatrix}, \tag{9.11}$$

where $u_1, u_2, \ldots, u_n$ are the orthogonal basis vectors of Q.

The computational steps for QR factorization are shown in Algorithm 9.4. This algorithm is illustrated through Example 9.5.

Algorithm 9.4. QR Factorization.

Given a symmetric tridiagonal matrix $A = [a_{ij}]$ for $i, j = 1, 2, \ldots, n$;

Let $Q_1 = A = \begin{bmatrix} u_1 & u_2 & \ldots & u_n \end{bmatrix}$;

For $i = 1$ to n

Compute $r_{ii} = \|u_i\|$;

Update $u_i = u_i / r_{ii}$;

For $j = i + 1$ to n

Compute $r_{ij} = u_i . u_j$;

Update $u_j = u_j / r_{ij}$;

Endfor

Endfor

Example 9.5. Find the QR factorization of matrix A below:

$$A = \begin{bmatrix} 2 & 4.2426 & 0 & 0 \\ 4.2426 & 1.6667 & 1.7795 & 0 \\ 0 & 1.7795 & 4.0292 & 0.9883 \\ 0 & 0 & 0.9883 & 1.3041 \end{bmatrix}.$$

Solution. Let $Q = A = \begin{bmatrix} u_1 & u_2 & u_3 & u_4 \end{bmatrix}$.

For $i = 1$,

$$r_{11} = 4.6904, \text{ update } u_1 = \frac{1}{r_{11}} u_1 = [0.4264 \quad 0.9045 \quad 0 \quad 0],$$

$$r_{12} = 3.3166, \text{ update } u_2 = [2.8284 \quad -1.3333 \quad 1.7795 \quad 0],$$

$$r_{13} = 1.6096, \text{ update } u_3 = [-0.6863 \quad 0.3235 \quad 4.0292 \quad 0.9883],$$

$$r_{14} = 0, \text{ update } u_4 = [0 \quad 0 \quad 0.9883 \quad 1.3041].$$

For $i = 2$,

$$r_{22} = 3.5978, \text{ update } u_2 = [0.7861 \quad -0.3706 \quad 0.4946 \quad 0],$$

$$r_{23} = 1.3334, \text{ update } u_3 = [-1.7346 \quad 0.8177 \quad 3.3697 \quad 0.9883],$$

$$r_{24} = 0.4888, \text{ update } u_4 = [-0.3843 \quad 0.1812 \quad 0.7465 \quad 1.3041].$$

For $i = 3$,

$$r_{33} = 4.0012, \text{ update } u_3 = [-0.4335 \quad 0.2044 \quad 0.8422 \quad 0.2470],$$

$$r_{34} = 1.1544, \text{ update } u_4 = [0.1162 \quad -0.0548 \quad -0.2257 \quad 1.0189].$$

For $i = 4$,

$$r_{44} = 1.0515, \text{ update } u_4 = [0.1105 \quad -0.0521 \quad -0.2147 \quad 0.9690].$$

Therefore,

$$
Q = \begin{bmatrix} u_1 & u_2 & u_3 & u_4 \end{bmatrix} = \begin{bmatrix} 0.4264 & 0.7861 & -0.4335 & 0.1105 \\ 0.9045 & -0.3706 & 0.2044 & -0.0521 \\ 0 & 0.4946 & 0.8422 & -0.2147 \\ 0 & 0 & 0.2470 & 0.9690 \end{bmatrix},
$$

$$
R = \begin{bmatrix} r_{11} & r_{12} & r_{13} & r_{14} \\ 0 & r_{22} & r_{23} & r_{24} \\ 0 & 0 & r_{33} & r_{34} \\ 0 & 0 & 0 & r_{44} \end{bmatrix} = \begin{bmatrix} 4.6904 & 3.3166 & 1.6096 & 0 \\ 0 & 3.5978 & 1.3334 & 0.4888 \\ 0 & 0 & 4.002 & 1.1544 \\ 0 & 0 & 0 & 1.0515 \end{bmatrix}.
$$

We obtain the factorization,

$$
A = \begin{bmatrix} 2 & 4.2426 & 0 & 0 \\ 4.2426 & 1.6667 & 1.7795 & 0 \\ 0 & 1.7795 & 4.0292 & 0.9883 \\ 0 & 0 & 0.9883 & 1.3041 \end{bmatrix}
$$

$$
= \begin{bmatrix} 0.4264 & 0.7861 & -0.4335 & 0.1105 \\ 0.9045 & -0.3706 & 0.2044 & -0.0521 \\ 0 & 0.4946 & 0.8422 & -0.2147 \\ 0 & 0 & 0.2470 & 0.9690 \end{bmatrix} \begin{bmatrix} 4.6904 & 3.3166 & 1.6096 & 0 \\ 0 & 3.5978 & 1.3334 & 0.4888 \\ 0 & 0 & 4.002 & 1.1544 \\ 0 & 0 & 0 & 1.0515 \end{bmatrix}.
$$

QR factorization is applied to find all the eigenvalues of a symmetrix matrix and their corresponding eigenvectors by factorizing the matrix as $A = QR$ and by reversing the resulting factorization through a series of iterations until all the errors from the eigenvalues are smaller than a tolerated value ε. The error is computed as $|\lambda_{k+1} - \lambda_k|$, where k is the iteration number. The iterations are shown as

$$
A = Q_0 R_0,
$$

$$
A_{k+1} = R_k Q_k = Q_{k+1} R_{k+1},
$$

$$
S_0 = Q_0 \quad S_k = S_{k-1} Q_k, \text{ for } k = 0, 1, 2, \ldots.
$$

Convergence to the solution is said to be achieved once all the errors are less than the tolerated value ε. The eigenvalues of A appear along the diagonal of final value of A_{k+1}, whereas the columns of S_k are their corresponding orthonormal eigenvectors basis, listed in the same order as eigenvalues.

Algorithm 9.5 summarizes the QR algorithm for finding the eigenvalues of a symmetric matrix A. The algorithm is further illustrated in Example 9.6.

Algorithm 9.5. QR Algorithm.

Given a symmetric tridiagonal matrix $A = [a_{ij}]$ for $i, j = 1, 2, \ldots, n$;

Given the error tolerance ε;

For $k = 1$ to max

Let $A_{k+1} = Q_k R_k$, and find Q_k and R_k using Algorithm 9.4;

Let $S_k = Q_k$;

Reverse multiply and compute $A_{k+1} = R_k Q_k$;

Compute the error of the diagonal elements in A_{k+1};

If error $< \varepsilon$

Determine the eigenvalues from the diagonal elements of A_{k+1};

Determine the eigenvectors from the columns of S_k;

Endif

Endfor

Example 9.6. Find all the eigenvalues and their corresponding eigenvectors of a symmetric matrix given by

$$
A = \begin{bmatrix} 2 & 4.2426 & 0 & 0 \\ 4.2426 & 1.6667 & 1.7795 & 0 \\ 0 & 1.7795 & 4.0292 & 0.9883 \\ 0 & 0 & 0.9883 & 1.3041 \end{bmatrix}.
$$

Solution. Iteration 0 :

Factorizing matrix $A = Q_0 R_0$, we have (from Example 9.5)

$$
S_0 = Q_0 = \begin{bmatrix} 0.4264 & 0.7861 & -0.4335 & 0.1105 \\ 0.9045 & -0.3706 & 0.2044 & -0.0521 \\ 0 & 0.4946 & 0.8422 & -0.2147 \\ 0 & 0 & 0.2470 & 0.9690 \end{bmatrix},
$$

$$
R_0 = \begin{bmatrix} 4.6904 & 3.3166 & 1.6096 & 0 \\ 0 & 3.5978 & 1.3334 & 0.4888 \\ 0 & 0 & 4.002 & 1.1544 \\ 0 & 0 & 0 & 1.0515 \end{bmatrix}.
$$

Reverse multiply these matrices to produce

$$A_1 = R_0 Q_0 = \begin{bmatrix} 5 & 3.2544 & 0 & 0 \\ 3.2544 & -0.6738 & 1.9790 & 0 \\ 0 & 1.9790 & 3.6549 & 0.2597 \\ 0 & 0 & 0.2597 & 1.0189 \end{bmatrix}.$$

Iteration 1:

Factorize $A_1 = Q_1 R_1$ to produce

$$Q_1 = \begin{bmatrix} 0.8381 & 0.4165 & -0.3515 & 0.0236 \\ 0.5455 & -0.6399 & 0.54 & -0.036 \\ 0 & 0.6458 & 0.7618 & -0.0512 \\ 0 & 0 & 0.0671 & 0.9977 \end{bmatrix},$$

$$R_1 = \begin{bmatrix} 5.9658 & 2.3599 & 1.0795 & 0 \\ 0 & 3.0646 & 1.0937 & 0.1677 \\ 0 & 0 & 3.8705 & 0.2662 \\ 0 & 0 & 0 & 1.0033 \end{bmatrix}.$$

Reverse multiply Q_1 and R_1 to produce A_2,

$$A_2 = R_1 Q_1 = \begin{bmatrix} 6.2874 & 1.6718 & 0 & 0 \\ 1.6718 & -1.2549 & 2.4993 & 0 \\ 0 & 2.4993 & 2.9665 & 0.0673 \\ 0 & 0 & 0.0673 & 1.0011 \end{bmatrix},$$

$$S_1 = S_0 Q_0 = \begin{bmatrix} 0.7862 & -0.6054 & -0.0482 & 0.1140 \\ 0.5559 & 0.7459 & -0.3658 & -0.0276 \\ 0.2698 & 0.2273 & 0.8943 & -0.2753 \\ 0 & 0.1595 & 0.2532 & 0.9542 \end{bmatrix}.$$

Continue the steps until iteration 13, where the error is less than ε, to produce

$$Q_{13} = \begin{bmatrix} 1 & -0.0003 & 0 & 0 \\ 0.0003 & 0.9986 & 0.0526 & 0 \\ 0 & 0.0526 & -0.9986 & 0 \\ 0 & 0 & 0 & 1 \end{bmatrix},$$

$$R_{13} = \begin{bmatrix} 6.7293 & 0.0033 & 0.0001 & 0 \\ 0 & 3.9441 & 0.0668 & 0 \\ 0 & 0 & 2.6751 & 0 \\ 0 & 0 & 0 & 1 \end{bmatrix},$$

$$S_{13} = \begin{bmatrix} 0.5892 & -0.4657 & 0.6479 & 0.1270 \\ 0.6566 & -0.1915 & -0.7289 & -0.0299 \\ 0.4633 & 0.8083 & 0.2169 & -0.2916 \\ 0.0843 & 0.3051 & -0.0431 & 0.9476 \end{bmatrix},$$

$$A_{14} = \begin{bmatrix} 6.7293 & 0.0012 & 0 & 0 \\ 0.0012 & 3.9421 & 0.1407 & 0 \\ 0 & 0.1407 & -2.6714 & 0 \\ 0 & 0 & 0 & 1 \end{bmatrix}.$$

We obtain all the eigenvalues and eigenvectors of matrix A from A_{14} and S_{13}, respectively, as follows:

$$\lambda_1 = 6.7283 \text{ and } v_1 = [0.5892 \quad 0.6566 \quad 0.4633 \quad 0.0843]^T,$$

$$\lambda_2 = 3.9421 \text{ and } v_2 = [-0.4657 \quad -0.1915 \quad 0.8083 \quad 0.3051]^T,$$

$$\lambda_3 = -2.6714 \text{ and } v_3 = [0.6479 \quad -0.7289 \quad 0.2169 \quad -0.0431]^T,$$

$$\lambda_4 = 1.0000 \text{ and } v_4 = [0.1270 \quad -0.0299 \quad -0.2916 \quad 0.9476]^T.$$

9.6 VISUAL SOLUTION: CODE9

Code9. User Manual.

1. Enter values for the matrix starting from the top left-hand corner.
2. Left-click the push button to see the results using the power and shifted-power methods.

Development files: Code9.cpp and Code9.h.

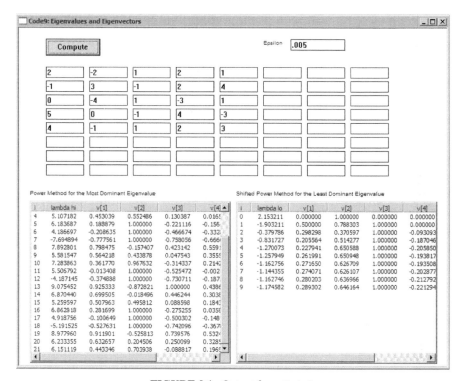

FIGURE 9.1. Output from Code9.

The eigenvalue problem is illustrated in a project called Code9. The project displays the power and the shifted power methods for finding the most dominant and least dominant eigenvalues of a given nonsymmetric matrix, respectively. For the case of a symmetric matrix, the program applies the Householder transformation and QR method automatically to compute all the eigenvalues.

Figure 9.1 shows an output from Code9. The figure shows the complete results for finding the most dominant and least dominant eigenvalues and their corresponding eigenvectors of the following matrix:

$$A = \begin{bmatrix} 2 & -2 & 1 & 2 & 1 \\ -1 & 3 & -1 & 2 & 4 \\ 0 & -4 & 1 & -3 & 1 \\ 5 & 0 & -1 & 4 & -3 \\ 4 & -1 & 1 & 2 & 3 \end{bmatrix}.$$

Input for the matrix is provided in the form of edit boxes. The results from the power series methods are displayed in two list view tables, the most dominant in the left and the least dominant in the right. The two tables display the m_k and v_k values for $k = 0, 1, \ldots, Stop$, where $Stop$ is the stopping number of the iterations that complies with the stopping criteria, given as $|m_{k+1} - m_k| < \varepsilon$. For the case of a symmetric matrix, the QR method displays the eigenvalues and their corresponding eigenvectors separately in two list view tables.

The maximum size of the matrix allowed in this interface is 8×8. The user has the option of selecting the matrix size by filling in the values starting from the top left-hand corner of the edit boxes. The actual size of the matrix is determined from the diagonal elements. An empty entry for the diagonal element $a_{i,i}$ indicates the size of matrix is $(i - 1) \times (i - 1)$. Therefore, the program is flexible in the sense that it allows the user to determine the size of the matrix freely by entering values in the edit boxes.

Figure 9.2 shows a schematic drawing of the computational steps in Code9. The progress in the runtime is monitored through fStatus whose initial value is 0. This value changes to 1 when either the power method or the QR method has been successfully applied to solve the given problem. Edit boxes for the input are created in

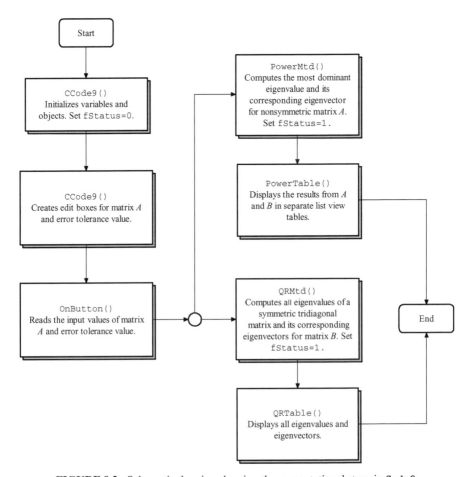

FIGURE 9.2. Schematic drawing showing the computational steps in Code9.

the constructor function, whereas their values are read in the OnButton() as soon as the *Compute* button is left-clicked.

The OnButton() performs a simple test to determine whether the input matrix is symmetric. If the matrix is symmetric, QRMtd() is called to execute the QR method. Otherwise, the processing is passed to PowerMtd(), which refers to the power method. A success in either method changes fStatus value to 1. The full eigenvalues and eigenvectors are then displayed in the list view tables through QRTable() for the QR method and PowerTable() for the power method.

Code9 consists of two files, Code9.h and Code9.cpp. A single class called CCode9 is used, and this class is derived from CFrameWnd. The main window from this class consists of two list view tables, 64 edit boxes for representing the 8×8 matrix, another edit box for the error stopping value ε, and a push button called *Compute*.

The elements in the iterations in the power and QR methods are represented by a structure called ITER, which are given by

```
typedef struct
{
      double lambda,v[M+1],error;
      double lambdaB,vB[M+1],errorB;
      double lambdaQR[M+1],vQR[M+1][M+1],errorQR[M+1];
} ITER;
ITER eigen[maxIter+1];
```

A maximum number of iterations allowed is a macro called maxIter. In the most dominant eigenvalue problem, an array called eigen stores the values of m_k and v_k for matrix A, for $k = 0, 1, \ldots, Stop$. Here, m_k is represented by lambda and v_k by the array v. In the least dominant eigenvalue problem, m_k and v_k of the matrix B are represented by lambdaB and vB, respectively. The errors in the iterations are error for the A matrix and errorB for the B matrix. For the QR method, lambdaQR and vQR are the eigenvalues and their corresponding eigenvectors, respectively. The error is errorQR.

The elements for the input boxes are represented by a structure called INPUT. The boxes form matrix A whose elements consist of the array input. The elements in the array are the edit boxes (ed), their home coordinates (hm), and their contents (item).

```
typedef struct
{
      CPoint hm;
      CString item;
      CEdit ed;
} INPUT;
INPUT input[nInputItems+1][M+1];
```

TABLE 9.3. Member functions in CCode6

Function	Description
CCode9()	Constructor.
~CCode9	Destructor.
PowerMtd()	The power method for finding the most dominant eigenvalue and its corresponding eigenvector.
QRMtd()	Computes all the eigenvalues and eigenvectors of a symmetrix matrix using the Householder transformation and QR method.
PowerTable()	Creates a list view table to display the results.
QRTable()	Displays all the eigenvalues and their corresponding eigenvectors in two separate list view tables.
OnButton()	Responds to ON_BN_CLICKED, which reads the input matrix from the user and calls the respective method to produce its solution.
OnPaint()	Displays and updates the output in the main window.

Table 9.3 lists the functions in CCode9. The key function in this class is PowerMtd(), which computes the most dominant and least dominant eigenvalues of the input matrix A and its corresponding eigenvectors. PowerTable() displays the detail results from the iterations in two list view tables, one for the most dominant eigenvalue and another for the least dominant eigenvalue.

Table 9.4 lists the main variables and objects in CCode9. The progress in execution is marked using fStatus, where, fStatus=0 indicates the stage before the power method is applied, and fStatus=1 indicates the power method has been successfully applied. The maximum matrix size shown is M, whereas its actual size is m. The number of iterations performed before convergence in matrix A is indicated by Stop1, whereas that in matrix B is indicated by Stop2.

There are two events, an update in the main window and the push button click, and they are handled by OnPaint() and OnButton(), respectively.

```
BEGIN_MESSAGE_MAP(CCode9,CFrameWnd)
    ON_WM_PAINT()
    ON_BN_CLICKED(IDC_BUTTON,OnButton)
END_MESSAGE_MAP()
```

The OnButton() responds to the push button click. There are three local arrays in the function: A, B, and I. A is the original matrix whose most dominant eigenvalue is the first subject in the problem. B is $A - \lambda I$, and its most dominant eigenvalue contributes to finding the least dominant eigenvalue for A. I is the identity matrix, which is needed in computing B. The function is given, as follows:

```
void CCode9::OnButton()
{
    int i,j,k;
    double **A,**B,**I;
```

TABLE 9.4. Main variables and objects in Code9

Variable/object	Type	Description
fStatus	bool	A flag whose values are fStatus=0 at the data input level and fStatus=1 when the power method has been successfully applied, respectively.
fMenu	int	The method assigned, with fMenu=1 for the power method and fMenu=2 for the QR method.
m, M	int	Actual size and maximum size of A matrix, respectively.
	macro	Maximum size of the A matrix.
StopA, StopB	int	Number of iterations before convergence to the most dominant and least dominant eigenvalues, respectively.
epsilon	double	ε or the stopping criteria for the iterations in the power method.
nInputItems	int	Number of input items in the selected method.
maxIter	int	Maximum number of iterations allowed for the power method.
idc	int	Id for the edit boxes.
btn	CButton	Push button object called *Compute*.

```
A=new double *[M+1];
B=new double *[M+1];
I=new double *[M+1];
for (i=1;i<=M;i++)
{
        A[i]=new double [M+1];
        B[i]=new double [M+1];
        I[i]=new double [M+1];
}
for (i=1;i<=M;i++)
        for (j=1;j<=M;j++)
                input[i][j].ed.GetWindowText
                        (input[i][j].item);
input[nInputItems][1].ed.GetWindowText
        (input[nInputItems][1].item);

// determine the actual size of matrix
for (i=1;i<=M;i++)
        if (input[i][i].item=="")
        {
                m=i-1; break;
        }
```

```
// read the contents of A
for (i=1;i<=m;i++)
      for (j=1;j<=m;j++)
      {
            I[i][j]=0;
            if (i==j)
                  I[i][j]=1;
            A[i][j]=atof(input[i][j].item);
      }

// test for matrix symmetry
k=0;
for (i=1;i<=m;i++)
      for (j=1;j<=m;j++)
            if (A[i][j]==A[j][i])
                  k++;
fMenu=((k==m*m)?2:1);
if (fMenu==1)
{
      PowerMtd(1,A);
      PowerTable(1);
      if (StopA!=-1)
      {
            for (i=1;i<=m;i++)
                  for (j=1;j<=m;j++)
                        B[i][j]=A[i][j]-eigen[StopA]
                              .lambda*I[i][j];
            PowerMtd(2,B);
            PowerTable(2);
      }
}
if (fMenu==2)
{
      QRMtd(A);
            QRTable(1);
            QRTable(2);
}
}
```

The OnButton() first reads the input values from *A* beginning with the top row, from left to right. The size of *A* is determined in this function by detecting the first null entry in its diagonal element. A null entry at $a_{i,i}$ implies the size of the matrix is $(i - 1) \times (i - 1)$, and this assigns $m = i - 1$. The code fragments for this task are

```
for (i=1;i<=M;i++)
      for (j=1;j<=M;j++)
```

```
            input[i][j].ed.GetWindowText
                 (input[i][j].item);
input[nInputItems][1].ed.GetWindowText
    (input[nInputItems][1].item);

// determine the actual size of matrix
for (i=1;i<=M;i++)
      if (input[i][i].item=="")
      {
            m=i-1; break;
      }
```

Once the actual size of the matrix is known, the corresponding identity matrix can now be created. At the same time, the values for *A* are determined by converting their corresponding strings in the edit boxes, as follows:

```
for (i=1;i<=m;i++)
      for (j=1;j<=m;j++)
      {
            I[i][j]=0;
            if (i==j)
                  I[i][j]=1;
            A[i][j]=atof(input[i][j].item);
      }
```

The OnButton() calls PowerMtd() twice for computing the most dominant eigenvalues and their corresponding eigenvectors in *A* and *B*. The results from the iterations are displayed in the list view tables through PowerTable().

```
if (fMenu==1)
{
      PowerMtd(1,A);
      PowerTable(1);
      if (StopA!=-1)
      {
            for (i=1;i<=m;i++)
                  for (j=1;j<=m;j++)
                        B[i][j]=A[i][j]-eigen[StopA].
                              lambda*I[i][j];
            PowerMtd(2,B);
            PowerTable(2);
      }
}
```

In the case where matrix A is symmetric, the flag fMenu=2 becomes active to indicate the problem is passed to the QR method in its solution. The code fragments are as follows:

```
if (fMenu==2)
{
      QRMtd(A);
      QRTable(1);
      QRTable(2);
}
```

Power Method

PowerMtd() computes the most dominant eigenvalue λ_1 and its corresponding eigenvectors v_1. There are two arguments in PowerMtd(). The first argument is fPower, and its value identifies the matrix to compute: 1 for A and 2 for B. The second argument is matrix c, which obtains its values through data passing from either A or B. The full code listing for PowerMtd() is given below:

```
void CCode9::PowerMtd(int fPower,double **c)
{
      double u[M+1],w[M+1],lamb[maxIter+1],error[maxIter+1];
      int i,j,k;
      for (i=1;i<=m;i++)
      {
            w[i]=0;
            if (i==2)
                  w[i]=1;
            eigen[0].v[i]=w[i];
            eigen[0].vB[i]=w[i];
      }
      epsilon=atof(input[nInputItems][1].item);
      for (k=0;k<=maxIter;k++)
      {
            for (i=1;i<=m;i++)
            {
                  u[i]=0;
                  for (j=1;j<=m;j++)
                        u[i] += c[i][j]*w[j];
            }
            lamb[k]=u[1];
            for (i=1;i<=m;i++)
                  if (fabs(lamb[k])<=fabs(u[i]))
                  {
                        lamb[k]=u[i];
                        if (fPower==1)
                              eigen[k+1].lambda=lamb[k];
```

```
                    if (fPower==2)
                            eigen[k+1].lambdaB=lamb[k];
            }
        for (i=1;i<=m;i++)
        {
            w[i]=u[i]/lamb[k];
            if (fPower==1)
                    eigen[k+1].v[i]=w[i];
            if (fPower==2)
                    eigen[k+1].vB[i]=w[i];
        }
        if (k>0)
        {
            error[k]=lamb[k]-lamb[k-1];
            if (fPower==1)
                    eigen[k].error=error[k];
            if (fPower==2)
                    eigen[k].errorB=error[k];
            if (fabs(error[k])<epsilon)
            {
                    if (fPower==1)
                    {
                        StopA=k;
                        eigen[StopA].lambda=lamb[StopA];
                        fStatus=1;
                        Invalidate();
                        break;
                    }
                    if (fPower==2)
                    {
                        StopB=k;
                        eigen[StopB].lambdaB=lamb[StopB];
                        break;
                    }
            }
            else
            {
                    if (k==maxIter && fPower==1)
                            StopA=-1;
                    if (k==maxIter && fPower==2)
                            StopB=-1;
            }
        }
    }
}
}
```

In PowerMtd(), the initial values consist of $v_0 = [0 \quad 1 \quad 0]^T$ for both matrix A and B, and ε. The code is given by

```
for (i=1;i<=m;i++)
{
      w[i]=0;
      if (i==2)
            w[i]=1;
      eigen[0].v[i]=w[i];
      eigen[0].vB[i]=w[i];
}
epsilon=atof(input[nInputItems][1].item);
```

Convergence to the solution is obtained from the error tolerance given by $|\lambda_{k+1} - \lambda_k| < \varepsilon$, and this condition is applied through fabs(error[k])<epsilon. The conditional test is performed at every iteration to determine whether the error complies with the given tolerance. The last iteration numbers before convergence, StopA for matrix A and StopB for matrix B, are stored in order to display the eigenvalues and eigenvectors in list view tables later. The code for this task is given by

```
error[k]=lamb[k]-lamb[k-1];
if (fPower==1)
      eigen[k].error=error[k];
if (fPower==2)
      eigen[k].errorB=error[k];
if (fabs(error[k])<epsilon)
{
      if (fPower==1)
      {
            StopA=k;
            eigen[StopA].lambda=lamb[StopA];
            fStatus=1;
            Invalidate();
            break;
      }
      if (fPower==2)
      {
            StopB=k;
            eigen[StopB].lambdaB=lamb[StopB];
            break;
      }
}
else
```

```
{
    if (k==maxIter && fPower==1)
        StopA=-1;
    if (k==maxIter && fPower==2)
        StopB=-1;
}
```

QR Method

The QR method is handled by QRMtd(). The function computes all the eigenvalues belonging to a symmetric matrix through a conditional check performed by OnButton(). The function is given by

```
void CCode9::QRMtd(double **c)
{
    int i,j,k,u,p;
    double a[maxIter+1][M+1][M+1],r[M+1][M+1];
    double q[M+1][M+1],s[M+1][M+1],w[M+1][M+1];
    double t[M+1],P[M+1][M+1],I[M+1][M+1],B[M+1][M+1];
    double Alpha,R,Sum;
    epsilon=atof(input[nInputItems][1].item);
    for (i=1;i<=m;i++)
        for (j=1;j<=m;j++)
        {
            r[i][j]=0;
            a[0][i][j]=c[i][j];
            I[i][j]=0;
            if (i==j)
                I[i][j]=1;
        }

    //Householder's Method
    for (k=1;k<=m-2;k++)
    {
        Sum=0;
        for (i=k+1;i<=m;i++)
            Sum+=pow(a[0][i][k],2);
        if (a[0][k+1][k]<0)
            Alpha=sqrt(Sum);
        else
            Alpha=(-1)*sqrt(Sum);
        R=sqrt((pow(Alpha,2)*1/2)-(Alpha*a[0][k+1][k]*1/2));
        for (i=1;i<=k;i++)
            t[i]=0;
        t[k+1]=(a[0][k+1][k]-Alpha)/(2*R);
```

```
        for (i=k+2;i<=m;i++)
             t[i]=a[0][i][k]/(2*R);
        for (i=1;i<=m;i++)
             for (j=1;j<=m;j++)
                  P[i][j]=I[i][j]-(2*t[i]*t[j]);
        for (i=1;i<=m;i++)
             for (j=1;j<=m;j++)
             {
                  Sum=0;
                  for (u=1;u<=m;u++)
                       Sum+=P[i][u]*a[0][u][j];
                  B[i][j]=Sum;
             }
        for (i=1;i<=m;i++)
             for (j=1;j<=m;j++)
             {
                  Sum=0;
                  for (u=1;u<=m;u++)
                       Sum += B[i][u]*P[u][j];
                  a[0][i][j]=Sum;
             }
}
for (i=1;i<=m;i++)
     for (j=1;j<=m;j++)
          q[i][j]=a[0][i][j];

//QR algorithm
StopA=-1;
for (p=0;p<=maxIter;p++)
{
     for (i=1;i<=m;i++)
          for (j=i;j<=m;j++)
          {
               if (i==j)
               {
                    Sum=0;
                    for (k=1;k<=m;k++)
                         Sum+=pow(q[k][i],2);
                    r[i][j]=pow(Sum,0.5);
                    for (k=1;k<=m;k++)
                         q[k][j]=(q[k][j]/r[i][j]);
               }
               if (i<j)
               {
                    Sum=0;
```

```
                    for (k=1;k<=m;k++)
                            Sum+=(q[k][i]*q[k][j]);
                    r[i][j]=Sum;
                    for (k=1;k<=m;k++)
                            q[k][j]=q[k][j]-(r[i][j]
                                    *q[k][i]);
            }
    }
if (p==0)
    for (i=1;i<=m;i++)
            for (j=1;j<=m;j++)
            {
                    s[i][j]=q[i][j];
                    eigen[p].vQR[i][j]=q[j][i];
            }
else
{
    for (i=1;i<=m;i++)
            for (j=1;j<=m;j++)
            {
                    Sum=0;
                    for (k=1;k<=m;k++)
                            Sum+=s[i][k]*q[k][j];
                    w[i][j]=Sum;
            }
    for (i=1;i<=m;i++)
            for (j=1;j<=m;j++)
            {
                    s[i][j]=w[i][j];
                    eigen[p].vQR[i][j]=w[j][i];
            }
}
for (i=1;i<=m;i++)
    for (j=1;j<=m;j++)
    {
            Sum=0;
            for (k=1;k<=m;k++)
                    Sum+=r[i][k]*q[k][j];
            a[p][i][j]=Sum;
            q[i][j]=a[p][i][j];
            eigen[p].lambdaQR[i]=a[p][i][i];
    }
if (p>0)
{
    j=0;
```

```
                    for (i=1;i<=m;i++)
                    {
                            eigen[p].errorQR[i]=eigen[p].lambdaQR[i]
                                    -eigen[p-1].lambdaQR[i];
                            if (fabs(eigen[p].errorQR[i])<epsilon)
                                    j++;
                    }
                    if (j==m)
                    {
                            fStatus=1;
                            Invalidate();
                            StopA=p;
                            break;
                    }
            }
        }
}
```

QRMtd() consists of two major steps. First, the symmetric matrix is converted into a symmetric tridiagonal matrix through the Householder transformation. This is achieved through

```
//Householder's Method
for (k=1;k<=m-2;k++)
{
        Sum=0;
        for (i=k+1;i<=m;i++)
                Sum+=pow(a[0][i][k],2);
        if (a[0][k+1][k]<0)
                Alpha=sqrt(Sum);
        else
                Alpha=(-1)*sqrt(Sum);
        R=sqrt((pow(Alpha,2)*1/2)-(Alpha*a[0][k+1][k]*1/2));
        for (i=1;i<=k;i++)
                t[i]=0;
        t[k+1]=(a[0][k+1][k]-Alpha)/(2*R);
        for (i=k+2;i<=m;i++)
                t[i]=a[0][i][k]/(2*R);
        for (i=1;i<=m;i++)
                for (j=1;j<=m;j++)
                        P[i][j]=I[i][j]-(2*t[i]*t[j]);
        for (i=1;i<=m;i++)
                for (j=1;j<=m;j++)
                {
                        Sum=0;
```

```
                    for (u=1;u<=m;u++)
                            Sum+=P[i][u]*a[0][u][j];
                    B[i][j]=Sum;
            }
    for (i=1;i<=m;i++)
            for (j=1;j<=m;j++)
            {
                    Sum=0;
                    for (u=1;u<=m;u++)
                            Sum += B[i][u]*P[u][j];
                    a[0][i][j]=Sum;
            }
    }
    for (i=1;i<=m;i++)
            for (j=1;j<=m;j++)
                    q[i][j]=a[0][i][j];
```

The second step consists of the application of the QR method in determining the eigenvalues and their corresponding eigenvectors. Iterations are performed using p until convergence, which is indicated by the integer variable, StopA. With the convergence, the flag status value is changed to fStatus=1, which means the method has been successfully applied. The eigenvalues are the last values of eigen[p].lambdaQR[i], whereas their eigenvectors are eigen[p].vQR[i]. The code fragments are given by

```
//QR algorithm
StopA=-1;
for (p=0;p<=maxIter;p++)
{
        for (i=1;i<=m;i++)
                for (j=i;j<=m;j++)
                {
                        if (i==j)
                        {
                                Sum=0;
                                for (k=1;k<=m;k++)
                                        Sum+=pow(q[k][i],2);
                                r[i][j]=pow(Sum,0.5);
                                for (k=1;k<=m;k++)
                                        q[k][j]=(q[k][j]/r[i][j]);
                        }
                        if (i<j)
                        {
                                Sum=0;
                                for (k=1;k<=m;k++)
```

```
                                        Sum+=(q[k][i]*q[k][j]);
                        r[i][j]=Sum;
                        for (k=1;k<=m;k++)
                                q[k][j]=q[k][j]-(r[i][j]*q[k][i]);
                }
        }
if (p==0)
        for (i=1;i<=m;i++)
                for (j=1;j<=m;j++)
                {
                        s[i][j]=q[i][j];
                        eigen[p].vQR[i][j]=q[j][i];
                }
        else
        {
                for (i=1;i<=m;i++)
                        for (j=1;j<=m;j++)
                        {
                                Sum=0;
                                for (k=1;k<=m;k++)
                                        Sum+=s[i][k]*q[k][j];
                                w[i][j]=Sum;
                        }
                for (i=1;i<=m;i++)
                        for (j=1;j<=m;j++)
                        {
                                s[i][j]=w[i][j];
                                eigen[p].vQR[i][j]=w[j][i];
                        }
        }
        for (i=1;i<=m;i++)
                for (j=1;j<=m;j++)
                {
                        Sum=0;
                        for (k=1;k<=m;k++)
                        Sum+=r[i][k]*q[k][j];
                        a[p][i][j]=Sum;
                        q[i][j]=a[p][i][j];
                        eigen[p].lambdaQR[i]=a[p][i][i];
                }
if (p>0)
{
        j=0;
        for (i=1;i<=m;i++)
{
```

```
                eigen[p].errorQR[i]=eigen[p].lambdaQR[i]
                      -eigen[p-1].lambdaQR[i];
                if (fabs(eigen[p].errorQR[i])<epsilon)
                      j++;
        }
        if (j==m)
        {
                fStatus=1;
                Invalidate();
                StopA=p;
                break;
        }
}
```

Output

Output in Code9 is produced in the form list view tables, which shows the results from every iteration. Two functions are created, PowerTable() for the results from power and shifted power methods and QRTable() for the QR method. The two functions are given as follows:

```
void CCode9::PowerTable(int c)
{
        int Stop,i,j,k;
        CString str;
        CRect rcTable[3];
        rcTable[c]=CRect(table[c].hm.x,table[c].hm.y,table[c].hm.x
                  +370,table[c].hm.y+290);
        table[c].list.DestroyWindow();
        table[c].list.Create(WS_VISIBLE | WS_CHILD | WS_DLGFRAME
                  | LVS_REPORT | LVS_NOSORTHEADER,rcTable[c],this,
                  idc++);
        table[c].list.InsertColumn(0,"i",LVCFMT_CENTER,25);
        table[c].list.InsertColumn(1,((c==1)?"lambda hi"
                  :"lambda lo"),LVCFMT_CENTER,70);
        for (i=2;i<=m+1;i++)
        {
                str.Format("v[%d]",i-1);
                table[c].list.InsertColumn(i,str,LVCFMT_CENTER,70);
        }
        table[c].list.InsertColumn(m+2,"error",LVCFMT_CENTER,70);
        Stop=((c==1)?StopA:StopB);
        for (k=0;k<=Stop;k++)
        {
                str.Format("%d",k);
```

```
                table[c].list.InsertItem(k,str,0);
                if (k>0)
                {
                        str.Format("%lf",((c==1)?eigen[k].lambda
                                :eigen[k].lambdaB+eigen[StopA].lambda));
                        table[c].list.SetItemText(k,1,str);
                        str.Format("%lf",((c==1)?eigen[k].error:
                                eigen[k].errorB));
                        table[c].list.SetItemText(k,m+2,str);
                }
                for (j=1;j<=m;j++)
                {
                        str.Format("%lf",((c==1)?eigen[k].v[j]:
                                eigen[k].vB[j]));
                        table[c].list.SetItemText(k,j+1,str);
                }
        }
}

void CCode9::QRTable(int c)
{
        int Stop,i,j,k,p;
        CString str;
        CRect rcTable[4];
        rcTable[c]=CRect(table[c].hm.x,table[c].hm.y,table[c].hm.x
                +370,table[c].hm.y+290);
        table[c].list.DestroyWindow();
        table[c].list.Create(WS_VISIBLE | WS_CHILD | WS_DLGFRAME
                | LVS_REPORT | LVS_NOSORTHEADER,rcTable[c],this,
                idc++);
        table[c].list.InsertColumn(0,"i",LVCFMT_CENTER,25);
        if (c==1)
                for (i=1;i<=m;i++)
                {
                        str.Format("lambda[%d]",i);
                        table[c].list.InsertColumn(i,str,
                                LVCFMT_CENTER,70);
                        str.Format("error[%d]",i);
                        table[c].list.InsertColumn(m+i,str,
                                LVCFMT_CENTER,70);
                }
        if (c==2)
        {
                k=0;
                for (i=1;i<=m;i++)
```

```
            for(j=1;j<=m;j++)
            {
                    str.Format("v[%d][%d]",i,j);
                    table[c].list.InsertColumn
                            (i+k,str,LVCFMT_CENTER,70);
                    k++;
            }
    }
    Stop=StopA;
    for (k=0;k<=Stop;k++)
    {
            str.Format("%d",k);
            table[c].list.InsertItem(k,str,0);
            if (k>0)
            {
                    if (c==1)
                            for (i=1;i<=m;i++)
                            {
                                    str.Format("%lf",eigen[k].
                                            lambdaQR[i]);
                                    table[c].list.SetItemText(k,i,
                                            str);
                                    str.Format("%lf",eigen[k].
                                            errorQR[i]);
                                    table[c].list.SetItemText(k,m+i,
                                            str);
                            }
                    if (c==2)
                    {
                            p=0;
                            for (i=1;i<=m;i++)
                            {
                                    for (j=1;j<=m;j++)
                                    {
                                            str.Format("%lf",eigen[k]
                                                    .vQR[i][j]);
                                            table[c].list.SetItemText
                                                    (k,j+p,str);
                                    }
                                    p += m;
                            }
                    }
            }
    }
}
```

9.7 SUMMARY

The chapter discusses the problem of finding eigenvalues of a given matrix. The power method has been applied in finding the most dominant eigenvalue, whereas its derivative called the shifted power method continues the step in determining the least dominant eigenvalue. Both methods also produce the corresponding eigenvectors for the computed eigenvalues through a series of iterations. In the case where the matrix is symmetric, a technique called the QR method is more practical as the method produces all the eigenvalues and their corresponding eigenvectors. The power, shifted power, and QR methods are illustrated in a project called Code9.

NUMERICAL EXERCISES

1. Find the most dominant and least dominant eigenvalues and their corresponding eigenvectors of the following matrices using the power and shifted power methods, with error tolerance of $\varepsilon = 0.05$:

   a. $A = \begin{bmatrix} 2 & 1 \\ 1 & 2 \end{bmatrix}$

   b. $A = \begin{bmatrix} 2 & 0 & 1 \\ 1 & -1 & 1 \\ 0 & 1 & 2 \end{bmatrix}$

   c. $A = \begin{bmatrix} 3 & -1 & 0 \\ 0 & 2 & 1 \\ 1 & -1 & 3 \end{bmatrix}$

   d. $A = \begin{bmatrix} 3 & -1 & 0 \\ -1 & 2 & -1 \\ 0 & -1 & 3 \end{bmatrix}$

   e. $A = \begin{bmatrix} 2 & -1 & 0 & 3 \\ 1 & 1 & 0 & 2 \\ 2 & -1 & 1 & 1 \\ 1 & -1 & -1 & 3 \end{bmatrix}$

   f. $A = \begin{bmatrix} 2 & -1 & 2 & 3 \\ 1 & 1 & 0 & 2 \\ 2 & 0 & 1 & 1 \\ 1 & -1 & 2 & 3 \end{bmatrix}$

   Check the results by running Code9.

2. Find all the eigenvalues and their corresponding eigenvectors of the following matrices using the QR method:

   a. $A = \begin{bmatrix} 2 & 1 \\ 1 & 2 \end{bmatrix}$

   b. $A = \begin{bmatrix} 2 & 0 & 1 \\ 0 & 2 & 1 \\ 1 & 1 & 2 \end{bmatrix}$

   c. $A = \begin{bmatrix} 3 & -1 & 0 \\ -1 & 2 & -1 \\ 0 & -1 & 3 \end{bmatrix}$

   d. $A = \begin{bmatrix} 2 & -1 & 0 & 3 \\ -1 & 1 & 0 & 2 \\ 0 & 0 & 3 & -1 \\ 3 & 2 & -1 & 3 \end{bmatrix}$

   Check the results by running Code9.

PROGRAMMING CHALLENGES

1. Modify Code9 by adding a new list view table to show the iterated values of the QR method.

2. Modify Code9 by adding the file read and save options. Data to be read are in matrix A, which can have a size from 2×2 to 8×8, whereas for storage are the eigenvalues and eigenvectors.

3. Modify Code9 by incorporating a method for detecting the singularity of the input matrix. A singular matrix A is one where $|A| = 0$, and this type of matrix does not have an inverse.

CHAPTER 10

Ordinary Differential Equations

10.1 INTRODUCTION

Problems involving the ordinary differential equation (ODE) arise in many areas, including engineering, natural sciences, medicine, economics, and anthropology. These problems are normally generally numeric-intensive and require fast computers in their implementation. Solutions to these problems are normally formulated as *models* that involve ordinary and partial differential equations. A mathematical model is the general solution to a given problem that is subject to some conditions and limitations. A typical model works best when all conditions and constraints in the problem are satisfied. At the same time, the model may not work if one or more of these conditions are not satisfied.

An *ordinary differential equation* is an equation that has one or more terms in the expression in the form of derivatives. A good understanding of problems involving ordinary differential equations is necessary in order to produce efficient mathematical models and their simulations on the computer. In engineering, models involving ordinary differential equations are commonly deployed as the fundamentals for the overall solution to a given problem.

Some well-known models for problems in engineering from ordinary differential equations are shown below:

Decay Equation: $\frac{dx}{dt} + \kappa x = 0$, where κ is a constant.

Damped harmonic oscillator equation: $a\frac{d^2x}{dt^2} + b\frac{dx}{dt} + cx = 0$, where a, b, c are constants.

Pendulum equation: $\frac{d^2\theta}{dt^2} + \frac{g}{l}\sin\theta = f(t)$, where g, l are constants.

Van der Pol equation: $\frac{d^2x}{dt^2} + \varepsilon(x^2 + 1)\frac{dx}{dt} + x = 0$, where ε is a constant.

The general form of an ordinary differential equation of order n with m variables, x_i, for $i = 1, 2, \ldots, m$, is stated as

$$g(x_1, x_2, \ldots, x_m, f(x_1, x_2, \ldots, x_m), f'(x_1, x_2, \ldots, x_m), \ldots, f^{(n)}(x_1, x_2, \ldots, x_m)) = 0.$$

(10.1)

In the above equation, $f^{(n)}$ denotes the nth derivative of the function $f(x_1, x_2, \ldots, x_m)$. Hence, the order of the differential equation is determined from the highest derivative in the given equation.

The *degree* of an ODE is the derivative with the highest power in the equation. We illustrate two examples to differentiate the concepts of degree and order, as follows:

$$5x^2 \left(\frac{dy}{dx}\right)^3 - 3\frac{dy}{dx}\sin xy = -1,$$

$$3x\frac{d^2y}{dx^2} - xe^{-y} + 5y = 3xy^2.$$

The first equation above is a first-order ODE, whereas the second is a second-order, ODE as determined from their highest derivatives. The degree of the first equation is three, whereas the second is one, as determined from their highest power.

An ODE is said to be *implicit* if the derivative term cannot be separated from other variables in the equation. Otherwise, if the differential equation can be separated from other variables, then the equation is said to be in an *explicit* form. It can be verified that the first equation above is implicit, whereas the second one can be written in an explicit form as follows:

$$\frac{d^2y}{dx^2} = \frac{3xy^2 + xe^{-y} - 5y}{3x}.$$

In this chapter, we will discuss several common numerical methods for solving the first- and second-order ordinary differential equation problems. A strong emphasis will be placed on the visual model using Visual C++ for each problem in the discussion.

10.2   INITIAL-VALUE PROBLEM FOR FIRST-ORDER ODE

In general, the first-order ordinary differential equation having two variables, x and $y = f(x)$, is expressed as

$$g(x, y, y') = 0.$$

(10.2)

Definition 10.1. The *initial-value problem* for the first-order ordinary differential equation consists of the equation

$$\frac{dy}{dx} = g(x, y), \tag{10.3a}$$

whose *initial value* is given by

$$y_0 = y(x_0). \tag{10.3b}$$

Equations (10.3a) and (10.3b) make up the initial value problem for a first-order ordinary differential equation. The initial value specifies the starting point where the solution to the first-order differential equation exists. The solution to the initial-value problem is obtained analytically according to the following steps:

$$\frac{dy}{dx} = g(x, y),$$

$$dy = g(x, y) \, dx,$$

$$y = f(x) = f(x_0) + \int g(x, y) \, dx. \tag{10.4}$$

It is obvious that the numerical solution requires computing the definite integral of $g(x, y)$ from $x = x_0 = a$ to $x = b$, assuming the function is continuous in this interval. The existence and uniqueness of the solution within a given domain is guaranteed if certain conditions are satisfied. We start with the following definition:

Definition 10.2. A function $g(x, y)$ at $y = y_1$ and $y = y_2$ is said to satisfy a *Lipschitz condition* in the variable y if a constant C called the Lipschitz constant exists in such a way that $|g(x, y_1) - g(x, y_2)| \le C|y_1 - y_2|$.

The above definition suggests the difference in $g(x, y)$ at two y locations is bounded by the Lipschitz constant. This definition paves the way for the uniqueness of the solution.

Theorem 10.1. Suppose $D = \{(x, y) | a \le x \le b, -\infty < y < \infty\}$ and $f(x, y)$ is continuous on D. If $f(x, y)$ satisfies the Lipschitz condition on D in the variable y, then the initial-value problem given by

$$y' = f(x, y), \quad a \le x \le b, \quad y(a) = y_0$$

has a unique solution in $a \le x \le b$.

Definition 10.2 and Theorem 10.1 together guarantee the uniqueness of a solution to the initial-value problem. Several exact methods for solving the initial-value problem have been studied and applied. However, we will not discuss these methods as our focus is on the numerical solutions to the problems.

There are also several numerical methods for solving the initial-value problem in the first-order ODE based on the general formulation in Equation (10.3). In this chapter, we will discuss some of these methods, including the Taylor series, Runge–Kutta of order 2, Runge–Kutta of order 4, and the Adams–Bashforth–Moulton multistep methods. We will also discuss the initial-value problem for a system of ODEs.

10.3 TAYLOR SERIES METHOD

The Taylor series method is based on the expansion of a given function based on the Taylor series equation. The nth-order expansion in the Taylor series is an approximation that produces a polynomial with $(n+1)$ terms.

Definition 10.3. The nth-order Taylor series of $y = f(x)$ at $y_{i+1} = y(x_{i+1})$ is the sum of $(n+1)$ terms at $y_i = y(x_i)$ for $i = 0, 1, \ldots, m$ and $x_0 \le x \le x_m$ in equal-width subintervals, each of size $h = \Delta x$. The Taylor series is defined as

$$y_{i+1} = y_i + \frac{h}{1!}y_i' + \frac{h^2}{2!}y_i'' + \frac{h^3}{3!}y_i''' + \cdots + \frac{h^n}{n!}y_i^{(n)} + O(n+1). \qquad (10.5)$$

In the above equation, $O(n + 1)$ is the sum of the terms involving $y^{(n+1)}$ onward. The Taylor series method for the initial-value problem is an approximation of the Taylor series taken up to the nth term expansion of Equation (10.5), as follows:

$$y_{i+1} \approx y_i + \frac{h}{1!}y_i' + \frac{h^2}{2!}y_i'' + \frac{h^3}{3!}y_i''' + \cdots + \frac{h^n}{n!}y_i^{(n)}. \qquad (10.6)$$

The initial value in the Taylor series method is $y_0 = y(x_0)$. It is obvious from the above equation that the nth-order expansion includes terms involving the derivatives from y_i' to $y_i^{(n)}$. The Taylor series method requires the subintervals in $x_0 \le x \le x_m$ to be uniform.

The implementation of the Taylor series method for solving the initial-value problem according to Equation (10.6) is very straightforward. Algorithm 10.1 shows how this method works.

Algorithm 10.1. Taylor series of order n method.
  Given $y' = g(x, y)$, $y_0 = y(x_0)$, and m equal-width subintervals $h = \Delta x$;
  Compute the derivatives $y'', y''', \ldots, y^{(n)}$;
  For $i = 0$ to $i = m - 1$
    Compute the discrete values of $y_i', y_i'', \ldots, y_i^{(n)}$;
    Compute y_{i+1} according to Equation (10.6);
  Endfor

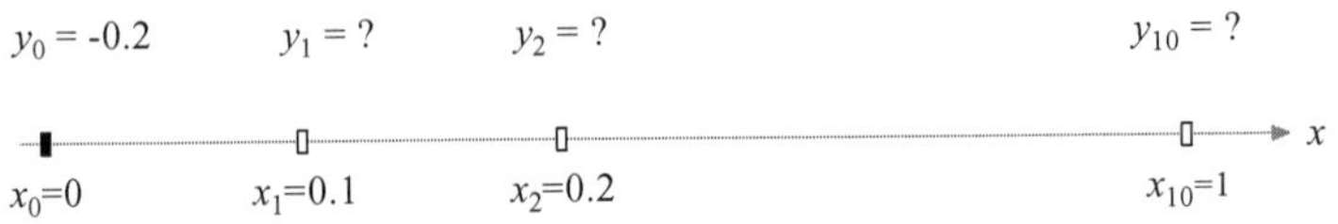

$y_0 = -0.2$ $y_1 = ?$ $y_2 = ?$ $y_{10} = ?$

$x_0 = 0$ $x_1 = 0.1$ $x_2 = 0.2$ $x_{10} = 1$

FIGURE 10.1. The solution diagram for the initial-value problem.

It can be verified from Equation (10.6) that the solution generated in the Taylor series method becomes more accurate with higher order expansion of the terms. This is obvious as a high-order value like $n = 5$ requires the computation of up to $y^{(5)}$ that results in six terms in the expansion. The expansion supports a higher precision value that arises from a large number of decimal places. Algorithm 10.1 is illustrated through the following example:

Example 10.1. Given $y' = x \cos 2y$ for $0 \le x \le 1$ with $h = \Delta x = 0.1$ and $y(0) = -0.2$, find $y(0.1), y(0.2), \ldots, y(1.0)$ using the Taylor series method of order 3.

Solution. First, sketch the domain of this problem, as shown in Figure 10.1. We obtain the Taylor series of order 3 as

$$y_{i+1} \approx y_i + \frac{h}{1!}y_i' + \frac{h^2}{2!}y_i'' + \frac{h^3}{3!}y_i''' = y_i + 0.1y_i' + 0.005y_i'' + 0.000167y_i'''.$$

The equation requires y, y', y'', and y''' in their discrete form. They are evaluated as

$$y' = x \cos 2y,$$
$$y'' = -2xy' \sin 2y + \cos 2y,$$
$$y''' = -2xy'' \sin 2y - 4y' \sin 2y - 4x(y')^2 \cos 2y.$$

The equations are expressed into their discrete forms as

$$y_i' = x_i \cos 2y_i,$$
$$y_i'' = -2x_i y_i' \sin 2y_i + \cos 2y_i,$$
$$y_i''' = -2x_i y_i'' \sin 2y_i - 4y_i' \sin 2y_i - 4x_i(y_i')^2 \cos 2y_i.$$

Starting at $i = 0$, we have $x_0 = 0$ and $y_0 = -0.2$. The discrete elements become

$$y_0' = x_0 \cos 2y_0 = 0,$$
$$y_0'' = -2x_0 y_0' \sin 2y_0 + \cos 2y_0 = 0.921061,$$
$$y_0''' = -2x_0 y_0'' \sin 2y_0 - 4y_0' \sin 2y_0 - 4x_0(y_0')^2 \cos 2y_0 = 0.$$

TABLE 10.1. Solution to the initial-value problem in Example 10.1

i	x_i	y_i	y_i'	y_i''	y_i'''
0	0	−0.2	0.000000	0.921061	0.000000
1	0.100000	−0.195395	0.092461	0.931653	0.208695
2	0.200000	−0.181456	0.186973	0.961417	0.375875
3	0.300000	−0.157888	0.285167	1.003691	0.448502
4	0.400000	−0.124279	0.387707	1.045571	0.354170
5	0.500000	−0.080221	0.493578	1.066008	0.004724
6	0.600000	−0.025532	0.599218	1.035399	−0.674864
7	0.700000	0.039454	0.697822	0.919879	−1.680773
8	0.800000	0.113555	0.779457	0.693513	−2.846112
9	0.900000	0.194494	0.832764	0.356803	−3.816957
10	1.000000	0.278919	0.848402	−0.049805	−4.186357

We obtain $y_1 = y(0.1)$

$$= y_0 + 0.1y_0' + 0.005y_0'' + 0.000167y_0'''$$
$$= (-0.2) + 0.1(0) + 0.005(0.921061) + 0.000167(0) = -0.195395.$$

By repeating the same step above for $i = 1$ with $x_1 = 0.1$ and $y_1 = -0.195395$, we get

$$y_1' = x_1 \cos 2y_1 = 0.092461,$$
$$y_1'' = -2x_1 y_1' \sin 2y_1 + \cos 2y_1 = 0.931653,$$
$$y_1''' = -2x_1 y_1'' \sin 2y_1 - 4y_1' \sin 2y_1 - 4x_1(y_1')^2 \cos 2y_1 = 0.208695.$$

Finally, we obtain the solution:

$$y_2 = y(0.2)$$

$$= y_1 + 0.1y_1' + 0.005y_1'' + 0.000167y_1'''$$

$$= -0.181456.$$

The complete results for other values in the range of $0 \leq x \leq 1.0$ are shown in Table 10.1. Figure 10.2 is the curve representing the solution to the initial-value problem.

Euler's Method

A classic technique called *Euler's method* is a special case of the Taylor series method whose order is $n = 1$. The method is derived from the fundamental theorem of calculus, stated as

$$\int_{x_i}^{x_{i+1}} g(x, f(x))\, dx = \int_{x_i}^{x_{i+1}} y(x)\, dx = y(x_{i+1}) - y(x_i). \tag{10.7}$$

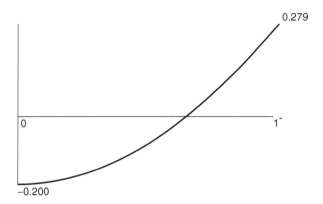

0.279

0

1‾

−0.200

FIGURE 10.2. Solution graph for Example 10.1.

Simplifying the above equation gives

$$y(x_{i+1}) = y(x_i) + \int_{x_i}^{x_{i+1}} g(x, f(x))\, dx$$

$$y_{i+1} = y_i + hy_i',$$

$$y_{i+1} = y_i + hg(x_i, y_i). \tag{10.8}$$

Euler's method is easy to implement as it involves only the first derivative in the Taylor series. It is not necessary to compute the derivatives of higher orders in this method. However, because of its two-term only expansion in the Taylor series, the results obtained using the Euler's method are not as accurate as the Taylor series method using higher orders.

10.4 RUNGE–KUTTA OF ORDER 2 METHOD

One difficulty with the Taylor series method is the necessity to evaluate one or higher derivatives of the given expression that may become very tedious and complicated. As a result, the method may not be practical for implementation on the computer as a special routine for evaluating the derivatives has to be developed along with the normal program. This special routine may involve symbolic computation, which is one area of study that requires a good understanding of data structure and numerical database knowledge. Therefore, the Taylor series method is seldom used in most applications for solving the initial-value problems in ODE.

A suitable alternative to the Taylor series method is the Runge–Kutta methods, which were first proposed by the German mathematicians, C. Runge and M. W. Kutta in 1900. Because of its simplicity, the Runge–Kutta method became very popular, and several variations to the original method were produced. The general solution for the explicit ODE is given by

$$y_{i+1} \approx y_0 + h \sum_{j=1}^{n} c_j k_j, \tag{10.9}$$

where n is the order of the Runge–Kutta equation, c_j is the weight, and k_j is the term in the Taylor series expansion of the equation $y' = g(x, y)$. We will discuss the Runge–Kutta of order 2 and 4 methods in this chapter, which has two and four terms in the summation of Equation (10.9), respectively.

Originally, the Runge–Kutta method of order 2 (RK2) is derived from the Taylor series method based on the equation given by

$$y_{i+1} \approx y_i + g(x_i, y_i)h + g'(x_i, y_i) \frac{h^2}{2}.$$

The above equation simplifies into

$$y_{i+1} \approx y_i + g(x_i, y_i)h + \left(\frac{\partial g}{\partial x} + \frac{\partial g}{\partial x} \frac{dy}{dx} \right) \frac{h^2}{2},$$

$$y_{i+1} \approx y_i + \left(1 - \frac{1}{2r} \right) k_1 + \frac{1}{2r} k_2, \tag{10.10}$$

where

$$k_1 = hg(x_i, y_i), \tag{10.11a}$$

$$k_2 = hg(x_i + rh, y_i + rk_1), \tag{10.11b}$$

and $0 < r \leq 1$. RK2 is stable and produces reasonably good solutions if the value of r is kept in the given range. The method assumes m equal-width subintervals with $h = \Delta x$ in $x_0 \leq x \leq x_m$.

Algorithm 10.2. Runge–Kutta of order 2 method (RK2).
  Given $y' = g(x, y)$, $y_0 = y(x_0)$, $h = \Delta x$, m and r;
  For $i = 0$ to m
    Compute $k_1 = hg(x_i, y_i)$;
    Compute $k_2 = hg(x_i + rh, y_i + rk_1)$;
    If $i < m$
      Compute y_{i+1} using Equation (10.10);
      Update $x_{i+1} \leftarrow x_i + h$;
    Endif
  Endfor

Example 10.2. Solve the problem in Example 10.1 using RK2 with $r = 0.8$.

Solution. Given $\frac{dy}{dx} = g(x, y) = x \cos 2y$. Hence,

$$k_1 = hg(x_i, y_i) = hx_i \cos 2y_i,$$

$$k_2 = hg(x_i + rh, y_i + rk_1) = h(x_i + rh)\cos 2(y_i + rk_1).$$

With $r = 0.8$, $y_{i+1} = y_i + (1 - \frac{1}{2r})k_1 + \frac{1}{2r}k_2 = y_i + 0.375k_1 + 0.625k_2$. At $i = 0$, $x_0 = 0$ and $y_0 = -0.2$. We obtain

$$k_1 = hx_0 \cos 2y_0 = 0,$$

$$k_2 = h(x_0 + rh)\cos 2(y_0 + rk_1) = 0.007368.$$

This produces

$$y(0.1) = y_1$$

$$= y_0 + 0.375k_1 + 0.625k_2$$

$$= -1 + 0.375(0) + 0.625(0.007368)$$

$$= -0.195394.$$

Continuing at $i = 1$, we have $x_1 = 0.1$ and $y_1 = -0.195394$:

$$k_1 = hx_1 \cos 2y_1 = 0.009246,$$

$$k_2 = h(x_1 + rh)\cos 2(y_1 + rk_1) = 0.016742.$$

And, finally,

$$y(0.2) = y_2$$

$$= y_1 + 0.375k_1 + 0.625k_2$$

$$= -0.195394 + 0.375(0.009246) + 0.625(0.016742)$$

$$= 0.181463.$$

Table 10.2 summarizes the results obtained for $i = 0, 1, \ldots, 10$.

Heun's Method

Heun's method is a special case of RK2 with $r = 1$. This produces

$$y_{i+1} = y_i + 0.5k_1 + 0.5k_2, \tag{10.12}$$

TABLE 10.2. RK2 solution to the initial-value problem

i	x_i	y_i	k_1	k_2
0	0	−0.2	0	0.007368
1	0.1	−0.195394	0.009246	0.016742
2	0.2	−0.181463	0.018697	0.026461
3	0.3	−0.157913	0.028516	0.036621
4	0.4	−0.124331	0.038769	0.047166
5	0.5	−0.080313	0.049356	0.057806
6	0.6	−0.025675	0.059920	0.067932
7	0.7	0.039252	0.069784	0.076593
8	0.8	0.113292	0.077955	0.082625
9	0.9	0.194166	0.083298	0.084967
10	1	0.278508	0.084883	0.083099

where $k_1 = hg(x_i, y_i)$ and $k_2 = hg(x_i + h, y_i + k_1)$. To implement this method, Algorithm 10.2 is used by setting $r = 1$.

Modified Euler–Cauchy Method

The modified Euler–Cauchy is another special case of RK2 with $r = 0.5$ to produce

$$y_{i+1} = y_i + k_2, \tag{10.13}$$

where $k_1 = hg(x_i, y_i)$ and $k_2 = hg(x_i + 0.5h, y_i + 0.5k_1)$. Although k_1 appears missing in Equation (10.13), the variable is still needed in the evaluation of k_2 and, therefore, will still need to be evaluated.

10.5 RUNGE–KUTTA OF ORDER 4 METHOD

The Runge–Kutta method of order 4 (RK4) is derived from the Taylor series of order 4 by approximating the second, third, and fourth derivatives. The method is given as

$$y_{i+1} \approx y_i + \frac{1}{6}[k_1 + 2k_2 + 2k_3 + k_4], \tag{10.14}$$

where

$$k_1 = hg(x_i, y_i), \tag{10.15a}$$

$$k_2 = hg(x_i + h/2, y_i + k_1/2), \tag{10.15b}$$

$$k_3 = hg(x_i + h/2, y_i + k_2/2), \tag{10.15c}$$

$$k_4 = hg(x_i + h, y_i + k_3). \tag{10.15d}$$

RK4 is a method of choice in many applications for solving the initial-value problem as it provides a more accurate solution compared with RK2. The method is also easy to implement, as summarized in Algorithm 10.3. Example 10.3 shows an example from this method.

Algorithm 10.3. Runge–Kutta of order 4 method (RK4).
  Given $y' = g(x, y)$, $y_0 = f(x_0)$, $h = \Delta x$ and m;
  For $i = 0$ to m
    Compute $k_1 = hg(x_i, y_i)$;
    Compute $k_2 = hg(x_i + h/2, y_i + k_1/2)$;
    Compute $k_3 = hg(x_i + h/2, y_i + k_2/2)$;
    Compute $k_4 = hg(x_i + h, y_i + k_3)$;
    If $i < m$
      Compute y_{i+1} using Equation (10.14);
      Update $x_{i+1} = x_i + h$;
    Endif
  Endfor

Example 10.3. Solve the problem in Example 10.1 using RK4.

Solution. Given $\frac{dy_i}{dx_i} = g(x_i, y_i) = x_i \cos 2y_i$.

At $i = 0$, $x_0 = 0$ and $y_0 = -1$. This produces

$$k_1 = x_0 \cos 2y_0 = 0,$$

$$k_2 = h(x_0 + h/2) \cos 2(y_0 + k_1/2) = 0.004605,$$

$$k_3 = h(x_0 + h/2) \cos 2(y_0 + k_2/2) = 0.004614,$$

$$k_4 = h(x_0 + h) \cos 2(y_0 + k_3) = 0.009246.$$

Therefore,

$$y(0.1) = y_1 = y_0 + \frac{1}{6} [k_1 + 2k_2 + 2k_3 + k_4] = -0.195386.$$

At $i = 1$, $x_0 = 0.1$ and $y_0 = -0.195386$. With similar steps, we get

$$k_1 = x_1 \cos 2y_1 = 0.009246,$$

$$k_2 = h(x_1 + h/2) \cos 2(y_1 + k_1/2) = 0.013921,$$

$$k_3 = h(x_1 + h/2) \cos 2(y_1 + k_2/2) = 0.013947,$$

$$k_4 = h(x_1 + h) \cos 2(y_1 + k_3) = 0.018698.$$

TABLE 10.3. Results using RK4 for Example 10.3

i	x_i	y_i	k_1	k_2	k_3	k_4
0	0	−0.2	0	0.004605	0.004614	0.009246
1	0.1	−0.195386	0.009246	0.013921	0.013947	0.018698
2	0.2	−0.181439	0.018697	0.023534	0.023574	0.028517
3	0.3	−0.157867	0.028517	0.033566	0.033616	0.038771
4	0.4	−0.124258	0.038771	0.044014	0.044062	0.049358
5	0.5	−0.080211	0.049357	0.054661	0.054693	0.059922
6	0.6	−0.025547	0.059921	0.064997	0.064994	0.069782
7	0.7	0.039401	0.069782	0.074174	0.074124	0.077947
8	0.8	0.113455	0.077949	0.081081	0.081000	0.083279
9	0.9	0.194353	0.083285	0.084613	0.084556	0.084841
10	1	0.278764	0.084856	0.084070	0.084120	0.082279

The above values produce

$$y(0.1) = y_1 = y_0 + \frac{1}{6}[k_1 + 2k_2 + 2k_3 + k_4] = -0.181439.$$

The full results of Example 10.3 are listed in Table 10.3.

10.6 PREDICTOR-CORRECTOR MULTISTEP METHOD

All the methods discussed in the earlier sections are single-step methods. The solutions to these methods are based on iterations starting from a single initial value. The results obtained are good, but they may not be precise because of factors such as error truncation and the approximated approach in the methods.

A refinement to the solution in a single-step method is the *multistep method*. A multistep method implements the predictor–corrector approach, which first deploys a *predictor function* to predict the solution using the Lagrange polynomial interpolation. The predicted value is then refined further using a *corrector function*.

We discuss the solution to the first-order ODE problem $\frac{dy}{dx} = g(x, y)$ with an initial value given by $y_0 = y(x_0)$ using this approach.

Adams–Bashforth–Moulton Method

The Adams–Bashforth–Moulton (ABM) method is the most popular multistep method for solving the initial-value problem. The method is based on the fundamental theorem of calculus from $y = f(x)$ given by

$$y_{i+1} = y_i + \int_{x_i}^{x_{i+1}} g(x, y)dx.$$

TABLE 10.4. Adams–Bashforth predictor functions

Adams–Bashforth	Predictor Function
Two-step	$p_{i+1} = y_i + \frac{h}{2}[-g(x_{i-1}, y_{i-1}) + 3g(x_i, y_i)]$
Three-step	$p_{i+1} = y_i + \frac{h}{12}[5g(x_{i-2}, y_{i-2}) - 16g(x_{i-1}, y_{i-1}) + 23g(x_i, y_i)]$
Four-step	$p_{i+1} = y_i + \frac{h}{24}[-9g(x_{i-3}, y_{i-3}) + 37g(x_{i-2}, y_{i-2}) - 59g(x_{i-2}, y_{i-1})$
	$\quad + 55g(x_i, y_i)]$

TABLE 10.5. Adams–Moulton corrector functions

Adams–Moulton	Corrector Function
Two-step	$y_{i+1} = y_i + \frac{h}{12}[-g(x_{i-1}, y_{i-1}) + 8g(x_i, y_i) + 5g(x_{i+1}, p_{i+1})]$
Three-step	$y_{i+1} = y_i + \frac{h}{24}[g(x_{i-2}, y_{i-2}) - 5g(x_{i-1}, y_{i-1}) + 19g(x_i, y_i)$
	$\quad + 9g(x_{i+1}, p_{i+1})]$

The Adams–Bashforth–Moulton method is a two-fold process consisting of the predictor and corrector functions, p_{i+1} and y_{i+1}. The predictor function is called the Adams–Bashforth function, whereas the corrector is the Adams–Moulton function. Several variations to the method have been documented, and they differ through the number of step sizes in both the predictor and the corrector functions.

Tables 10.4 and 10.5 show some of the most common Adams–Bashforth predictor and Adams–Moulton corrector functions. We discuss the implementation of the Adams–Bashforth–Moulton method using the four-step Adams–Bashforth method as the predictor function and the three-step Adams–Moulton function as the corrector function. The functions are given as

Four-step Adams–Bashforth Equation (Predictor):

$$p_{i+1} = y_i + \frac{h}{24}[-9g(x_{i-3}, y_{i-3}) + 37g(x_{i-2}, y_{i-2}) - 59g(x_{i-2}, y_{i-1}) + 55g(x_i, y_i)].$$
$$(10.16a)$$

Three-step Adams–Moulton Equation (Corrector):

$$y_{i+1} = y_i + \frac{h}{24}[g(x_{i-2}, y_{i-2}) - 5g(x_{i-1}, y_{i-1}) + 19g(x_i, y_i) + 9g(x_{i+1}, p_{i+1})].$$
$$(10.16b)$$

The Adams–Bashforth predictor is based on the Lagrange polynomial approximation on four points, (x_{i-3}, y_{i-3}), (x_{i-2}, y_{i-2}), (x_{i-1}, y_{i-1}), and (x_i, y_i) to produce the extrapolated point (x_{i+1}, p_{i+1}) using Equation (10.16a). Starting with $i = 3$, the values of $g(x_0, y_0)$, $g(x_1, y_1)$, $g(x_2, y_2)$ and $g(x_3, y_3)$ are first evaluated to produce (x_4, p_4). Any single-step method discussed earlier can be used to evaluate these three values.

The Adams–Moulton corrector involves a Lagrange interpolation over the points (x_{i-2}, y_{i-2}), (x_{i-1}, y_{i-1}), (x_i, y_i) and the new point (x_{i+1}, p_{i+1}) starting at $i = 3$, to produce the corrected value, (x_{i+1}, y_{i+1}). With $i = 3$, the corrected value of (x_4, y_4) is obtained using Equation (10.16b) from the values of $g(x_1, y_1), g(x_2, y_2)$, and $g(x_3, y_3)$.

Algorithm 10.4. Adams–Bashforth–Moulton method.
  Given $y' = g(x, y)$, $y_0 = f(x_0)$, $h = \Delta x$ and m;
  For $i = 0$ to 2
    Evaluate y_{i+1} using RK2, RK4, or Taylor methods;
  Endfor
  For $i = 3$ to m
    If $i < m$
      Evaluate p_{i+1} using Equation (10.16a);
      Evaluate y_{i+1} using Equation (10.16b);
      Update $x_{i+1} = x_i + h$;
  Endfor

Algorithm 10.4 outlines the computational steps for the Adams–Bashforth–Moulton method. This algorithm is illustrated using Example 10.4.

Example 10.4. Solve the problem in Example 10.1 using the Adams–Bashforth–Moulton method, starting with RK4 for the first three values.

Solution. Starting with $x_0 = 0$ from the initial value, we apply RK4 to produce the values of y_1, y_2, and y_3. The values are obtained from Example 10.3, as

$$y_1 = -0.195386, \quad y_2 = -0.181439 \text{ and } y_3 = -0.157867.$$

We get the predictor value at $i = 3$:

$$p_4 = y_3 + \frac{h}{24}[-9g(x_0, y_0) + 37g(x_1, y_1) - 59g(x_2, y_2) + 55g(x_3, y_3)]$$

$$= -0.157867 + \frac{.1}{24}[-9(0) + 37(-0.092462) - 59(0.186976) + 55(0.285171)]$$

$$= -0.124226.$$

The corrected value follows:

$$y_4 = y_3 + \frac{h}{24}[g(x_1, y_1) - 5g(x_2, y_2) + 19g(x_3, y_3) + 9g(x_4, p_4)]$$

$$= -0.172063 + \frac{0.1}{24}[-0.092462 - 5(0.186976) + 19(-0.157867) + 9(0.387718)]$$

$$= -0.124262.$$

TABLE 10.6. Results from Example 10.4

i	x_i	p_i	y_i
0	0		−0.200000
1	0.1		−0.195386
2	0.2		−0.181439
3	0.3		−0.157867
4	0.4	−0.124226	−0.124262
5	0.5	−0.080158	−0.080220
6	0.6	−0.025469	−0.025564
7	0.7	0.039497	0.039375
8	0.8	0.113541	0.113423
9	0.9	0.194379	0.194324
10	1	0.278675	0.278753

Similar calculations for $i = 4$ produce p_5 and y_5, as follows:

$$p_5 = y_4 + \frac{h}{24}[-9g(x_1, y_1) + 37g(x_2, y_2) - 59g(x_3, y_3) + 55g(x_4, y_4)] = -0.80158,$$

$$y_5 = y_4 + \frac{h}{24}[g(x_2, y_2) - 5g(x_3, y_3) + 19g(x_4, y_4) + 9g(x_5, p_5)] = -0.080220.$$

The full results from Example 10.4 are shown in Table 10.6.

10.7 SYSTEM OF FIRST-ORDER ODEs

An initial-value problem also arises from a system of ordinary differential equations. A *system of first-order ordinary differential equations* consists of two or more differential equations that share the same set of variables. As in the system of linear equations, a system with n differential equations requires n initial values before it can be solved with unique solutions.

Definition 10.4. In general, the initial-value problem for a system of n ordinary differential equations involving $y_1(x)$, $y_2(x)$, ..., $y_n(x)$ can be written as

$$\frac{dy_1}{dx} = g_1(x, y_1, y_2, \ldots, y_n),$$

$$\frac{dy_2}{dx} = g_2(x, y_1, y_2, \ldots, y_n),$$

$$\cdots$$

$$\frac{dy_n}{dx} = g_n(x, y_1, y_2, \ldots, y_n),$$

(10.17a)

where $g_1(x, y_1, y_2, \ldots, y_n)$, $g_2(x, y_1, y_2, \ldots, y_n), \ldots, g_n(x, y_1, y_2, \ldots, y_n)$ are the given functions in $x_0 \leq x \leq x_m$. The initial values in this problem are given by

$$y_1(x_0), \quad y_2(x_0), \ldots, y_n(x_0). \tag{10.17b}$$

The domain for this problem consists of the successive points $x_0, x_1, \ldots, x_m$ in m equal-width subintervals that are $h = \Delta x$ apart.

Primarily, the solution to an initial-value problem involving a system of differential equations is derived from the same methods used in the single-equation case. Therefore, any method discussed earlier can be applied to solve this problem.

We discuss the Runge–Kutta of order 4 method for solving a system of two differential equations having three variables each. A system of two differential equations with three variables x, $y(x)$, and $z(x)$ in $x_0 \leq x \leq x_m$ has the following form:

$$\frac{dy}{dx} = g_1(x, y, z),$$

$$\frac{dz}{dx} = g_2(x, y, z).$$

The initial values are given by $y_0 = y(x_0)$ and $z_0 = z(x_0)$. The discrete values of x are expressed as x_i for $i = 0, 1, 2, \ldots, m$. Since the intervals are uniform with width $h = \Delta x$, the terms in x can be expressed as

$$x_{i+1} = x_i + h = x_0 + ih.$$

The RK4 method for a system of two differential equations is expressed as

$$y_{i+1} = y_i + \frac{1}{6}[k_1 + 2k_2 + 2k_3 + k_4], \tag{10.18a}$$

$$z_{i+1} = z_i + \frac{1}{6}[K_1 + 2K_2 + 2K_3 + K_4]. \tag{10.18b}$$

In Equations (10.18a) and (10.18b), k_i and K_i for $i = 1, 2, 3, 4$ are the RK4 parameters defined in Equations (10.15a), (10.15b), (10.15c), and (10.15d). They are

$k_1 = hg_1(x_i, y_i, z_i), \qquad\qquad\quad K_1 = hg_2(x_i, y_i, z_i),$

$k_2 = hg_1(x_i + h/2, y_i + k_1/2, z_i + K_1/2), \quad K_2 = g_2(x_i + h/2, y_i + k_1/2, z_i + K_1/2),$

$k_3 = hg_1(x_i + h/2, y_i + k_2/2, z_i + K_2/2), \quad K_3 = g_2(x_i + h/2, y_i + k_2/2, z_i + K_2/2),$

$k_4 = hg_1(x_i + h, y_i + k_3, z_i + K_3), \qquad K_4 = g_2(x_i + h, y_i + k_3, z_i + K_3).$

Algorithm 10.5 summarizes the steps in solving the initial-value problem from a system of two ordinary differential equations using RK4.

Algorithm 10.5. System of first-order ODE with variables using RK4.

Given $\frac{dy}{dx} = g_1(x, y, z)$, $\frac{dz}{dx} = g_2(x, y, z)$, $h = \Delta x$ and m;

Given the initial values, $y_0 = y(x_0)$ and $z_0 = z(x_0)$;

For $i = 0$ to m

    Evaluate $k_1 = hg_1(x_i, y_i, z_i)$;

    Evaluate $K_1 = hg_2(x_i, y_i, z_i)$;

    Evaluate $k_2 = hg_1(x_i + h/2, y_i + k_1/2, z_i + K_1/2)$;

    Evaluate $K_2 = hg_2(x_i + h/2, y_i + k_1/2, z_i + K_1/2)$;

    Evaluate $k_3 = hg_1(x_i + h/2, y_i + k_2/2, z_i + K_2/2)$;

    Evaluate $K_3 = hg_2(x_i + h/2, y_i + k_2/2, z_i + K_2/2)$;

    Evaluate $k_4 = hg_1(x_i + h, y_i + k_3, z_i + K_3)$;

    Evaluate $K_4 = hg_2(x_i + h, y_i + k_3, z_i + K_3)$.

    If $i < m$

        Compute y_{i+1} using Equation (10.18a);

        Compute z_{i+1} using Equation (10.18b);

        Update $x_{i+1} \leftarrow x_i + h$;

    Endif

Endfor

Example 10.5. Solve the differential equations $\frac{dy}{dx} = y - 2zx$ and $\frac{dz}{dx} = xyz$ with the initial values given by $x_0 = 0$, $y_0 = 2$, and $z_0 = -1$, and $h = \Delta x = 0.1$ in $0 \le x \le 1$ using the RK4 method.

Solution. The system consists of

$$g_1(x, y, z) = y - 2zx,$$

$$g_2(x, y, z) = xyz.$$

We obtain $m = (x_m - x_0)/h = (1 - 0)/0.1 = 10$. The parameters in RK4 are evaluated as

$$k_1 = y_0 - 2z_0x_0 = 0.2, \quad K_1 = x_0y_0z_0 = 0,$$

$$k_2 = (y_0 + k_1/2) - 2(z_0 + K_1/2)(x_0 + h/2) = 0.220,$$

$$K_2 = (x_0 + h/2)(y_0 + k_1/2)(z_0 + K_1/2) = -0.0105,$$

$$k_3 = (y_0 + k_2/2) - 2(z_0 + K_2/2)(x_0 + h/2) = 0.221053,$$

$$K_3 = (x_0 + h/2)(y_0 + k_2/2)(z_0 + K_2/2) = -0.010605,$$

$$k_4 = (y_0 + k_3) - 2(z_0 + K_3)(x_0 + h) = 0.242317,$$

$$K_4 = (x_0 + h)(y_0 + k_3)(z_0 + K_3) = -0.022446,$$

TABLE 10.7. RK4 Solution to the system of ODE problem in Example 10.5

i	x_i	y_i	z_i	k_1, K_1	k_2, K_2	k_3, K_3	k_4, K_4
0	0	2.000000	−1.000000	0.200000	0.220000	0.221053	0.242317
				0.000000	−0.010500	−0.010605	−0.022446
1	0.1	2.220737	−1.010776	0.242289	0.264848	0.266178	0.290575
				−0.022447	−0.035901	−0.036311	−0.052080
2	0.2	2.486556	−1.047268	0.242289	0.264848	0.266178	0.290575
				−0.022447	−0.035901	−0.036311	−0.052080
3	0.3	2.752376	−1.083760	0.242289	0.264848	0.266178	0.290575
				−0.022447	−0.035901	−0.036311	−0.052080
4	0.4	3.018195	−1.120252	0.242289	0.264848	0.266178	0.290575
				−0.022447	−0.035901	−0.036311	−0.052080
5	0.5	3.284015	−1.156744	0.242289	0.264848	0.266178	0.290575
				−0.022447	−0.035901	−0.036311	−0.052080
6	0.6	3.549834	−1.193236	0.242289	0.264848	0.266178	0.290575
				−0.022447	−0.035901	−0.036311	−0.052080
7	0.7	3.815653	−1.229728	0.242289	0.264848	0.266178	0.290575
				−0.022447	−0.035901	−0.036311	−0.052080
8	0.8	4.081473	−1.266220	0.242289	0.264848	0.266178	0.290575
				−0.022447	−0.035901	−0.036311	−0.052080
9	0.9	4.347292	−1.302712	0.242289	0.264848	0.266178	0.290575
				−0.022447	−0.035901	−0.036311	−0.052080
10	1	4.613112	−1.339204	0.242289	0.264848	0.266178	0.290575
				−0.022447	−0.035901	−0.036311	−0.052080

Therefore,

$$y_1 = y(0.1) = y_0 + \frac{h}{6}[k_1 + 2k_2 + 2k_3 + k_4]$$

$$= 2 + \frac{0.1}{6}[0.2 + 2(0.22) + 2(0.221053) + 0.242317]$$

$$= 2.220737.$$

$$z_1 = z(0.1) = z_0 + \frac{h}{6}[K_1 + 2K_2 + 2K_3 + K_4]$$

$$= 2 + \frac{0.1}{6}[0 + 2(-0.0105) + 2(-0.010606) - 0.022446]$$

$$= -1.010776.$$

Table 10.7 summarizes the results for (x_i, y_i, z_i) for $i = 0, 1, \ldots, 10$.

10.8 SECOND-ORDER ODE

The second-order ordinary differential equation is an equation that has the second derivative as its highest derivative. The general form of a second-order ODE involving

$y(x)$ is

$$g(x, y, y', y'') = 0. \tag{10.19}$$

Two problems arise in the second-order ODE, and they are called the initial-value problem and the boundary-value problem. For an interval defined as $a \leq x \leq b$, the initial-value problem involves Equation (10.19) with its initial values at $x = a$ given. A boundary-value problem has its boundary values given at $x = a$ and $x = b$ for solving Equation (10.19).

In solving a second-order ODE problem, conditions from either its initial values or boundary values are needed. A solution obtained from the second-order differential equations may exist uniquely or infinitely. The following theorems describe the cases of unique solution.

Theorem 10.2. Given $y'' = f(x, y, y')$ for $x_0 \leq x \leq x_m$ with $y(x_0) = y_0$ and $y(x_m) = y_m$, which is continuous in $D = \{(x, y, y')|x_0 \leq x \leq x_m, -\infty < y < \infty, -\infty < y' < \infty\}$. If f_y and $\partial f/\partial y'$ are also continuous in D, and $|\frac{\partial f}{\partial y'}(x, y, y')| \leq M$, then the solution is unique.

Theorem 10.3. If $y'' = p(x)y' + q(x)y + r(x)$ for $x_0 \leq x \leq x_m$, with $y(x_0) = y_0$ and $y(x_m) = y_m$; $p(x)$, $q(x)$, and $r(x)$ are continuous in $[x_0, x_m]$; and $q(x) > 0$ in $[x_0, x_m]$, then the solution is unique.

We discuss the solution to the initial-value problem in the second-order ordinary differential equation using a technique of reducing its order to a system of first-order equations. This is followed by the boundary-value problem involving a method called finite-difference.

10.9 INITIAL-VALUE PROBLEM FOR SECOND-ORDER ODE

The initial-value problem for a second-order ordinary differential equation is defined as follows:

Definition 10.5. The initial-value problem for a continuous second-order ordinary differential equation in an interval defined as $x_0 \leq x \leq x_m$ with the variables $y = f(x)$ consists of solving Equation (10.19) with the initial conditions given by

$$y(x_0) = y_0, \tag{10.20a}$$

$$y'(x_0) = y'_0. \tag{10.20b}$$

A second-order ODE requires two initial conditions, one from the normal starting value and another involving the first derivative. One good strategy for solving the initial-value problem for the second-order ODE is to reduce its order to a system of first-order ODEs. The same technique discussed in Section 10.8 can then be applied to solve the problem once the system of first-order ODEs has been obtained. To achieve this objective, any single-step method discussed in this chapter can be used to generate the solution.

We discuss the reduction of the second-order ODE into a system of first-order ODEs. The numerical solution to the initial-value problem involving the second-order ODE can be modeled as the discrete points (x_i, y_i) over m uniform intervals in $x_0 \leq x \leq x_m$ whose width is given by $h = \Delta x$.

A second-order ODE with two variables, x and $y = f(x)$, can be reduced into a system of two first-order ODEs. This is possible by setting $z = y'$, and this transforms y'' into z'. Hence, $g(x, y, y', y'') = 0$ is reduced to the following system of first-order ODEs:

$$g_1(x, y, z) = z,$$

$$g_2(x, y, z) = z'.$$

The initial-value conditions for this problem become $y(x_0) = y_0$ and $z_0 = y'(x_0)$. The solutions are then obtained by applying the same method discussed in Section 10.8. Any suitable first-order method such as RK2 and RK4 can be used to solve the system of differential equations.

Algorithm 10.6 summarizes the steps in solving the initial-value problem. The method applies RK4 for solving the system of first-order ODEs. The algorithm is illustrated using Example 10.6.

Algorithm 10.6. Second-order ODE reduction to first-order ODE system.
   Given $g(x, y, y', y'') = 0$, $y_0 = y(x_0)$, $y'(x_0) = z_0$, $h = \Delta x$ and m;
   Let $z = y'$ and $z' = y''$;
   Form a system with $z = y' = g_1(x, y, z)$ and $z' = y'' = g_2(x, y, z)$;
   For $i = 0$ to m
      Evaluate $k_1 = hg_1(x_i, y_i, z_i)$;
      Evaluate $K_1 = hg_2(x_i, y_i, z_i)$;
      Evaluate $k_2 = hg_1(x_i + h/2, y_i + k_1/2, z_i + K_1/2)$;
      Evaluate $K_2 = hg_2(x_i + h/2, y_i + k_1/2, z_i + K_1/2)$;
      Evaluate $k_3 = hg_1(x_i + h/2, y_i + k_2/2, z_i + K_2/2)$;
      Evaluate $K_3 = hg_2(x_i + h/2, y_i + k_2/2, z_i + K_2/2)$;
      Evaluate $k_4 = hg_1(x_i + h, y_i + k_3, z_i + K_3)$;
      Evaluate $K_4 = hg_2(x_i + h, y_i + k_3, z_i + K_3)$;
      If $i < m$
         Compute y_{i+1} using Equation (10.18a);
         Compute z_{i+1} using Equation (10.18b);
         Update $x_{i+1} \leftarrow x_i + h$;
      Endif
   Endfor

Example 10.6. Solve the equation $y'' + 4y' + 5y = 0$, with the initial conditions given by $y(0) = -1$ dan $y'(0) = 2$, for $0 \leq x \leq 1$ and $h = 0.1$.

Solution. The number of subintervals is $m = (1 - 0)/0.1 = 10$. Let $z = y'$, and this reduces $y'' + 4y' + 5y = 0$ into $z' + 4z + 5y = 0$. We obtain a system of first-order ODEs, as follows:

$$y' = g_1(x, y, z) = z,$$
$$z' = g_2(x, y, z) = -4z - 5y$$

The initial values are $y_0 = y(0) = -1$ and $z_0 = z(0) = 2$. The parameters in RK4 are obtained as follows:

$$k_1 = hz_i, \quad K_1 = h(-4z_i - 5y_i),$$

$$k_2 = h(z_i + K_1/2), \quad K_2 = h[-4(z_i + K_1/2) - 5(y_i + k_1/2)],$$

$$k_3 = h(z_i + K_2/2), \quad K_3 = h[-4(z_i + K_2/2) - 5(y_i + k_2/2)],$$

$$k_4 = h(z_i + K_3), \quad K_1 = h[-4(z_i + K_3) - 5(y_i + k_3)].$$

At $i = 0$, $x_0 = 0$, $y_0 = -1$, and $z_0 = 2$. This gives

$$k_1 = hz_0 = 0.2, \quad K_1 = h(-4z_0 - 5y_0) = -0.3,$$

$$k_2 = h(z_0 + K_1/2) = 0.185, \quad K_2 = h[-4(z_0 + K_1/2) - 5(y_0 + k_1/2)] = -0.29,$$

$$k_3 = h(z_0 + K_2/2) = 0.1855, \quad K_3 = h[-4(z_0 + K_2/2) - 5(y_0 + k_2/2)] = -0.288250,$$

$$k_4 = h(z_0 + K_3/2) = 0.171175, \quad K_4 = h[-4(z_0 + K_3/2) - 5(y_0 + k_3/2)] = -0.277450.$$

We obtain the first solution as

$$y_1 = y(0.1) = y_0 + \frac{h}{6}[k_1 + 2k_2 + 2k_3 + k_4]$$

$$= -1 + \frac{0.1}{6}[0.2 + 2(0.185) + 2(0.1855) + 0.171175]$$

$$= -0.814638.$$

$$z_1 = z(0.1) = z_0 + \frac{h}{6}[K_1 + 2K_2 + 2K_3 + K_4]$$

$$= 2 + \frac{0.1}{6}[-0.3 + 2(-0.29) + 2(-0.288250) - 0.277450]$$

$$= 1.711008.$$

TABLE 10.8. RK4 Solution to the second-order ODE boundary value problem in Example 10.6

i	x_i	y_i	z_i	k_1, K_1	k_2, K_2	k_3, K_3	k_4, K_4
0	0	−1.000000	2.000000	0.200000	0.185000	0.185500	0.171175
				−0.300000	−0.290000	−0.288250	−0.277450
1	0.1	−0.814638	1.711008	0.171101	0.157247	0.157879	0.144750
				−0.277085	−0.264443	−0.263508	−0.250621
2	0.2	−0.656954	1.447074	0.171101	0.157247	0.157879	0.144750
				−0.277085	−0.264443	−0.263508	−0.250621
3	0.3	−0.499270	1.183139	0.171101	0.157247	0.157879	0.144750
				−0.277085	−0.264443	−0.263508	−0.250621
4	0.4	−0.341587	0.919205	0.171101	0.157247	0.157879	0.144750
				−0.277085	−0.264443	−0.263508	−0.250621
5	0.5	−0.183903	0.655271	0.171101	0.157247	0.157879	0.144750
				−0.277085	−0.264443	−0.263508	−0.250621
6	0.6	−0.026220	0.391336	0.171101	0.157247	0.157879	0.144750
				−0.277085	−0.264443	−0.263508	−0.250621
7	0.7	0.131464	0.127402	0.171101	0.157247	0.157879	0.144750
				−0.277085	−0.264443	−0.263508	−0.250621
8	0.8	0.289148	−0.136533	0.171101	0.157247	0.157879	0.144750
				−0.277085	−0.264443	−0.263508	−0.250621
9	0.9	0.446831	−0.400467	0.171101	0.157247	0.157879	0.144750
				−0.277085	−0.264443	−0.263508	−0.250621
10	1	0.604515	−0.664401	0.171101	0.157247	0.157879	0.144750
				−0.277085	−0.264443	−0.263508	−0.250621

The full results from this problem are shown in Table 10.8.

10.10   FINITE-DIFFERENCE METHOD FOR SECOND-ORDER ODE

The boundary-value problem for a second-order ordinary differential equation has the boundary conditions given. In an interval defined as $a \leq x \leq b$, the boundaries for the continuous function $y = f(x)$ in the interval are the left and right points, $(a, f(a))$ and $(b, f(b))$. To solve the differential equation, the boundary values must be given.

A *boundary* is an end point in the given interval or domain of the problem. There are two types of boundaries in an ordinary differential equation:

a. Dirichlet boundary conditions, which are stated as the given values at the ends of one of the intervals, for example, $y(a) = \alpha$ and $y(b) = \beta$ in $a \leq x \leq b$.

b. Neumann boundary conditions, which are stated as the given values of the first derivatives at the ends of one of the intervals. For example, $y'(a) = \lambda$ and $y'(b) = \mu$ in $a \leq x \leq b$ are the Neumann boundary conditions.

We will deal with the first type of boundary condition in this section and with the second type in the next section.

Definition 10.6. The *boundary-value problem* involving a second-order ordinary differential equation is Equation (10.19) in $x_0 \leq x \leq x_m$ with boundary conditions given by $y(x_0) = y_0$ and $y(x_m) = y_m$. In this interval, the width is given by $h = \Delta x$ and there are m uniform subintervals.

We restrict our discussion on the boundary-value problems to the case of linear second-order ODEs. A second-order differential equation is said to *linear* if it can be expressed into the following form:

$$p(x)y'' + q(x)y' + r(x)y = w(x), \tag{10.21}$$

where $p(x)$, $q(x)$, $r(x)$, and $w(x)$ are continuous functions of x in the interval $x_0 \leq x \leq x_m$.

A common approach for solving a linear second-order differential equation with boundary conditions is the *finite-difference method*. The method is based on the approximation of the derivatives of y at several finite points in the interval to yield a *finite-difference formula*. The points are distributed at equal-width subintervals so that the derivatives y' and y'' can be replaced by their approximated discrete values.

The solution to the boundary-value problem for ODE2 consists of two main steps, as depicted in Figure 10.3. First, the differential equation is discretized where the terms involving y' and y'' are replaced by their approximated values using the central-difference rules. This step leads the way to the formation of the finite-difference formula for the problem.

The second step starts by applying the finite-difference formula to the m subintervals in $x_0 \leq x \leq x_m$. The finite-difference formula is applied at each of the $m - 1$ interior points in the interval to produce a system of $(m - 1) \times (m - 1)$ linear equations. A technique from Chapter 5, such as the Gaussian elimination method, is then applied to solve this system to produce the final solution to the boundary-value problem.

Algorithm 10.7 outlines the implementation of the finite-difference method for solving the boundary-value problem for a linear second-order ODE.

Algorithm 10.7. Boundary-value problem for the linear second-order ODE.

Given $p(x)y'' + q(x)y' + r(x)y = w(x)$;

Given $y(x_0) = y_0$, $y(x_m) = y_m$ and $h = \Delta x$;

Discretize the variables into $p(x_i)y_i'' + q(x_i)y_i' + r(x_i)y_i = w(x_i)$;

Obtain the finite-difference equation by substituting y_i' and y_i'';

Form a system of linear equations from the finite-difference equation;

Solve the system of linear equations to get $y_1, y_2, \ldots, y_{m-1}$;

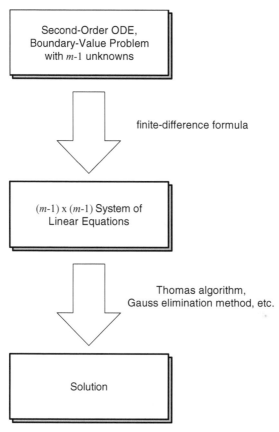

FIGURE 10.3. Two-step solution to the boundary value problem in the second-order ODE.

We discuss the general solution to Equation (10.21) using Algorithm 10.7. The discrete form of this equation is given by

$$p(x_i)y_i'' + q(x_i)y_i' + r(x_i)y_i = w(x_i).$$

Let $p_i = p(x_i)$, $q_i = q(x_i)$, $r_i = r(x_i)$, and $w_i = w(x_i)$, and the above equation becomes

$$p_i y_i'' + q_i y_i' + r_i y_i = w_i.$$

Finite-difference values are obtained by employing the central difference rules discussed in Chapter 8 to approximate the first and second derivatives, given as

$$y_i' \approx \frac{y_{i+1} - y_{i-1}}{2h},$$

$$y_i'' \approx \frac{y_{i+1} - 2y_i + y_{i-1}}{h^2}.$$

Substituting y' and y'', we get

$$p_i \frac{y_{i+1} - 2y_i + y_{i-1}}{h^2} + q_i \frac{y_{i+1} - y_{i-1}}{2h} + r_i y_i = w_i.$$

Rearranging the terms in the order of y_{i-1}, y_i and y_{i+1}, we obtain the finite-difference formula for the given problem:

$$\left(\frac{p_i}{h^2} - \frac{q_i}{2h}\right) y_{i-1} + \left(\frac{-2p_i}{h^2} + r_i\right) y_i + \left(\frac{p_i}{h^2} + \frac{q_i}{2h}\right) y_{i+1} = w_i. \qquad (10.22)$$

There are $m - 1$ unknowns in the system, y_i for $i = 1, 2, \ldots, m - 1$, which can be solved from the $(m - 1) \times (m - 1)$ system of linear equations. This system of linear equations is obtained by first substituting $i = 1, 2, \ldots, m - 1$ into the above equation. The process starts with $i = 1$:

$$\left(\frac{p_1}{h^2} - \frac{q_1}{2h}\right) y_0 + \left(\frac{-2p_1}{h^2} + r_1\right) y_1 + \left(\frac{p_1}{h^2} + \frac{q_1}{2h}\right) y_2 = w_1.$$

Since the value of y_0 is given, the first term above is moved to the right-hand side of the equation to give

$$\left(\frac{-2p_1}{h^2} + r_1\right) y_1 + \left(\frac{p_1}{h^2} + \frac{q_1}{2h}\right) y_2 = w_1 - \left(\frac{p_1}{h^2} - \frac{q_1}{2h}\right) y_0.$$

At $i = m - 1$, the equation becomes

$$\left(\frac{p_{m-1}}{h^2} - \frac{q_{m-1}}{2h}\right) y_{m-2} + \left(\frac{-2p_{m-1}}{h^2} + r_{m-1}\right) y_{m-1} + \left(\frac{p_{m-1}}{h^2} + \frac{q_{m-1}}{2h}\right) y_m = w_{m-1}.$$

Since the value of y_m is given, the last term in the left-hand side is moved to the right-hand side, and this produces the last equation in the system as

$$\left(\frac{p_{m-1}}{h^2} - \frac{q_{m-1}}{2h}\right) y_{m-2} + \left(\frac{-2p_{m-1}}{h^2} + r_{m-1}\right) y_{m-1} = w_{m-1} - \left(\frac{p_{m-1}}{h^2} + \frac{q_{m-1}}{2h}\right) y_m.$$

Regrouping all the above equations into a matrix form, we obtain a tridiagonal system of linear equations of size $(m-1) \times (m-1)$, as follows:

$$
\begin{bmatrix}
\left(\frac{-2p_1}{h^2}+r_1\right) & \left(\frac{p_1}{h^2}+\frac{q_1}{2h}\right) & 0 & \cdots & 0 & 0 & 0 \\
\left(\frac{p_2}{h^2}-\frac{q_2}{2h}\right) & \left(\frac{-2p_2}{h^2}+r_2\right) & \left(\frac{p_2}{h^2}+\frac{q_2}{2h}\right) & \cdots & 0 & 0 & 0 \\
0 & \left(\frac{p_3}{h^2}-\frac{q_3}{2h}\right) & \left(\frac{-2p_3}{h^2}+r_3\right) & \cdots & 0 & 0 & 0 \\
\cdots & \cdots & \cdots & \cdots & \cdots & \cdots & \cdots \\
0 & 0 & 0 & \cdots & \left(\frac{-2p_{m-3}}{h^2}+r_{m-2}\right) & \left(\frac{p_{m-3}}{h^2}+\frac{q_{m-3}}{2h}\right) & 0 \\
0 & 0 & 0 & \cdots & \left(\frac{p_{m-2}}{h^2}-\frac{q_{m-2}}{2h}\right) & \left(\frac{-2p_{m-2}}{h^2}+r_{m-2}\right) & \left(\frac{p_{m-2}}{h^2}+\frac{q_{m-2}}{2h}\right) \\
0 & 0 & 0 & \cdots & 0 & \left(\frac{p_{m-1}}{h^2}-\frac{q_{m-1}}{2h}\right) & \left(\frac{-2p_{m-1}}{h^2}+r_{m-1}\right)
\end{bmatrix}
\begin{bmatrix}
y_1 \\ y_2 \\ y_3 \\ \cdots \\ y_{m-3} \\ y_{m-2} \\ y_{m-1}
\end{bmatrix}
$$

$$
= \begin{bmatrix}
w_1 - \left(\frac{p_1}{h^2}-\frac{q_1}{2h}\right)y_0 \\
w_2 \\
w_3 \\
\cdots \\
w_{m-3} \\
w_{m-2} \\
w_{m-1} - \left(\frac{p_{m-1}}{h^2}+\frac{q_{m-1}}{2h}\right)y_m
\end{bmatrix}. \tag{10.23}
$$

From the above system of linear equations, we obtain the algorithm for the nonzero entries of $A = [a_{i,j}]$ and $b = [b_i]$ in $Ay = b$, as follows:

$$
a_{i,j} = \begin{cases}
\dfrac{-2p_i}{h^2}+r_i & \text{for } j = i \text{ and } i = 1, 2, \ldots, m-1, \\[2mm]
\dfrac{p_i}{h^2}+\dfrac{q_i}{2h} & \text{for } j = i+1 \text{ and } i = 1, 2, \ldots, m-2, \\[2mm]
\dfrac{p_i}{h^2}-\dfrac{q_i}{2h} & \text{for } j = i-1 \text{ and } i = 2, \ldots, m-1,
\end{cases} \tag{10.24a}
$$

$$
b_i = \begin{cases}
w_1 - \left(\dfrac{p_1}{h^2}-\dfrac{q_1}{2h}\right)y_0 & \text{if } i = 1, \\[2mm]
w_i & \text{if } i = 2, 3, \ldots, m-2, \\[2mm]
w_{m-1} - \left(\dfrac{p_{m-1}}{h^2}+\dfrac{q_{m-1}}{2h}\right)y_m & \text{if } i = m-1.
\end{cases} \tag{10.24b}
$$

Since Equation (10.23) is tridiagonal, the most practical approach for solving this system is the Thomas algorithm as the computational steps required in this method are not as massive as in other methods. We discuss an example that shows the method for solving this problem.

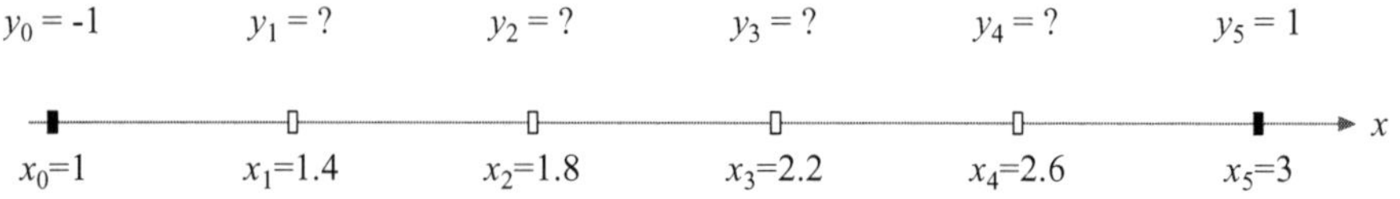

$y_0 = -1$ $y_1 = ?$ $y_2 = ?$ $y_3 = ?$ $y_4 = ?$ $y_5 = 1$

$x_0=1$ $x_1=1.4$ $x_2=1.8$ $x_3=2.2$ $x_4=2.6$ $x_5=3$

FIGURE 10.4. The solution diagram for Example 10.7.

Example 10.7. Given $(\cos x)y'' + (\sin(2x - 1)) y' + (\sin(1 - 5x))y = x \cos x$ in $1 < x < 3$ whose width is $h = \Delta x = 0.4$, the boundary values in this problem are given as $y(1) = -1$ and $y(3) = 1$. Find the values of $y(1.4)$, $y(1.8)$, $y(2.2)$, and $y(2.6)$.

Solution. Figure 10.4 shows the solution diagram for the problem. There are four unknowns in this problem, y_i for $i = 1, 2, 3, 4$, since $m = 5$. We start with

$$(\cos x_i)y_i'' + (\sin(2x_i - 1))y_i' + (\sin(1 - 5x_i))y_i = x_i \cos x_i.$$

Applying the central-difference rules for substituting y_i' and y_i'',

$$[\cos x_i]\frac{y_{i+1} - 2y_i + y_{i-1}}{h^2} + [\sin(2x_i - 1)]\frac{y_{i+1} - y_{i-1}}{2h} + [\sin(1 - 5x_i)]y_i = x_i \cos x_i.$$

Simplifying the terms in the above equation, we obtain the following finite-difference formula:

$$\left[\frac{\cos x_i}{h^2} - \frac{\sin(2x_i - 1)}{2h}\right]y_{i-1} + \left[\frac{-2\cos x_i}{h^2} + \sin(1-5x_i)\right]y_i$$

$$+ \left[\frac{\cos x_i}{h^2} + \frac{\sin(2x_i - 1)}{2h}\right]y_{i+1} = x_i \cos x_i.$$

The next step is to form the system of linear equations. A system of four linear equations is to be formed since there are four unknowns, y_1, y_2, y_3, and y_4. The equations are found by setting $i = 1, 2, 3, 4$ into the finite-difference formula to form a 4×4 system of linear equations, as follows:

$i = 1: -0.155015(-1) - 1.845174y_1 + 2.279604y_2 = 0.237954$

    $-1.845174y_1 + 2.279604y_2 = 0.082939,$

$i = 2: -2.064390y_1 + 1.850668y_2 - 0.775636y_3 = -0.4089638,$

$i = 3: -3.358706y_2 + 7.900285y_3 - 3.997558y_4 = -1.294702,$

$i = 4: -4.266085y_3 + 11.247682y_4 - 6.445024(1) = -2.227911$

    $-4.266085y_3 + 11.247682y_4 = 4.217114.$

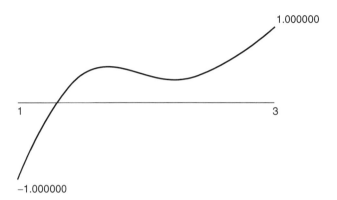

FIGURE 10.5. Solution curve for Example 10.7.

In matrix form, the linear equations become

$$
\begin{bmatrix}
-1.845174 & 2.279604 & 0 & 0 \\
-2.064390 & 1.850668 & -0.775636 & 0 \\
0 & -3.358706 & 7.900285 & -3.997558 \\
0 & 0 & -4.266085 & 11.247682
\end{bmatrix}
\begin{bmatrix}
y_1 \\ y_2 \\ y_3 \\ y_4
\end{bmatrix}
=
\begin{bmatrix}
0.082939 \\ -0.408964 \\ -1.294702 \\ 4.217114
\end{bmatrix}.
$$

The above system of linear equations is solved to produce the following solution:

$$
y = [\, y_1 \quad y_2 \quad y_3 \quad y_4 \,]^T.
$$

$$
= [\, 0.486949 \quad 0.430533 \quad 0.258478 \quad 0.472969 \,]^T.
$$

Figure 10.5 shows the solution graph for Example 10.7.

10.11 DIFFERENTIATED BOUNDARY CONDITIONS

Under certain circumstances, the boundary conditions for the second-order ODE in Equation (10.21) may be given in the form of derivatives, as follows:

$$
y'(x_0) = \alpha \text{ and } y'(x_m) = \beta,
$$

where α and β are constants and m is the number of subintervals in $x_0 \le x \le x_m$.

In solving this problem, similar steps as in the previous case are applied to obtain the finite-difference formula in Equation 10.22.

$$
\left(\frac{p_i}{h^2} - \frac{q_i}{2h} \right) y_{i-1} + \left(\frac{-2p_i}{h^2} + r_i \right) y_i + \left(\frac{p_i}{h^2} + \frac{q_i}{2h} \right) y_{i+1} = w_i.
$$

The left boundary is $y_0' = \alpha$. Applying the central-difference rule from Equation (8.5a),

$$\frac{y_1 - y_{-1}}{2h} = \alpha.$$

We obtain a virtual value, y_{-1} as this quantity is not inside $y_0 \le y \le y_m$. Expressing y_{-1} as a subject of the equation, we obtain

$$y_{-1} = y_1 - 2h\alpha.$$

The right boundary condition consists of $y_m' = \beta$, which becomes

$$\frac{y_{m+1} - y_{m-1}}{2h} = \beta.$$

Again, another virtual value y_{m+1} is obtained as it is outside of $y_0 \le y \le y_m$. This value is made the subject, as follows:

$$y_{m+1} = y_{m-1} + 2h\beta.$$

It can be verified that the virtual values are only applicable in the finite-difference formula in cases of $i = 0$ and $i = m$. At $i = 0$, Equation (10.22) produces:

$$\left(\frac{p_0}{h^2} - \frac{q_0}{2h}\right) y_{-1} + \left(\frac{-2p_0}{h^2} + r_0\right) y_0 + \left(\frac{p_0}{h^2} + \frac{q_0}{2h}\right) y_1 = w_0.$$

Substituting the value of y_{-1}, the above equation simplifies to

$$\left(\frac{p_0}{h^2} - \frac{q_0}{2h}\right)(y_1 - 2h\alpha) + \left(\frac{-2p_0}{h^2} + r_0\right) y_0 + \left(\frac{p_0}{h^2} + \frac{q_0}{2h}\right) y_1 = w_0$$

$$\left(\frac{-2p_0}{h^2} + r_0\right) y_0 + \frac{2p_0}{h^2} y_1 = w_0 + 2h\alpha \left(\frac{p_0}{h^2} - \frac{q_0}{2h}\right). \tag{10.25}$$

Similarly, applying Equation (10.22) at $i = m$:

$$\left(\frac{p_m}{h^2} - \frac{q_m}{2h}\right) y_{m-1} + \left(\frac{-2p_m}{h^2} + r_m\right) y_m + \left(\frac{p_m}{h^2} + \frac{q_m}{2h}\right) y_{m+1} = w_m.$$

Substituting the value of y_{m+1}:

$$\left(\frac{p_m}{h^2} - \frac{q_m}{2h}\right) y_{m-1} + \left(\frac{-2p_m}{h^2} + r_m\right) y_m + \left(\frac{p_m}{h^2} + \frac{q_m}{2h}\right)(y_{m-1} + 2h\beta) = w_m,$$

$$\frac{2p_m}{h^2} y_{m-1} + \left(\frac{-2p_m}{h^2} + r_m\right) y_m = w_m - 2h\beta \left(\frac{p_m}{h^2} + \frac{q_m}{2h}\right). \tag{10.26}$$

For $i = 1, 2, \ldots, m - 1$, the entries in each row of the coefficient matrix consist of the diagonal, and its left and right terms, as follows:

$i = 1:$
$$\left(\frac{p_1}{h^2} - \frac{q_1}{2h}\right) y_0 + \left(\frac{-2p_1}{h^2} + r_1\right) y_1 + \left(\frac{p_1}{h^2} + \frac{q_1}{2h}\right) y_2 = w_1,$$

$i = 2:$
$$\left(\frac{p_2}{h^2} - \frac{q_2}{2h}\right) y_1 + \left(\frac{-2p_2}{h^2} + r_2\right) y_2 + \left(\frac{p_2}{h^2} + \frac{q_2}{2h}\right) y_3 = w_2,$$

$\ldots$

$i = m - 2:$ $\left(\dfrac{p_{m-2}}{h^2} - \dfrac{q_{m-2}}{2h}\right) y_{m-3} + \left(\dfrac{-2p_{m-2}}{h^2} + r_{m-2}\right) y_{m-2} + \left(\dfrac{p_{m-2}}{h^2} + \dfrac{q_{m-2}}{2h}\right) y_{m-1} = w_{m-2},$

$i = m - 1:$ $\left(\dfrac{p_{m-1}}{h^2} - \dfrac{q_{m-1}}{2h}\right) y_{m-2} + \left(\dfrac{-2p_{m-1}}{h^2} + r_{m-1}\right) y_{m-1} + \left(\dfrac{p_{m-1}}{h^2} + \dfrac{q_{m-1}}{2h}\right) y_m = w_{m-1}.$

We obtain a $(m + 1) \times (m + 1)$ tridiagonal system of linear equations:

$$
\begin{bmatrix}
\frac{-2p_0}{h^2} + r_0 & \frac{2p_0}{h^2} & 0 & \cdots & 0 & 0 & 0 \\
\frac{p_1}{h^2} - \frac{q_1}{2h} & \frac{-2p_1}{h^2} + r_1 & \frac{p_1}{h^2} + \frac{q_1}{2h} & \cdots & 0 & 0 & 0 \\
0 & \frac{p_2}{h^2} - \frac{q_2}{2h} & \frac{-2p_2}{h^2} + r_2 & \cdots & 0 & 0 & 0 \\
\cdots & \cdots & \cdots & \cdots & \cdots & \cdots & \cdots \\
0 & 0 & 0 & \cdots & \frac{-2p_{m-2}}{h^2} + r_{m-2} & \frac{p_{m-2}}{h^2} + \frac{q_{m-2}}{2h} & 0 \\
0 & 0 & 0 & \cdots & \frac{p_{m-1}}{h^2} - \frac{q_{m-1}}{2h} & \frac{-2p_{m-1}}{h^2} + r_{m-1} & \frac{p_{m-1}}{h^2} + \frac{q_{m-1}}{2h} \\
0 & 0 & 0 & \cdots & 0 & \frac{2p_m}{h^2} & \frac{-2p_m}{h^2} + r_m
\end{bmatrix}
\begin{bmatrix}
y_0 \\ y_1 \\ y_2 \\ \cdots \\ y_{m-2} \\ y_{m-1} \\ y_m
\end{bmatrix}
$$

$$
=
\begin{bmatrix}
w_0 + 2h\alpha \left(\frac{p_0}{h^2} - \frac{q_0}{2h}\right) \\
w_1 \\
w_2 \\
\cdots \\
w_{m-2} \\
w_{m-1} \\
w_m - 2h\beta \left(\frac{p_m}{h^2} + \frac{q_m}{2h}\right)
\end{bmatrix}.
\tag{10.27}
$$

Equation (10.27) is summarized as $A\mathbf{y} = \mathbf{b}$, where $A = [a_{ij}]$ is the coefficient matrix in the left-hand side, $\mathbf{b} = [b_i]$ is the vector in the right side of the equation, and $\mathbf{y} = [y_i]$ is the unknown vector, for $i, j = 0, 1, 2, \ldots, m$. The tridiagonal elements of A are obtained as

$$a_{ii} = \frac{-2p_i}{h^2} + r_i \quad \text{for } i = 0, 1, 2, \ldots, m, \tag{10.28a}$$

$$a_{i,i+1} = \begin{cases} \dfrac{p_i}{h^2} + \dfrac{q_i}{2h} & \text{for } i = 1, 2, \ldots, m - 1, \\[2mm] \dfrac{2p_0}{h^2} & \text{for } i = 0, \end{cases} \tag{10.28b}$$

$$a_{i+1,i} = \begin{cases} \dfrac{p_i}{h^2} - \dfrac{q_i}{2h} & \text{for } i = 0, 1, \ldots, m - 2, \\[2mm] \dfrac{2p_m}{h^2} & \text{for } i = m - 1. \end{cases} \tag{10.28c}$$

From the same equation, we obtain the representation for b, as follows:

$$b_i = \begin{cases} w_0 + 2h\alpha \left(\dfrac{p_0}{h^2} - \dfrac{q_0}{2h} \right) & \text{for } i = 0, \\[2mm] w_i & \text{for } i = 1, 2, \ldots, m - 1, \\[2mm] w_m - 2h\beta \left(\dfrac{p_m}{h^2} + \dfrac{q_m}{2h} \right) & \text{for } i = m. \end{cases} \tag{10.28d}$$

Collectively, Equations (10.28a), (10.28b), (10.28c), and (10.28d) are sufficient to solve for y. Since the coefficient matrix in Equation (10.27) is tridiagonal, the most suitable choice for solving the system of linear equations is the Thomas algorithm method.

Example 10.8. Given $(2 \cos x)y'' + (5 \cos 2x)y' - (9 \sin 3x)y = -7 \sin 2x$ with differentiated boundary values given as $y'(0) = 0.4$ and $y'(1) = 0.9$ in $0 \le x \le 1$, find y_i for $i = 0, 1, \ldots, m$ on m equal-width intervals with $h = \Delta x = 0.2$.

Solution. There are $m = \frac{x_m - x_0}{h} = \frac{1-0}{0.2} = 5$ subintervals in the domain, and this gives the graphical representation of the problem, as shown in Figure 10.6. The discrete

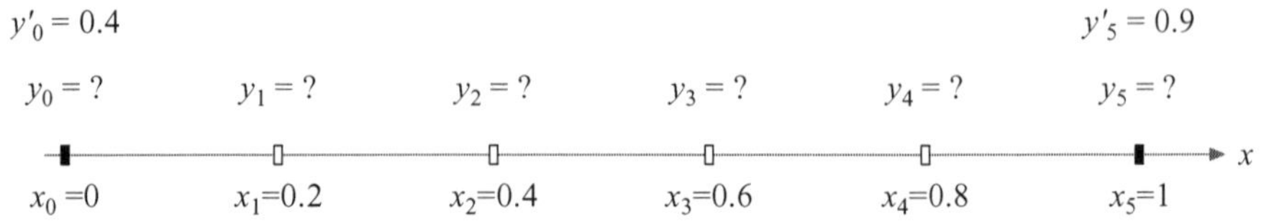

FIGURE 10.6. Graphical representation of the problem.

solution of the second-order differential equation is given by

$$(2\cos x_i)y_i'' + (5\cos 2x_i)y_i' - (9\sin 3x_i)y_i = -7\sin 2x_i.$$

Replacing y_i' and y_i'' in the above equation using central-difference rules, we get

$$(2\cos x_i)\frac{y_{i+1} - 2y_i + y_{i-1}}{h^2} + (5\cos 2x_i)\frac{y_{i+1} - y_{i-1}}{2h} - (9\sin 3x_i)y_i = -7\sin 2x_i.$$

The above equation is simplified to produce the finite-difference formula given by

$$\left[\frac{2\cos x_i}{h^2} - \frac{5\cos 2x_i}{2h}\right]y_{i-1} + \left[\frac{-4\cos x_i}{h^2} - 9\sin 3x_i\right]y_i$$

$$+ \left[\frac{2\cos x_i}{h^2} + \frac{5\cos 2x_i}{2h}\right]y_{i+1} = -7\sin 2x_i.$$

The boundary values at $x_0 = 0$ and x_5 in this problem are given in the form of first derivatives at these points. This implies y_0 and y_5 are also the unknowns in this problem along with y_1, y_2, y_3, and y_4. Therefore, there are six unknowns that require the reduction of the boundary-value problem to a 6×6 system of linear equations.

The given first derivative at $x_0 = 0$ is simplified using the central-difference rule to produce a virtual value, y_{-1}. This is obtained as follows:

$$y'(0) = 0.4,$$

$$\frac{y_1 - y_{-1}}{2h} = 0.4,$$

$$y_{-1} = y_1 - 0.8h.$$

The right boundary value is simplified in a similar fashion to produce another virtual value, y_6, as follows:

$$y'(1) = 0.9,$$

$$\frac{y_6 - y_4}{2h} = 0.9,$$

$$y_6 = y_4 + 1.8h.$$

The first virtual value and the finite-difference formula are applied at $i = 0$ to produce

$$\left[\frac{2\cos x_0}{h^2} - \frac{5\cos 2x_0}{2h}\right] y_{-1} + \left[\frac{-4\cos x_0}{h^2} - 9\sin 3x_0\right] y_0$$

$$+ \left[\frac{2\cos x_0}{h^2} + \frac{5\cos 2x_0}{2h}\right] y_1 = -7\sin 2x_0,$$

$$\left[\frac{2\cos x_0}{h^2} - \frac{5\cos 2x_0}{2h}\right](y_1 - 0.8h) + \left[\frac{-4\cos x_0}{h^2} - 9\sin 3x_0\right] y_0$$

$$+ \left[\frac{2\cos x_0}{h^2} + \frac{5\cos 2x_0}{2h}\right] y_1 = -7\sin 2x_0,$$

$$\left[\frac{-4\cos x_0}{h^2} - 9\sin 3x_0\right] y_0 + \left[\frac{4\cos x_0}{h^2}\right] y_1 = -7\sin 2x_0$$

$$+ 0.8h\left[\frac{4\cos x_0}{h^2} - \frac{5\cos 2x_0}{2h}\right].$$

Applying the formula at the other interior points produce

At $i = 1$, $\left[\dfrac{2\cos x_1}{h^2} - \dfrac{5\cos 2x_1}{2h}\right] y_0 + \left[\dfrac{-4\cos x_1}{h^2} - 9\sin 3x_1\right] y_1$

$$+ \left[\frac{2\cos x_1}{h^2} + \frac{5\cos 2x_1}{2h}\right] y_2 = -7\sin 2x_1.$$

At $i = 2$, $\left[\dfrac{2\cos x_2}{h^2} - \dfrac{5\cos 2x_2}{2h}\right] y_1 + \left[\dfrac{-4\cos x_2}{h^2} - 9\sin 3x_2\right] y_2$

$$+ \left[\frac{2\cos x_2}{h^2} + \frac{5\cos 2x_2}{2h}\right] y_3 = -7\sin 2x_2.$$

At $i = 3$, $\left[\dfrac{2\cos x_3}{h^2} - \dfrac{5\cos 2x_3}{2h}\right] y_2 + \left[\dfrac{-4\cos x_3}{h^2} - 9\sin 3x_3\right] y_3$

$$+ \left[\frac{2\cos x_3}{h^2} + \frac{5\cos 2x_3}{2h}\right] y_4 = -7\sin 2x_3.$$

At $i = 4$, $\left[\dfrac{2\cos x_4}{h^2} - \dfrac{5\cos 2x_4}{2h}\right] y_3 + \left[\dfrac{-4\cos x_4}{h^2} - 9\sin 3x_4\right] y_4$

$$+ \left[\frac{2\cos x_4}{h^2} + \frac{5\cos 2x_4}{2h}\right] y_5 = -7\sin 2x_4.$$

The second virtual value is applied to the right boundary point, at $i = 5$, to produce

$$\left[\frac{2\cos x_5}{h^2} - \frac{5\cos 2x_5}{2h}\right] y_4 + \left[\frac{-4\cos x_5}{h^2} - 9\sin 3x_5\right] y_5$$

$$+ \left[\frac{2\cos x_5}{h^2} + \frac{5\cos 2x_5}{2h}\right] y_6 = -7\sin 2x_5,$$

$$\left[\frac{2\cos x_5}{h^2} - \frac{5\cos 2x_5}{2h}\right] y_4 + \left[\frac{-4\cos x_5}{h^2} - 9\sin 3x_5\right] y_5$$

$$+ \left[\frac{2\cos x_5}{h^2} + \frac{5\cos 2x_5}{2h}\right] (y_4 + 1.8h) = -7\sin 2x_5,$$

$$\frac{4\cos x_5}{h^2} y_4 + \left[\frac{-4\cos x_5}{h^2} - 9\sin 3x_5\right] y_5 = -7\sin 2x_5$$

$$-1.8h\left[\frac{2\cos x_5}{h^2} + \frac{5\cos 2x_5}{2h}\right].$$

We obtain the following system of linear equations:

$$\begin{bmatrix} -100 & 4 & 0 & 0 & 0 & 0 \\ 37.490066 & -103.088440 & 59.320024 & 0 & 0 & 0 \\ 0 & 37.344216 & -100.494451 & 50.582522 & 0 & 0 \\ 0 & 0 & 36.737309 & -91.298190 & 38.426755 & 0 \\ 0 & 0 & 0 & 35.200329 & -75.749840 & 25.617914 \\ 0 & 0 & 0 & 0 & 54.030231 & -55.300311 \end{bmatrix} \begin{bmatrix} y_0 \\ y_1 \\ y_2 \\ y_3 \\ y_4 \\ y_5 \end{bmatrix}$$

$$= \begin{bmatrix} 1.2 \\ -2.725928 \\ -5.021493 \\ -6.524274 \\ -6.997015 \\ -14.217863 \end{bmatrix}.$$

The above system is solved to produce the final solutions, given by

$$y = -0.002208 \quad 0.244800 \quad 0.380864 \quad 0.476674 \quad 0.598627 \quad 0.841981]^T.$$

The solution curve for this problem is shown in Figure 10.7.

FIGURE 10.7. Solution curve for Example 10.8 in $0 \le x \le 1$.

10.12   VISUAL SOLUTION: CODE10

Code10. User Manual.

1. Select a method from the menu.
2. Enter the input according to the selected method.
3. Click the push button to view the results.

   Development files: `Code10.cpp`, `Code10.h`, and `MyParser.obj`.

We discuss the visual interface for the ordinary differential equations problems. The project is called `Code10`, and it consists of all the methods discussed in this chapter. `Code10` is menu-driven, and this provides friendliness to problems that are generally considered difficult.

Figure 10.8 shows the output from `Code10`, which consists of a menu with eight items that represent the methods in order from top to bottom, as follows: Taylor, Runge–Kutta of order 2, Runge–Kutta of order 4, Adams–Bashforth–Moulton, ODE1 system, ODE2 initial-value problem, finite-difference 1, and finite-difference 2. The figure shows the solution to a sample problem from the fifth item in the menu (ODE 1 System), which is an initial-value problem on a system of first-order ODEs given by

$$y' = f(x, y) = 3 \sin x \sin yz \text{ and } z' = g(x, y) = 3 \cos x \cos yz,$$

whose initial values are $x_0 = 0$, $y_0 = 0.5$, and $z_0 = 0.5$, for $0 \le x \le 8$. The results are shown in the table with their corresponding graphs of $y = q(x)$ and $z = r(x)$ generated.

`Code10` has been designed to allow the user a full control of the input as well as the output. To allow this flexibility, the output from Figure 10.8 is divided into four regions. The first region is the menu displayed as shaded rectangles on the top left. The second region is the input area, which becomes active when an item in the menu is selected. The third region is the list view table for displaying the results from the

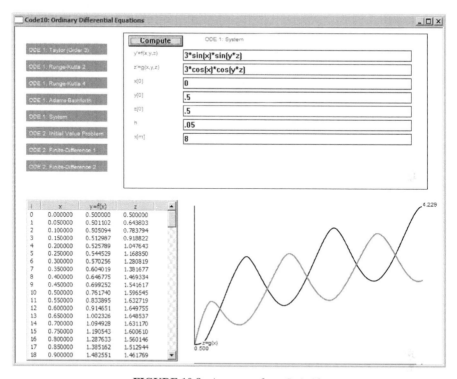

FIGURE 10.8. An output from Code10.

calculations. The fourth region is the graphical area that displays the solution graph for the problem.

Code10 has a single class called CCode10, and this class is derived from CFrameWnd. The development files in Code10 are Code10.cpp, Code10.h, and MyParser.obj.

Figure 10.9 is the schematic drawing of Code10 that shows the development stages of Code10. In this diagram, a status flag called fStatus monitors the execution progress whose updated value represents a step in the execution. Initially, fStatus has a value of 0. When an item in the menu is selected, fStatus changes its value to 1. At the same time, a variable called fMenu is assigned with a number that represents the order of the item from top to bottom. The selection also creates edit boxes for collecting input from the user and a push button called *Compute* that, when activated, calls the function that corresponds to the selected method.

When input has been completed and confirmed with a click at the *Compute* button, fStatus value is updated to 2. The click causes a call to be made to a function corresponding to the selected method in the menu. This function represents a method for solving the given problem. The function solves the problem according to the input values, and the results are displayed in the list view table through ShowTable(). The solution graph for the problem is also displayed through DrawCurve().

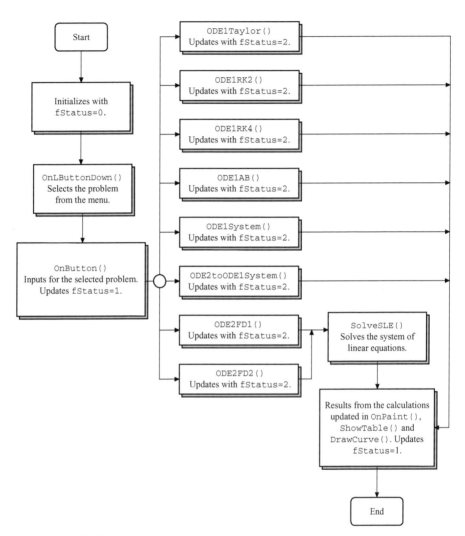

FIGURE 10.9. Schematic drawing showing the development stages of Code10.

Changes in the input values in the edit boxes are allowed by resetting fStatus to 1 in DrawCurve() once the results have been obtained and displayed. This option is necessary as part of the user-friendliness features for the given problem. With this update, any small changes in the input values will cause the whole data to be reevaluated, and the results are immediately updated both in the table and in the solution graph.

The data structure in Code10 consists of four structures. The first structure is PT, which represents the variables x_i, y_i, p_i, and z_i as the array pt in the methods. The structure also defines the left, right, maximum, and minimum points in the solution curve in the problem. The structure is declared as follows:

```
typedef struct
{
        double x,y,p,z;      // xᵢ, yᵢ, pᵢ, and zᵢ components
} PT;
PT *pt;
```

A structure called INPUT declares the input objects, and these objects are linked to an array called input. The structure is declared as

```
typedef struct
{
        CString item, label,;  // input string and its label
        CPoint hm;             // home coordinates
        CEdit ed;              // edit box
        CRect rc;              // rectangular object
} INPUT;
INPUT input[maxInput+1];
```

The third structure is MENU, which declares objects for the items in the menu. This structure is declared as

```
typedef struct
{
        CString item;        // menu item
        CPoint hm;           // home coordinates
        CRect rc;            // rectangular region
} MENU;
MENU menu[nMenuItems+1];
```

The last structure is CURVE, which provides objects for creating the solution curve. The structure defines the rectangular region for displaying the solution graph.

```
typedef struct
{
        CRect rc;            // rectangular region
        CPoint hm,end;       // starting and end
                             //      coordinates
} CURVE;
CURVE curve;
```

Three events are mapped in Code10, namely, the display update, left-button click, and push button click.

```
BEGIN_MESSAGE_MAP(CCode10,CFrameWnd)
    ON_WM_PAINT()
    ON_WM_LBUTTONDOWN()
```

TABLE 10.9. Functions in Code10

Function	Description
ODE1Taylor()	Taylor series method for solving the first-order initial-value problem.
ODE1RK2()	Runge–Kutta of order 2 method for solving the first-order initial-value problem.
ODE1RK4()	Runge–Kutta of order 4 method for solving the first-order initial-value problem.
ODE1AB()	Adams–Bashforth multistep method for solving the first-order initial-value problem.
ODE1System()	Runge–Kutta of order 4 method for solving the initial-value problem in a system of first-order equations.
ODE2toODE1System()	Reduction of the initial-value problem from the second-order ODE to first-order ODE, and its solution using the Runge–Kutta of order 4 method.
ODE2FD1()	Finite-difference equation method for the boundary-value problem in the second-order ODE.
ODE2FD2()	Finite-difference equation method for the boundary-value problem in the second-order ODE having differentiated boundary conditions.

```
ON_BN_CLICKED(IDC_BUTTON,OnButton)
END_MESSAGE_MAP()
```

Eight items in the menu represent eight different methods for solving the initial- and boundary-value problems. Each method is represented by a function as described in Table 10.9.

Taylor Series of Order 3 Solution

The Taylor series method is represented by ODE1Taylor(). This function supports the Taylor series method of order three only. It would be a good challenge for the reader to modify the item to make it more flexible by supporting the method using any high order. In the given problem, $y' = g(x, y)$ is the input function. Because of the non-symbolic nature of the application, the program does not evaluate the second or third derivatives automatically from the input function. The user needs to enter the derivatives in the edit boxes, and the string will then be passed to parse() for processing.

In reading the input string, the derivatives are represented as single characters in parse(), as follows:

$$u = y', \quad v = y'', \text{ and } w = y'''.$$

The representation is necessary so as to use the character codes defined in Table 4.1 where the corresponding codes for x, y, u, v, w, and z are 23, 24, 20, 21, 22, and 25, respectively . For example, $y''' = 3xy' - 4(y'')^2 \cos xy''$ is written as $v = 3*x*u - 4v \wedge 2*\cos(x*v)$.

The Taylor series method in ODE1Taylor() is written based on Algorithm 10.1 and Equation (10.6). The function is given as follows:

```
void CCode10::ODE1Taylor()
{
        double psv[6],tmp,max;
        int psi[6];
        int i;
        double h,u,v,w,z;

        pt[0].x=atof(input[2].item);
        pt[0].y=atof(input[5].item);
        h=atof(input[4].item);
        tmp=atof(input[3].item);
        m=(tmp-pt[0].x)/h; m=((m<M)?m:M);
        max=pt[0].x+(double)m*h;
        tmp=(tmp<max)?tmp:max;
        pt[m].x=tmp;
        psi[1]=23; psi[2]=24;
        psi[3]=20; psi[4]=21;
        for (i=0;i<=m;i++)
        {
                psv[1]=pt[i].x;
                psv[2]=pt[i].y;
                u=parse(input[1].item,2,psv,psi);
                psv[3]=u;
                v=parse(input[6].item,3,psv,psi);
                psv[4]=v;
                w=parse(input[7].item,4,psv,psi);
                if (i<m)
                {
                        pt[i+1].y=pt[i].y+h*u+pow(h,2)/
                                2*v+pow(h,3)/6*w;
                        pt[i+1].x=pt[i].x+h;
                }
        }
}
```

There are m subintervals that require m iterations for computing y_i, for $i = 1, 2, \ldots, m$. The input strings for the equations are read as input[1].item, input[6].item, and input[7].item. These strings are passed to parse() for processing. The returned values are stored as u, v, and w, which represent y', y'', and y''', respectively.

```
psv[1]=pt[i].x;
psv[2]=pt[i].y;
u=parse(input[1].item,2,psv,psi);
psv[3]=u;
v=parse(input[6].item,3,psv,psi);
psv[4]=v;
w=parse(input[7].item,4,psv,psi);
```

The Taylor series method solves the initial-value problem according to Equation (10.6). The code for this equation is written as

```
if (i<m)
{
    pt[i+1].y=pt[i].y+h*u+pow(h,2)/2*v+pow(h,3)/6*w;
    pt[i+1].x=pt[i].x+h;
}
```

Runge–Kutta of Order 2 Solution

RK2 is easier to implement than the Taylor series method as it does not require the evaluation of high-order derivatives. In Code10, RK2 is handled by ODE1RK2(). The method is very straightforward, as shown below:

```
void CCode10::ODE1RK2()
{
    int i,psi[6];
    double h,r,psv[6],tmp,max;
    double k1,k2;
    pt[0].x=atof(input[2].item);
    pt[0].y=atof(input[5].item);
    h=atof(input[4].item);
    tmp=atof(input[3].item);
    m=(tmp-pt[0].x)/h; m=((m<M)?m:M);
    max=pt[0].x+(double)m*h;
    tmp=(tmp<max)?tmp:max;
    pt[m].x=tmp;
    r=atof(input[6].item);
    psi[1]=23; psi[2]=24;
    for (i=0;i<=m;i++)
    {
```

```
            psv[1]=pt[i].x;
            psv[2]=pt[i].y;
            k1=h*parse(input[1].item,2,psv,psi);
            psv[1]=pt[i].x+r*h;
            psv[2]=pt[i].y+r*k1;
            k2=h*parse(input[1].item,2,psv,psi);
            if (i<m)
            {
                    pt[i+1].y=pt[i].y+(1-1/(2*r))*k1+1/(2*r)*k2;
                    pt[i+1].x=pt[i].x+h;
            }
        }
}
```

ODE1RK2() solves the initial-value problem using Equations (10.10) and (10.11), whose computational steps have been outlined in Algorithm 10.2. Equation (10.11a) and (10.11b) are processed using

```
psv[1]=pt[i].x;
psv[2]=pt[i].y;
k1=h*parse(input[1].item,2,psv,psi);
psv[1]=pt[i].x+r*h;
psv[2]=pt[i].y+r*k1;
k2=h*parse(input[1].item,2,psv,psi);
```

The values of y_i, for $i = 1, 2, \ldots, m$, are updated using Equation (10.10), and they are written in ODE1RK2() as

```
if (i<m)
{
        pt[i+1].y=pt[i].y+(1-1/(2*r))*k1+1/(2*r)*k2;
        pt[i+1].x=pt[i].x+h;
}
```

Runge–Kutta of Order 4 Solution

The function ODE1RK4() represents the solution to the initial-value problem based on Algorithm 10.3 and Equation (10.14). The code segment is given as

```
void CCode10::ODE1RK4()
{
        int i,psi[6];
        double h,psv[6],tmp,max;
        double k1,k2,k3,k4;
        pt[0].x=atof(input[2].item);
```

```
        pt[0].y=atof(input[5].item);
        h=atof(input[4].item);
        tmp=atof(input[3].item);
        m=(tmp-pt[0].x)/h;
        m=((m<M)?m:M);
        max=pt[0].x+(double)m*h;
        tmp=(tmp<max)?tmp:max;
        pt[m].x=tmp;
        psi[1]=23; psi[2]=24;
        for (i=0;i<=m;i++)
        {
                psv[1]=pt[i].x;
                psv[2]=pt[i].y;
                k1=h*parse(input[1].item,2,psv,psi);

                psv[1]=pt[i].x+h/2;
                psv[2]=pt[i].y+k1/2;
                k2=h*parse(input[1].item,2,psv,psi);

                psv[1]=pt[i].x+h/2;
                psv[2]=pt[i].y+k2/2;
                k3=h*parse(input[1].item,2,psv,psi);

                psv[1]=pt[i].x+h;
                psv[2]=pt[i].y+k3;
                k4=h*parse(input[1].item,2,psv,psi);
                if (i<m)
                {
                        pt[i+1].y=pt[i].y+(k1+2*k2+2*k3+k4)/6;
                        pt[i+1].x=pt[i].x+h;
                }
        }
}
```

The parameters k_1, k_2, k_3, and k_4 in Equations (10.15a), (10.15b), (10.15c), and (10.15d), respectively, are written in the following code fragments:

```
psv[1]=pt[i].x;
psv[2]=pt[i].y;
k1=h*parse(input[1].item,2,psv,psi);

psv[1]=pt[i].x+h/2;
psv[2]=pt[i].y+k1/2;
k2=h*parse(input[1].item,2,psv,psi);
```

```
psv[1]=pt[i].x+h/2;
psv[2]=pt[i].y+k2/2;
k3=h*parse(input[1].item,2,psv,psi);

psv[1]=pt[i].x+h;
psv[2]=pt[i].y+k3;
k4=h*parse(input[1].item,2,psv,psi);
```

RK4 solution in Equation (10.14) is written compactly as

```
if (i<m)
{
        pt[i+1].y=pt[i].y+(k1+2*k2+2*k3+k4)/6;
        pt[i+1].x=pt[i].x+h;
}
```

Adams–Bashforth–Moulton Multistep Solution

The Adams–Bashforth–Moulton method is a multistep method that requires a single-step method in its first few iterations to produce the predictor values. These values are read and inserted into the corrector function to produce solutions in the subsequent iterations. In Code10, the function ODE1AB() implements this method based on Algorithm 10.4.

```
void CCode10::ODE1AB()
{
        int i,psi[6];
        double h,f0,f1,f2,f3,fp,psv[6],tmp,max;
        double k1,k2,k3,k4;
        pt[0].x=atof(input[2].item); pt[0].y=atof(input[5].item);
        h=atof(input[4].item);
        tmp=atof(input[3].item);
        m=(tmp-pt[0].x)/h; m=((m<M)?m:M);
        max=pt[0].x+(double)m*h;
        tmp=(tmp<max)?tmp:max;
        pt[m].x=tmp;
        psi[1]=23; psi[2]=24;
        for (i=0;i<=m;i++)
        {
                if (i<3)
                {
                        psv[1]=pt[i].x;
                        psv[2]=pt[i].y;
                        k1=h*parse(input[1].item,2,psv,psi);

                        psv[1]=pt[i].x+h/2;
```

```
                    psv[2]=pt[i].y+k1/2;
                    k2=h*parse(input[1].item,2,psv,psi);

                    psv[1]=pt[i].x+h/2;
                    psv[2]=pt[i].y+k2/2;
                    k3=h*parse(input[1].item,2,psv,psi);

                    psv[1]=pt[i].x+h;
                    psv[2]=pt[i].y+k3;
                    k4=h*parse(input[1].item,2,psv,psi);

                    pt[i+1].y=pt[i].y+(k1+2*k2+2*k3+k4)/6;
            }
            if (i<m)
            {
                    pt[i+1].x=pt[i].x+h;
                    if (i>=3)
                    {
                            psv[1]=pt[i-3].x;
                            psv[2]=pt[i-3].y;
                            f0=parse(input[1].item,2,psv,psi);

                            psv[1]=pt[i-2].x;
                            psv[2]=pt[i-2].y;
                            f1=parse(input[1].item,2,psv,psi);

                            psv[1]=pt[i-1].x;
                            psv[2]=pt[i-1].y;
                            f2=parse(input[1].item,2,psv,psi);

                            psv[1]=pt[i].x;
                            psv[2]=pt[i].y;
                            f3=parse(input[1].item,2,psv,psi);
                            pt[i+1].p=pt[i].y+h/24*(-9*f0+37*f1
                                    -59*f2+55*f3);

                            psv[1]=pt[i+1].x;
                            psv[2]=pt[i+1].p;
                            fp=parse(input[1].item,2,psv,psi);
                            pt[i+1].y=pt[i].y+h/24*(f1-5*f2+19*f3
                                    +9*fp);
                    }
            }
        }
    }
}
```

The Adams–Bashforth predictor function is Equation (10.16a), and it is based on RK4. The function is represented by the following code segment:

```
if (i<3)
{
        psv[1]=pt[i].x;
        psv[2]=pt[i].y;
        k1=h*parse(input[1].item,2,psv,psi);

        psv[1]=pt[i].x+h/2;
        psv[2]=pt[i].y+k1/2;
        k2=h*parse(input[1].item,2,psv,psi);

        psv[1]=pt[i].x+h/2;
        psv[2]=pt[i].y+k2/2;
        k3=h*parse(input[1].item,2,psv,psi);

        psv[1]=pt[i].x+h;
        psv[2]=pt[i].y+k3;
        k4=h*parse(input[1].item,2,psv,psi);

        pt[i+1].y=pt[i].y+(k1+2*k2+2*k3+k4)/6;
}
```

The Adams–Moulton corrector function is based on Equation (10.16b). The code segment consists of

```
if (i<m)
{
        pt[i+1].x=pt[i].x+h;
        if (i>=3)
        {
                psv[1]=pt[i-3].x;
                psv[2]=pt[i-3].y;
                f0=parse(input[1].item,2,psv,psi);

                psv[1]=pt[i-2].x;
                psv[2]=pt[i-2].y;
                f1=parse(input[1].item,2,psv,psi);

                psv[1]=pt[i-1].x;
                psv[2]=pt[i-1].y;
                f2=parse(input[1].item,2,psv,psi);

                psv[1]=pt[i].x;
```

```
                psv[2]=pt[i].y;
                f3=parse(input[1].item,2,psv,psi);
                pt[i+1].p=pt[i].y+h/24*(-9*f0+37*f1-59*f2+55*f3);

                psv[1]=pt[i+1].x;
                psv[2]=pt[i+1].p;
                fp=parse(input[1].item,2,psv,psi);
                pt[i+1].y=pt[i].y+h/24*(f1-5*f2
                        +19*f3+9*fp);
        }
}
```

ODE System Solution

The initial-value problem can be extended into a system of ordinary differential equations by having an equivalent number of initial values. A system with three independent variables requires two equations and two initial values in order to produce unique solutions.

In Code10, a system with two ordinary differential equations is solved in ODE1System(). The code fragments are written based on Algorithm 10.5, and they are given as

```
void CCode10::ODE1System()
{
        int i,psi[6];
        double h,psv[6],tmp,max;
        double k1,k2,k3,k4;
        double K1,K2,K3,K4;
        pt[0].x=atof(input[3].item);
        pt[0].y=atof(input[4].item);
        pt[0].z=atof(input[5].item);
        h=atof(input[6].item);
        tmp=atof(input[7].item);
        m=(tmp-pt[0].x)/h; m=((m<M)?m:M);
        max=pt[0].x+(double)m*h;
        tmp=(tmp<max)?tmp:max;
        pt[m].x=tmp;
        psi[1]=23; psi[2]=24; psi[3]=25;
        for (i=0;i<=m;i++)
        {
                psv[1]=pt[i].x;
                psv[2]=pt[i].y;
                psv[3]=pt[i].z;
                k1=h*parse(input[1].item,3,psv,psi);
```

```
K1=h*parse(input[2].item,3,psv,psi);

psv[1]=pt[i].x+h/2;
psv[2]=pt[i].y+k1/2;
psv[3]=pt[i].z+K1/2;
k2=h*parse(input[1].item,3,psv,psi);
K2=h*parse(input[2].item,3,psv,psi);

psv[1]=pt[i].x+h/2;
psv[2]=pt[i].y+k2/2;
psv[3]=pt[i].z+K2/2;
k3=h*parse(input[1].item,3,psv,psi);
K3=h*parse(input[2].item,3,psv,psi);

psv[1]=pt[i].x+h;
psv[2]=pt[i].y+k3;
psv[3]=pt[i].z+K3;
k4=h*parse(input[1].item,3,psv,psi);
K4=h*parse(input[2].item,3,psv,psi);

if (i<m)
{
        pt[i+1].y=pt[i].y+(k1+2*k2+2*k3+k4)/6;
        pt[i+1].z=pt[i].z+(K1+2*K2+2*K3+K4)/6;
        pt[i+1].x=pt[i].x+h;
}
    }
}
```

RK4 is deployed in ODE1System() based on the solutions provided by Equations (10.18a) and (10.18b).

ODE2 to ODE1 Solution

The solution to the initial-value problem for the second-order ODE consists of reducing the equation into a system of two first-order ODEs. The two systems are then solved using RK4, using Algorithm 10.6. In Code10, a function called ODE2toODE1System() performs this task.

```
void CCode10::ODE2toODE1System()
{
        int i,psi[6];
        double psv[6];
        double h,tmp;
```

```
double k1,k2,k3,k4;
double K1,K2,K3,K4;
pt[0].x=atof(input[2].item);
pt[0].y=atof(input[3].item);
pt[0].z=atof(input[4].item);
h=atof(input[5].item);
tmp=atof(input[6].item);
m=(tmp-pt[0].x)/h; m=((m<M)?m:M);
pt[m].x=tmp;
psi[1]=23; psi[2]=24; psi[3]=25;
for (i=0;i<=m;i++)
{
        psv[1]=pt[i].x;
        psv[2]=pt[i].y;
        psv[3]=pt[i].z;
        k1=h*pt[i].z; K1=h*parse(input[1].item,3,psv,psi);

        psv[1]=pt[i].x+h/2;
        psv[2]=pt[i].y+k1/2;
        psv[3]=pt[i].z+K1/2;
        k2=h*(pt[i].z+K1/2); K2=h*parse(input[1].item,3,
                psv,psi);

        psv[1]=pt[i].x+h/2;
        psv[2]=pt[i].y+k2/2;
        psv[3]=pt[i].z+K2/2;
        k3=h*(pt[i].z+K2/2); K3=h*parse(input[1].item,
                3,psv,psi);

        psv[1]=pt[i].x+h;
        psv[2]=pt[i].y+k3;
        psv[3]=pt[i].z+K3;
        k4=h*(pt[i].z+K3); K4=h*parse(input[1].item,
                3,psv,psi);

        if (i<m)
        {
                pt[i+1].y=pt[i].y+(k1+2*k2+2*k3+k4)/6;
                pt[i+1].z=pt[i].z+(K1+2*K2+2*K3+K4)/6;
                pt[i+1].x=pt[i].x+h;
        }
}
}
```

ODE2 Finite-Difference 1 Solution

The boundary-value problem for the second-order ODE is solved using the finite-difference method. The solution is provided based on Algorithm 10.7. The operating function is ODE2FD1(), and this function represents the method in item 7 of the menu. The code fragments for this function are given as

```
void CCode10::ODE2FD1()
{
      int i,j,psi[6];
      double psv[6],h,tmp;
      double **a,*b;
      double *p,*q,*r,*w;
      b=new double [M+1];
      p=new double [M+1];
      q=new double [M+1];
      r=new double [M+1];
      w=new double [M+1];
      a=new double *[M+1];
      h=atof(input[5].item);
      m=(int)(atof(input[7].item)-atof(input[6].item))/h;
            m=((m<M)?m:M);
      pt[0].x=atof(input[6].item);
      pt[m].x=pt[0].x+(double)m*h;
      pt[0].y=atof(input[8].item);
      pt[m].y=atof(input[9].item);
      psi[1]=23;
      for (i=1;i<=m-1;i++)
      {
            pt[i].x=pt[i-1].x+h;
            psv[1]=pt[i].x;
            p[i]=parse(input[1].item,1,psv,psi);
            q[i]=parse(input[2].item,1,psv,psi);
            r[i]=parse(input[3].item,1,psv,psi);
            w[i]=parse(input[4].item,1,psv,psi);
      }
      for (i=0;i<=M;i++)
            a[i]=new double [M+1];
      for (i=1;i<=m-1;i++)
      {
            for (j=1;j<=m-1;j++)
                  a[i][j]=0;
            b[i]=0;
      }
      b[1]=w[1]-(p[1]/(h*h)-q[1]/(2*h))*pt[0].y;
```

```
        b[m-1]=w[m-1]-(p[m-1]/(h*h)+q[m-1]/(2*h))*pt[m].y;
        for (i=1;i<=m-1;i++)
        {
                a[i][i]=-2*p[i]/(h*h)+r[i];
                if (i<m-1)
                        a[i+1][i]=p[i+1]/(h*h)-q[i+1]/(2*h);
                if (i>1)
                        a[i-1][i]=p[i-1]/(h*h)+q[i-1]/(2*h);
                if (i>1 && i<m-1)
                        b[i]=w[i];
        }
        SolveSLE(a,b);
        for (i=0;i<=M;i++)
                delete a[i];
        delete p,q,r,w,a,b;
}
```

The input for the ordinary differential equation is obtained from Equation (10.21) in the form of $p(x), q(x), r(x)$, and $s(x)$. Their values are read as input strings from the edit boxes, and these strings are processed into numerical values through parse(). The following code segment implements this idea:

```
psi[1]=23;
for (i=1;i<=m-1;i++)
{
        pt[i].x=pt[i-1].x+h;
        psv[1]=pt[i].x;
        p[i]=parse(input[1].item,1,psv,psi);
        q[i]=parse(input[2].item,1,psv,psi);
        r[i]=parse(input[3].item,1,psv,psi);
        w[i]=parse(input[4].item,1,psv,psi);
}
```

Two main steps are involved in solving the boundary-value problem. First, a system of linear equations $Ay = b$ is to be formed using the finite-difference equation in Equation (10.22). Here, $A = [a_{ij}]$ and $b = [b_i]$ are determined using Equations (10.24a) and (10.24b), respectively. The code segment is given as

```
for (i=1;i<=m-1;i++)
{
        for (j=1;j<=m-1;j++)
                a[i][j]=0;
        b[i]=0;
}
```

```
b[1]=w[1]-(p[1]/(h*h)-q[1]/(2*h))*pt[0].y;
b[m-1]=w[m-1]-(p[m-1]/(h*h)+q[m-1]/(2*h))*pt[m].y;
for (i=1;i<=m-1;i++)
{
        a[i][i]=-2*p[i]/(h*h)+r[i];
        if (i<m-1)
                a[i+1][i]=p[i+1]/(h*h)-q[i+1]/(2*h);
        if (i>1)
                a[i-1][i]=p[i-1]/(h*h)+q[i-1]/(2*h);
        if (i>1 && i<m-1)
                b[i]=w[i];
}
```

The second step is to solve the system of linear equations. We apply the Gaussian elimination method by calling a function called SolveSLE(). Two arguments are supplied as input to this function, namely, $A = [a_{ij}]$ and $b = [b_i]$. SolveSLE() solves the problem and produces the solution as $y = [y_i]$, which is pt[i].y in the program. The function is shown as follows:

```
void CCode10::SolveSLE(double **a,double *b)
{
        int i,j,k,lo,hi;
        double m1,Sum;
        if (fMenu==7)
        {
                lo=1; hi=m-1;
        }
        if (fMenu==8)
        {
                lo=0; hi=m;
        }
        for (k=lo;k<=hi-1;k++)
                for (i=k+1;i<=hi;i++)
                {
                        m1=a[i][k]/a[k][k];
                        for (j=lo;j<=hi;j++)
                                a[i][j] -= m1*a[k][j];
                        b[i] -= m1*b[k];
                }
        for (i=hi;i>=lo;i--)
        {
                Sum=0;
                pt[i].y=0;
                for (j=i;j<=hi;j++)
```

```
                     Sum += a[i][j]*pt[j].y;
              pt[i].y=(b[i]-Sum)/a[i][i];
        }
}
```

SolveSLE() is shared by items 7 and 8 in the menu. The size of the matrices in the systems of linear equations in the two items are not the same. Two local variables called lo and hi have been introduced to represent the starting and ending indices of the elements in the matrices. The current system has a size of $(m - 1) \times (m - 1)$, whose rows and columns start with $i = 1$ to $i = m - 1$. Therefore, lo and hi are 1 and $m - 1$, respectively.

ODE2 Finite-Difference 2 Solution

Item 8 in the menu is represented by ODE2FD2(). This function solves the boundary-value problem in the second-order ODE whose boundary conditions are given in the form of first-order derivatives. Basically, ODE2FD2() is a little bit different from ODE2FD1() as it has to support the differentiated boundary conditions, which results in a $(m + 1) \times (m + 1)$ system of linear equations. The complete code for ODE2FD2() is given below:

```
void CCode10::ODE2FD2()
{
        int i,j,psi[6];
        double psv[6],tmp;
        double h,alpha,beta;
        double **a,*b;
        double *p,*q,*r,*w;
        b=new double [M+1];
        p=new double [M+1];
        q=new double [M+1];
        r=new double [M+1];
        w=new double [M+1];
        a=new double *[M+1];
        for (i=0;i<=M;i++)
                a[i]=new double [M+1];
        h=atof(input[5].item);
        m=(int)(atof(input[7].item)-atof(input[6].item))/h;
                m=((m<M)?m:M);
        pt[0].x=atof(input[6].item);
        pt[m].x=pt[0].x+(double)m*h;
        alpha=atof(input[8].item); beta=atof(input[9].item);
        for (i=0;i<=m;i++)
        {
                if (i<m)
                        pt[i+1].x=pt[i].x+h;
```

```
            psv[1]=pt[i].x; psi[1]=23;
            p[i]=parse(input[1].item,1,psv,psi);
            q[i]=parse(input[2].item,1,psv,psi);
            r[i]=parse(input[3].item,1,psv,psi);
            w[i]=parse(input[4].item,1,psv,psi);
    }
    for (i=0;i<=m;i++)
    {
            for (j=0;j<=m;j++)
                    a[i][j]=0;
            b[i]=0;
    }
    for (i=0;i<=m;i++)
    {
            a[i][i]=-2*p[i]/(h*h)+r[i];
            if (i>0 && i<m)
                    a[i][i+1]=p[i]/(h*h)+q[i]/(2*h);
            if (i<m-1)
                    a[i+1][i]=p[i+1]/(h*h)-q[i+1]/(2*h);
            if (i>0 && i<m)
                    b[i]=w[i];
    }
    a[0][1]=2*p[0]/m;
    a[m][m-1]=2*p[m]/(h*h);
    b[0]=w[0]+(p[0]/(h*h)-q[0]/(2*h))*2*h*alpha;
    b[m]=w[m]-(p[m]/(h*h)+q[m]/(2*h))*2*h*beta;
    SolveSLE(a,b);
    for (i=0;i<=M;i++)
            delete a[i];
    delete p,q,r,w,a,b;
}
```

As in item 7, the solution to the boundary-value problem consists of two main steps. First, a system of linear equations is created using the finite-difference equation of Equation (10.22). We have to solve $Ay = b$, and $A = [a_{ij}]$ and $b = [b_i]$ are obtained from Equations (10.28a), (10.28b), (10.28c), and (10.28d). The boundary conditions are given in the form of first-order derivatives with α and β as the left and right derivative values in the interval. Therefore, y_0 and y_m become the unknowns along with y_i for $i = 1, 2, \ldots, m - 1$.

There are $m + 1$ unknowns, and this creates a system of linear equations of size $(m + 1) \times (m + 1)$. From Equations (10.28a), (10.28b), (10.28c), and (10.28d), we obtain the code segment for creating $A = [a_{ij}]$ and $b = [b_i]$:

```
for (i=0;i<=m;i++)
```

```
{
    a[i][i]=-2*p[i]/(h*h)+r[i];
    if (i>0 && i<m)
        a[i][i+1]=p[i]/(h*h)+q[i]/(2*h);
    if (i<m-1)
        a[i+1][i]=p[i+1]/(h*h)-q[i+1]/(2*h);
    if (i>0 && i<m)
        b[i]=w[i];
}
a[0][1]=2*p[0]/m;
a[m][m-1]=2*p[m]/(h*h);
b[0]=w[0]+(p[0]/(h*h)-q[0]/(2*h))*2*h*alpha;
b[m]=w[m]-(p[m]/(h*h)+q[m]/(2*h))*2*h*beta;
```

The system of linear equations is solved using `SolveSLE()`. The unknowns in the $(m + 1) \times (m + 1)$ system are y_i for $i = 0, 1, \ldots, m$, and this prompts `lo=0` and `hi=m` in `SolveSLE()`. The solution is provided as `pt[i].y`, which represents $y = [y_i]$.

10.13 SUMMARY

The chapter discusses the numerical solutions to the first- and second-order ordinary differential equations. The first-order differential equations involve initial-value problems, and their numerical solutions consist of the Taylor series, Runge–Kutta of order 2, Runge–Kutta of order 4, and the Adam–Bashforth–Moulton methods. Problems in the second-order ordinary differential equations include both the initial-value and boundary-value problems. Their initial-value solutions are derived from the first-order methods, whereas the boundary-value problems are solved using the finite-difference methods.

Several creative projects can be embarked from our discussion. Problems in ordinary differential equations are commonly encountered in science and engineering. The problems are expressed in the form of models involving ordinary differential equations, whereas simulations are performed to support these theoretical models. In most cases, visualization from the solution is necessary to support the work. Therefore, programming with a friendly graphical user interface is needed for visualizing the results from the applications.

NUMERICAL EXERCISES

The following exercises are intended to test your understanding of the materials discussed in this chapter. Use six decimal places for calculations.

1. Solve and compare the results from the following initial-value problems using the Taylor series method of order 2 and 3:

   a. $y' = 3 \sin x - 4 \cos x$, for $0 \leq x \leq 1$, $h = \Delta x = 0.25$ and $y(0) = -1$.

b. $y' = 3x^2y - 4y$, for $0 \le x \le 1$, $h = \Delta x = 0.25$ and $y(0) = -1$.

c. $y' = 3 \sin xy$, for $0 \le x \le 1$, $h = \Delta x = 0.25$ and $y(0) = -1$.

Run Code10 to check the results from your work.

2. Solve and compare the results from the following initial-value problems using RK2 (with $r = 0.8$) and RK4:

a. $y' = 3 \sin x - 4 \cos x$, for $0 \le x \le 1$, $h = \Delta x = 0.25$ and $y(0) = -1$.

b. $y' = 3x^2y - 4y$, for $0 \le x \le 1$, $h = \Delta x = 0.25$ and $y(0) = -1$.

c. $y' = 3 \sin xy$, for $0 \le x \le 1$, $h = \Delta x = 0.25$ and $y(0) = -1$.

d. $3xy' - 2y^2 + 4x = -1$, for $0 \le x \le 1$, $h = \Delta x = 0.25$ and $y(0) = -1$.

Run Code10 to check the results from your work.

3. Solve the following initial-value problems using the Adams–Bashforth–Moulton method with RK4 as their starting predictor values:

a. $y' = 3 \sin x - 4 \cos x$, for $0 \le x \le 1$, $h = \Delta x = 0.25$ and $y(0) = -1$.

b. $y' = 3x^2y - 4y$, for $0 \le x \le 1$, $h = \Delta x = 0.25$ and $y(0) = -1$.

c. $y' = 3 \sin xy$, for $0 \le x \le 1$, $h = \Delta x = 0.25$ and $y(0) = -1$.

d. $3xy' - 2y^2 + 4x = -1$, for $0 \le x \le 1$, $h = \Delta x = 0.25$ and $y(0) = -1$.

Run Code10 to check the results from your work.

4. Solve the following initial-value problems for the system of first-order ordinary differential equations using RK4:

a. $y' = 2x - 3y$ and $z' = 3z + 2x$, for $0 \le x \le 1$. The initial values are $y(0) = -1$ and $z(0) = -2$.

b. $y' = 2 \cos y - 3 \sin x$ and $z' = 1 - 2 \sin xyz$, for $0 \le x \le 1$. The initial values are $y(0) = -0.7$ and $z(0) = 0.2$.

Run Code10 to check the results from your work.

5. Solve the following initial-value problems for the second-order ordinary differential equations using RK4 by reducing the equations to a system of first-order ordinary differential equations:

a. $y'' = 3y' + y - x - 1$, for $0 \le x \le 1$ and $h = \Delta x = 0.25$. The initial values are $y(0) = 0.5$ and $y'(0) = -0.5$.

b. $y'' = 3x^2y' + 2xy - 3x - 2$, for $0 \le x \le 1$ and $h = \Delta x = 0.25$. The initial values are $y(0) = 0.5$ and $y'(0) = -0.5$.

c. $y'' = (3 \sin x)y' + (2 \cos x)y - x - 1$, for $0 \le x \le 1$ and $h = \Delta x = 0.25$. The initial values are $y(0) = 0.5$ and $y'(0) = -0.5$.

Run Code10 to check the results from your work.

6. Solve the boundary-value problems for the second-order ordinary differential equations using the finite-difference method, given by

a. $y'' = 3y' + y - x - 1$, for $0 < x < 1$, and $h = \Delta x = 0.25$, $y(0) = -1$ and $y(1) = -0.5$.

b. $y'' = 3x^2y' + 2xy - 3x - 2$, for $0 < x < 2$, and $h = \Delta x = 0.5$, $y(0) = 0$ and $y(2) = -2$.

c. $2y'' + 3x^2y' - 3xy - 4x = 1$, for $-1 < x < 1$, and $h = \Delta x = 0.5$, $y(-1) = -1$ and $y(1) = -0.5$.

d. $y'' = (3\sin x)y' + (2\cos x)y - x - 1$, for $0 < x < 1$, and $h = \Delta x = 0.25$, $y'(0) = -0.5$ and $y'(1) = 1$.

e. $y'' = (3\sin x)y' + (2\cos x)y - x - 1$, for $0 < x < 1$, and $h = \Delta x = 0.25$, $y'(0) = -1$ and $y(1) = -0.5$.

f. $y'' = (3\sin x)y' + (2\cos x)y - x - 1$, for $0 < x < 1$, and $h = \Delta x = 0.25$, $y(0) - y'(0) = -1$ and $y(1) + 2y'(1) = -0.5$.

Run `Code10` to check the results from your work in problems (a), (b), (c), and (d).

PROGRAMMING CHALLENGES

1. Modify the `Code10` project by improving on the module in the Taylor series method to allow the user to determine the order of the series from two to five. This improvement requires new edit boxes for collecting the additional derivatives as input to the method.

2. Modify the `Code10` project by improving on the module in the Adams–Bashforth–Moulton method to allow the user to determine the step size in both the predictor and the corrector functions according to the given equations in Tables 10.4 and 10.5.

3. Modify the `Code10` project by improving on the finite-difference method for the second-order ordinary differential equations to support mixed boundary-value conditions. For example, the boundary values may be given in the form of Problems 6(e) and (f) in the above numerical exercises.

4. Modify the `Code10` project to add some important features, such as file open and retrieve options. These features are important because data are normally placed in separate files from the program files.

CHAPTER 11

Partial Differential Equations

11.1 INTRODUCTION

Partial differential equation (PDE) is an equation that has one or more partial derivatives as independent variables in its terms. Some typical examples are

$$3\sin x \frac{\partial f}{\partial x} + 4y^2 x \left(\frac{\partial f}{\partial y}\right)^3 = \cos xy,$$

$$3\frac{\partial^2 f}{\partial x^2} - 2\frac{\partial^2 f}{\partial y^2} - 3x\frac{\partial f}{\partial x} + 4y\frac{\partial f}{\partial y} = -1.$$

The *order* of a partial differential equation is defined as the highest partial derivative of the terms in the equation. Therefore, the first example above is the first-order PDE, whereas the second is the second-order PDE. The *degree* of a partial differential equation is defined as the power of the highest derivative term in the equation. It can be verified from the definition that the two equations above have the degrees of three and one, respectively.

In general, a partial differential equation of order n having m variables x_i for $i = 1, 2, \ldots, m$ is expressed as

$$f\left(x_1, x_2, \ldots, x_m, u(x_1, x_2, \ldots, x_m), \frac{\partial u}{\partial x_1}, \ldots, \frac{\partial u}{\partial x_m}, \frac{\partial^2 u}{\partial x_1^2}, \ldots, \frac{\partial^2 u}{\partial x_m^2}, \ldots, \frac{\partial^n u}{\partial x_1^n}, \ldots, \frac{\partial^n u}{\partial x_m^n}\right) = 0.$$
(11.1)

In the above equation, $\frac{\partial u}{\partial x}$ is simply u_x in its compact form, whereas $\frac{\partial^2 u}{\partial x^2}$ and $\frac{\partial^2 u}{\partial x \partial y}$ are u_{xx} and u_{xy}, respectively.

In this chapter, we will concentrate on numerical problems involving second-order partial differential equations only. A second-order partial differential equation with variables x_1, x_2, and $u(x_1, x_2)$ has the following general form:

$$f\left(x_1, x_2, u(x_1, x_2), \frac{\partial u}{\partial x_1}, \frac{\partial u}{\partial x_2}, \frac{\partial^2 u}{\partial x_1^2}, \frac{\partial^2 u}{\partial x_1 \partial x_2}, \frac{\partial^2 u}{\partial x_2^2}\right) = 0. \tag{11.2}$$

Table 11.1 lists some very common second-order partial differential equations. The equations are common in many engineering applications such as heat distribution in a plate, wave propagation, and electromagnetic fields. A second-order PDE $u(x, y)$ is said to be *linear* if it can be written into the following form:

$$A\frac{\partial^2 u}{\partial x^2} + B\frac{\partial^2 u}{\partial x \partial y} + C\frac{\partial^2 u}{\partial y^2} + D\frac{\partial u}{\partial x} + E\frac{\partial u}{\partial y} + Fu + G = 0, \tag{11.3}$$

where A, B, C, D, E, F, and G are constants. A linear second-order PDE can further be classified as elliptic, parabolic, and hyperbolic according to the following rules:

1. If $B^2 - 4AC < 0$ or $Z \equiv \begin{bmatrix} A & B \\ B & C \end{bmatrix}$ is positive definite, then the equation is *elliptic*.
2. If $B^2 - 4AC = 0$, then the equation is *parabolic*.
3. If $B^2 - 4AC > 0$, then the equation is *hyperbolic*.

TABLE 11.1. Some common second-order PDEs

Equation	Description
$\frac{\partial^2 u}{\partial x^2} + \frac{\partial^2 u}{\partial y^2} = 0$	Laplace's equation
$\frac{\partial^2 u}{\partial x^2} + \frac{\partial^2 u}{\partial y^2} = w(x, y)$	Poisson's equation
$\frac{\partial^2 u}{\partial x^2} - \kappa\frac{\partial^2 u}{\partial t^2} = 0$, for $\kappa > 0$	Wave equation
$\sigma\frac{\partial^2 u}{\partial x^2} - \frac{\partial u}{\partial t} = 0$	Diffusion equation
$\frac{\partial u}{\partial t} - \alpha^2\frac{\partial^2 u}{\partial x^2} = 0$	Heat equation
$\frac{\partial^2 u}{\partial x^2} + \frac{\partial^2 u}{\partial y^2} + f(x, y)u = g(x, y)$	Helmholtz's equation
$\frac{\partial^2 u}{\partial t^2} = a^2\frac{\partial^2 u}{\partial x^2} - bu$	Klein-Gordon's equation
$\frac{\partial^2 u}{\partial t^2} = a^2\frac{\partial^2 u}{\partial x^2} + be^{\beta u}$	Modified Loiuville's equation
$\frac{1}{k^2}u_{tt} = u_{xx} + u_{yy}$	Vibrating membrane

It can be easily verified from the rules that Laplace's, Poissson's, and Helmholtz's equations from Table 11.1 are good examples of elliptic equations. Similarly, heat equation is parabolic, whereas Klein–Gordon's and wave equations are hyperbolic.

Boundary-Value Problem

Mathematical models using partial differential equations are mostly based on their boundary conditions. A typical boundary-value problem for a given function $u(x, y)$ that is continuous in a rectangular domain bounded by $a \leq x \leq b$ and $c \leq y \leq d$ has its values defined in at least one pair of its sides. There are two types of boundary conditions for this problem:

   a. Dirichlet boundary conditions, which are stated as the given values at the ends of one of the intervals, for example, $u(a, y) = f_1(y)$ and $u(b, y) = f_2(y)$ in $a \leq x \leq b$.

   b. Neumann boundary conditions, which are stated as the given values of the first derivatives at the ends of one of the intervals. For example, $u_x(a, y) = f_3(y)$ and $u_x(b, y) = f_4(y)$ in $a \leq x \leq b$ are the Neumann boundary conditions.

The initial conditions at one of the parameters may also arise in a boundary-value problem. For example, a two-dimensional heat equation has $u(x, t)$ defined in $a \leq x \leq b$ and $c \leq t \leq d$, where t represents time. The boundary conditions for this problem are stated as $u(a, t) = f_1(t)$ and $u(b, t) = f_2(t)$, where $f_1(t)$ and $f_2(t)$ are functions governing the left and right sides of the domain, respectively. The initial condition is $u(x, c) = g(x)$, which suggests the initial heat value depends on its location x from the origin.

The numerical solution to a partial differential equation problem is an approximated approach based on some finite points in the domain of the problem. The most common approaches include finite-difference, finite element, and boundary element methods. Finite difference applies to cases where the boundaries are fixed, and the subintervals within the boundaries have uniform width. The finite-element method is more flexible as it can be applied on non-uniform elements in the problem. The boundary element method applies to linear PDEs, which are formulated as integral equations. In this method, calculations are made to evaluate the values along the boundaries only as defined by its governing partial differential equation.

The easiest approach for solving the boundary-value problems involving partial differential equations is the finite-difference method. This method is based on the same concept as in the ordinary differential equation, that is, by dividing the given interval into several equal-width subintervals and substituting the derivatives in order to yield the finite-difference formula for the problem. We will focus on the finite-difference methods for several boundary-value problems in this chapter.

The solution to a boundary-value problem from the partial differential equations can be expressed as an explicit or implicit form. The explicit form of the solution is a straightforward method that has its values evaluated directly from the finite-difference equation. The implicit form is more difficult as it requires the reduction of the problem into a system of linear equations before its solution is obtained.

Central-Difference Rules

In the finite-difference method, solutions are obtained by replacing the given first and second derivatives into their approximated form using the central-difference rules. The rules for partial derivatives are merely the extension from the full derivatives derived from the Taylor series expansion of the terms.

Figure 11.1 shows the finite points in the rectangular grids. In approximating the partial derivatives, the grids are divided into horizontal and vertical subintervals whose widths are h and k, respectively. A point $u(x, y)$ is written as $u(x_i, y_j)$, where the subscripts i and j denote its x and y positions, respectively. $u(x_{i+1}, y_j)$ is one unit in front of $u(x_i, y_j)$, whereas $u(x_{i-1}, y_j)$ is one unit behind. Similarly, $u(x_i, y_{j+1})$ refers to a point above $u(x_i, y_j)$, whereas $u(x_i, y_{j-1})$ is one unit below. For simplicity, $u(x_i, y_j)$ is referred simply as $u_{i,j}$, $u(x_{i+1}, y_j)$ as $u_{i+1,j}$, and so on.

For $u(x, y)$ continuous and defined in a rectangular domain whose horizontal and vertical grids have the widths of h and k, respectively, the derivatives at the finite point (x_i, y_j) are approximated using the central-difference rules from Equations (8.5a) and (8.5b) as

$$\left. \frac{\partial u}{\partial x} \right|_{(x_i, y_j)} \approx \frac{u(x_{i+1}, y_j) - u(x_{i-1}, y_j)}{2h}, \tag{11.4a}$$

$$\left. \frac{\partial u}{\partial y} \right|_{(x_i, y_j)} \approx \frac{u(x_i, y_{j+1}) - u(x_i, y_{j-1})}{2k}, \tag{11.4b}$$

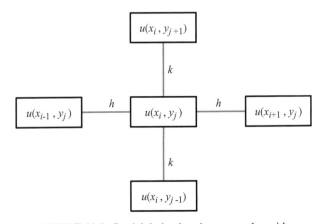

FIGURE 11.1. Partial derivatives in rectangular grids.

$$\frac{\partial^2 u}{\partial x^2}\bigg|_{(x_i,y_j)} \approx \frac{u(x_{i+1}, y_j) - 2u(x_i, y_j) + u(x_{i-1}, y_j)}{h^2}, \qquad (11.4c)$$

$$\frac{\partial^2 u}{\partial y^2}\bigg|_{(x_i,y_j)} \approx \frac{u(x_i, y_{j+1}) - 2u(x_i, y_j) + u(x_i, y_{j-1})}{k^2}. \qquad (11.4d)$$

The partial derivatives given by Equations (11.4a), (11.4b), (11.4c), and (11.4d) are merely the two-dimensional extension of the full derivatives discussed earlier in Chapter 8. The partial derivative of $u(x, y)$ with respect to the variable x is obtained by differentiating that variable only. This means only the i subscript is involved. Similarly, the partial derivative with respect to y involves the subscript j only.

11.2 POISSON'S EQUATION

The elliptic partial differential equation has many applications in areas such as harmonic analysis in potential theory. The most common types of elliptic equations are the Poisson's and Laplace's equations. The Poisson's equation has many applications, such as in the two-dimensional heat flow modeling, given by

$$\nabla^2 u = -\frac{1}{\kappa} q(x, y),$$

where the constant κ is the conductivity and $q(x, y)$ is the rate of heat distribution per unit area.

Definition 11.1. Poisson's equation is stated as $\nabla^n u = w(x_1, x_2, \dots, x_n)$, where ∇^n is the sum of the nth partial derivative with respect to the variables.

In the above definition, ∇ is an operator, pronounced as *del*, which denotes the *potential function*. In the two-variable case of x and y, ∇ is the first partial derivative defined as

$$\nabla = \frac{\partial}{\partial x} + \frac{\partial}{\partial y}.$$

The one-dimensional form of the Poisson's equation with variables x and $u(x)$ in the interval $a < x < b$ is given by

$$\frac{\partial^2 u}{\partial x^2} = \frac{d^2 u}{dx^2} = w(x).$$

Similarly, $\nabla^2 = \frac{\partial^2}{\partial x^2} + \frac{\partial^2}{\partial y^2}$ and $\nabla^2 = \frac{\partial^2}{\partial x^2} + \frac{\partial^2}{\partial y^2} + \frac{\partial^2}{\partial z^2}$ are the Poisson's equations in two and three variables, respectively.

The Poisson's equation with two independent variables, x and y, and two dependent variables, $u(x, y)$ and $w(x, y)$, is given by

$$\nabla^2 u = \frac{\partial^2 u}{\partial x^2} + \frac{\partial^2 u}{\partial y^2} = w(x, y). \tag{11.5}$$

The domain Ω in the Poisson's equation problem consists of a rectangular region defined as $\{\Omega : a < x < b, c < y < d\}$. Both $u(x, y)$ and $w(x, y)$ are assumed to be continuous in this rectangular region. The objective in the boundary-value problem for the Poisson's equation is to find the values of $u(x, y)$ inside Ω, given its values along the sides of the rectangular region.

Figure 11.2 shows Ω, which is made from $m \times n$ rectangular grids with $m = 4$ and $n = 5$ in the boundary-value problem. Discrete coordinates are used to label the points in the grids with $u_{i,j}$ representing the value of $u(x_i, y_j)$. The points are labeled starting from the bottom left-hand corner as their origin whose coordinates are given by $u_{0,0} = u(x_0, y_0)$. This coordinates system is slightly different from the normal Cartesian representation as the subscripts i and j in this system are assumed to be positive integers, or zero only.

The points in Figure 11.2 are labeled according to their coordinates with black squares denoting the given boundary and initial values and white squares for the unknowns in the problem. The widths of the intervals are determined, as follows:

$$\text{horizontal interval, } h = \Delta x = \frac{x_n - x_0}{n},$$

$$\text{vertical interval, } k = \Delta y = \frac{y_m - y_0}{m}.$$

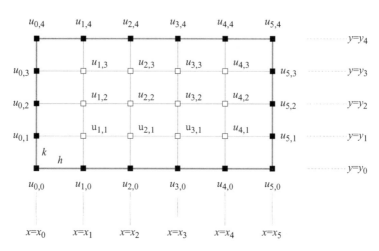

FIGURE 11.2. The values of $u(x, y)$ in the rectangular grids.

The boundaries for Equation (11.5) are the left and right sides of the rectangles, as shown in Figure 11.2. The Dirichlet boundary conditions are clearly functions of x, given by

$$u(x, y_0) = f_1(x), \quad \text{for } x_0 < x < x_n,$$

$$u(x, y_m) = f_2(x), \quad \text{for } x_0 < x < x_n.$$

The vertical sides of the rectangle in Figure 11.2 are the initial conditions. The bottom side of the rectangle denotes the initial values, whereas the top side indicates the end values. In the Poisson's equation, these values are functions of y represented by $g_1(y)$ and $g_2(y)$, as follows:

$$u(x_0, y) = g_1(y), \quad \text{for } y_0 \le y \le y_m,$$

$$u(x_n, y) = g_2(y), \quad \text{for } y_0 \le y \le y_m.$$

These initial values cannot be mistakenly assumed as the boundary conditions because the quantity y may be replaced by t, which represents time. A time-dependent variable normally starts at a given value and ends at another given value.

We discuss the finite-difference method for solving the boundary-value problem whose domain is $\{\Omega : x_0 < x < x_n, \ y_0 < y < y_m\}$. The numerical solution to the boundary-value problem involving the Poisson's equation consists of two main steps, as depicted in Figure 11.3. First, the partial differential equation is reduced to its finite-difference form by eliminating its first and second derivatives using the central-difference rules. From the finite-difference formula, a system of linear equations is obtained by substituting the initial and boundary values. A problem from m rows and n columns in the rectangular grid results in $(m - 1)(n - 1)$ unknowns, which generate a system of linear equations of size $(m - 1)(n - 1) \times (m - 1)(n - 1)$.

The second step is solving the system of linear equations using any suitable method as discussed in Chapter 5, such as the Gaussian elimination and Crout methods. This step produces the values of $u_{i,j} = u(x_i, y_j)$ for $1 \le j \le m - 1$ and $1 \le i \le n - 1$.

We start with the first step by finding the finite-difference formula for the problem. This is achieved by eliminating the partial derivatives using the central-difference differentiation rules:

$$\left.\frac{\partial^2 u}{\partial x^2}\right|_{(x_i, y_j)} + \left.\frac{\partial^2 u}{\partial y^2}\right|_{(x_i, y_j)} = w_{i,j},$$

$$\frac{u_{i+1,j} - 2u_{i,j} + u_{i-1,j}}{h^2} + \frac{u_{i,j+1} - 2u_{i,j} + u_{i,j-1}}{k^2} = w_{i,j},$$

$$k^2\left(u_{i+1,j} - 2u_{i,j} + u_{i-1,j}\right) + h^2\left(u_{i,j+1} - 2u_{i,j} + u_{i,j-1}\right) = h^2 k^2 w_{i,j}.$$

Rearranging the terms, we arrive at the finite-difference formula:

$$h^2 u_{i,j-1} + k^2 u_{i-1,j} - 2(h^2 + k^2)u_{i,j} + k^2 u_{i+1,j} + h^2 u_{i,j+1} = h^2 k^2 w_{i,j}. \quad (11.6)$$

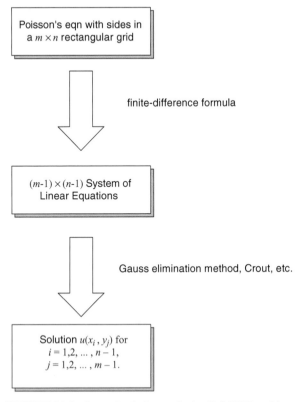

FIGURE 11.3. General solution to the implicit PDE problem.

Graphically, Equation (11.6) can be represented as a molecule, as shown in Figure 11.4. This molecule can be mapped into the rectangular grids starting at the bottom left-hand corner of the rectangle to generate a system of linear equations. The initial position for this mapping is achieved by setting $i = 1$ and $j = 1$. The molecule is moved from left to right until it reaches the right side of the rectangular, then continues to the next level, and so on until the right side of the topmost side is reached. At each point in (i, j), an equation is generated. Therefore, a total of $(m - 1)(n - 1)$ equations are generated, and the number matches the number of unknowns in the problem.

In forming the equations, the terms involving the sides of the rectangle can be evaluated using the boundary and initial conditions. Therefore, these terms are moved to the right-hand side of the equations. Referring to the rectangular grids in Figure 11.2, the following results are obtained in the case of a $m \times n$ rectangular domain:

At level $j = 1$,

Left node, $i = 1$:

$$h^2 u_{1,0} + k^2 u_{0,1} - 2(h^2 + k^2)u_{1,1} + k^2 u_{2,1} + h^2 u_{1,2} = h^2 k^2 w_{1,1},$$

$$-2(h^2 + k^2)u_{1,1} + k^2 u_{2,1} + h^2 u_{1,2} = h^2 k^2 w_{1,1} - h^2 u_{1,0} - k^2 u_{0,1}.$$

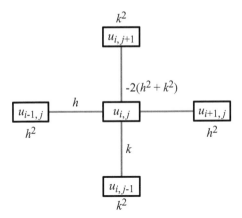

FIGURE 11.4. Finite-difference molecule for the Poisson's equation.

Interior node:

$$h^2 u_{i,0} + k^2 u_{i-1,1} - 2(h^2 + k^2)u_{i,1} + k^2 u_{i+1,1} + h^2 u_{i,2} = h^2 k^2 w_{i,1},$$
$$k^2 u_{i-1,1} - 2(h^2 + k^2)u_{i,1} + k^2 u_{i+1,1} + h^2 u_{i,2} = h^2 k^2 w_{i,1} - h^2 u_{i,0}.$$

Right node, $i = n - 1$:

$$h^2 u_{n-1,0} + k^2 u_{n-2,1} - 2(h^2 + k^2)u_{n-1,1} + k^2 u_{n,1} + h^2 u_{n-1,2} = h^2 k^2 w_{n-1,1},$$
$$k^2 u_{n-2,1} - 2(h^2 + k^2)u_{n-1,1} + h^2 u_{n-1,2} = h^2 k^2 w_{n-1,1} - h^2 u_{n-1,0} - k^2 u_{n,1}.$$

At the middle level,
Left node, $i = 1$:

$$h^2 u_{1,j-1} + k^2 u_{0,j} - 2(h^2 + k^2)u_{1,j} + k^2 u_{2,j} + h^2 u_{1,j+1} = h^2 k^2 w_{1,j},$$
$$h^2 u_{1,j-1} - 2(h^2 + k^2)u_{1,j} + k^2 u_{2,j} + h^2 u_{1,j+1} = h^2 k^2 w_{1,j} - k^2 u_{0,j}.$$

Interior node:

$$h^2 u_{i,j-1} + k^2 u_{i-1,j} - 2(h^2 + k^2)u_{i,j} + k^2 u_{i+1,j} + h^2 u_{i,j+1} = h^2 k^2 w_{i,j}.$$

Right node, $i = n - 1$:

$$h^2 u_{n-1,j-1} + k^2 u_{n-2,j} - 2(h^2 + k^2)u_{n-1,j} + k^2 u_{n,j} + h^2 u_{n-1,j+1} = h^2 k^2 w_{n-1,j},$$
$$h^2 u_{n-1,j-1} + k^2 u_{n-2,j} - 2(h^2 + k^2)u_{n-1,j} + h^2 u_{n-1,j+1} = h^2 k^2 w_{n-1,j} - k^2 u_{n,j}.$$

At level $j = m - 1$,

Left node, $i = 1$:

$$h^2 u_{1,m-2} + k^2 u_{0,m-1} - 2(h^2 + k^2)u_{1,m-1} + k^2 u_{2,m-1} + h^2 u_{1,m} = h^2 k^2 w_{1,m-1},$$
$$h^2 u_{1,m-2} - 2(h^2 + k^2)u_{1,m-1} + k^2 u_{2,m-1} = h^2 k^2 w_{1,m-1} - k^2 u_{0,m-1} - h^2 u_{1,m}.$$

Interior node:

$$h^2 u_{i,m-2} + k^2 u_{i-1,m-1} - 2(h^2 + k^2)u_{i,m-1} + k^2 u_{i+1,m-1} + h^2 u_{i,m} = h^2 k^2 w_{i,m-1},$$
$$h^2 u_{i,m-2} + k^2 u_{i-1,m-1} - 2(h^2 + k^2)u_{i,m-1} + k^2 u_{i+1,m-1} = h^2 k^2 w_{i,m-1} - h^2 u_{i,m}.$$

Right node, $i = n - 1$:

$$h^2 u_{n-1,m-2} + k^2 u_{n-2,m-1} - 2(h^2 + k^2)u_{n-1,m-1} + k^2 u_{n,m-1} + h^2 u_{n-1,m}$$
$$= h^2 k^2 w_{n-1,m-1},$$
$$h^2 u_{n-1,m-2} + k^2 u_{n-2,m-1} - 2(h^2 + k^2)u_{n-1,m-1}$$
$$= h^2 k^2 w_{n-1,m-1} - k^2 u_{n,m-1} - h^2 u_{n-1,m}.$$

Applying the approach to a special case with $m = 3$ and $n = 3$ results in a system of linear equations of size 9×9, as follows:

$$
\begin{bmatrix}
-2(h^2+k^2) & k^2 & 0 & h^2 & 0 & 0 & 0 & 0 & 0 \\
k^2 & -2(h^2+k^2) & k^2 & 0 & h^2 & 0 & 0 & 0 & 0 \\
0 & k^2 & -2(h^2+k^2) & 0 & 0 & h^2 & 0 & 0 & 0 \\
h^2 & 0 & 0 & -2(h^2+k^2) & k^2 & 0 & h^2 & 0 & 0 \\
0 & h^2 & 0 & k^2 & -2(h^2+k^2) & k^2 & 0 & h^2 & 0 \\
0 & 0 & h^2 & 0 & k^2 & -2(h^2+k^2) & 0 & 0 & h^2 \\
0 & 0 & 0 & h^2 & 0 & 0 & -2(h^2+k^2) & k^2 & 0 \\
0 & 0 & 0 & 0 & h^2 & 0 & k^2 & -2(h^2+k^2) & k^2 \\
0 & 0 & 0 & 0 & 0 & h^2 & 0 & k^2 & -2(h^2+k^2)
\end{bmatrix}
\begin{bmatrix}
u_{1,1} \\ u_{2,1} \\ u_{3,1} \\ u_{1,2} \\ u_{2,2} \\ u_{3,2} \\ u_{1,3} \\ u_{2,3} \\ u_{3,3}
\end{bmatrix}
$$

$$
=
\begin{bmatrix}
h^2 k^2 w_{1,1} - h^2 u_{1,0} - k^2 u_{0,1} \\
h^2 k^2 w_{2,1} - h^2 u_{2,0} \\
h^2 k^2 w_{3,1} - h^2 u_{3,0} - k^2 u_{4,1} \\
h^2 k^2 w_{1,2} - k^2 u_{0,2} \\
h^2 k^2 w_{2,2} \\
h^2 k^2 w_3 - k^2 u_{4,2} \\
h^2 k^2 w_{1,3} - k^2 u_{0,3} - h^2 u_{1,4} \\
h^2 k^2 w_{2,3} - h^2 u_{2,4} \\
h^2 k^2 w_{3,3} - k^2 u_{4,3} - h^2 u_{3,4}
\end{bmatrix}.
$$

From the above results, we obtain the formulation for the system of linear equations $A\mathbf{u} = \mathbf{b}$, where $A = [a_{ij}]$ is the coefficient matrix, $\mathbf{b} = [b_i]$, and $\mathbf{u}$ is the vector for the unknowns. The values of the elements in A are obtained by substituting the boundary and initial conditions into the finite-difference equation of Equation (11.6). They are determined, as follows:

$$a_{i,j} = \begin{cases} -2(h^2 + k^2) & \text{for } i = j, \\ h^2 & \text{for } a_{i,n-1+i} = a_{n-1+j,j}, \\ k^2 & \text{if } i \bmod (n-1) \neq 0 \text{ and } i < (m-1)(n-1), \\ 0 & \text{elsewhere,} \end{cases} \tag{11.7a}$$

$$b_r = h^2 k^2 w_{ij}, \tag{11.7b}$$

where $r = i + (j-1)(n-1)$. Algorithm 11.1 summarizes the finite-difference method for the Poisson's equation problem. This algorithm is illustrated using Example 11.1.

Algorithm 11.1. Finite-difference method for the Poisson's equation.
Given $m, n, h = \Delta x$ and $k = \Delta y$;
Given $\nabla^2 u = w(x, y)$ in $x_0 < x < x_n$ and $y_0 < y < y_m$;
Given the boundary values, $u(x_0, y) = g_1(y)$ and $u(x_n, y) = g_2(y)$;
Given the initial values, $u(x, y_0) = f_1(x)$ and $u(x, y_m) = f_2(x)$;
Evaluate $m = (y_m - y_0)/k$ and $n = (x_n - x_0)/h$;
Evaluate $u_{0,j}$ and $u_{n,j}$ for $j = 1, 2, \ldots, m-1$;
Evaluate $u_{i,0}$ and $u_{i,m}$ for $i = 1, 2, \ldots, n-1$;
Find the finite-difference formula using Equation (11.6);
Find matrix A in $Au = b$, as follows:
  For $r = 1$ to $(m-1)(n-1)$
    $a_{rr} = -2(h^2 + k^2)$;
    If $r \leq 2(n-1)$
      $a_{r,n-1+r} = a_{n-1+r,r} = h^2$;
    Endif
    If $r\%(n-1)! = 0$ and $r < (m-1)(n-1)$
      $a_{r,r+1} = a_{r+1,r} = k^2$;
    Endif
  Endfor
Find vector b in $Au = b$ as follows:
  For i=1 to n-1
    For j=1 to m-1
      $r = i + (j-1)(n-1)$;
      $b_r = h^2 k^2 w_{ij}$;
    Endfor
  Endfor
Solve for u in $Au = b$;

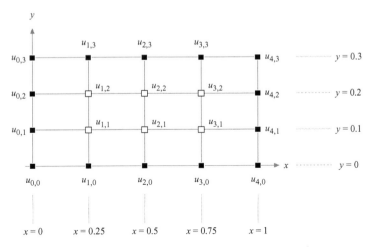

FIGURE 11.5. The rectangular domain in the example.

Example 11.1. Solve the Poisson's equation given by $u_{xx} + u_{yy} = 10xy$, for $0 < x < 1$ and $0 < y < 0.3$, with $h = \Delta x = 0.25$ and $k = \Delta y = 0.1$. The boundary values in this problem are $u(0, y) = y \sin y$, and $u(1, y) = e^{-y}$, for $0 < y < 0.3$, whereas the initial values are $u(x, 0) = xe^{-x}$ and $u(x, 0.3) = 1 - x$, for $0 \le x \le 1$.

Solution. From the given values, we obtain $m = (0.3 - 0)/0.1 = 3$ and $n = (1 - 0)/0.25 = 4$. These values result in a rectangular domain, as shown in Figure 11.5. The values in the left boundaries are satisfied from $u(0, y) = y \sin y$:

$$u_{0,1} = u(0, 0.1) = 0.1 \sin 0.1 = 0.009983,$$

$$u_{0,2} = u(0, 0.2) = 0.2 \sin 0.2 = 0.0397339.$$

The values in the right boundaries are determined from $u(1, y) = e^{-y}$:

$$u_{4,1} = u(1, 0.1) = e^{-0.1} = 0.904837,$$

$$u_{4,2} = u(1, 0.2) = e^{-0.2} = 0.818731.$$

From the initial values, $u(x, 0) = xe^{-x}$:

$$u_{1,0} = u(0.25, 0) = 0.194700,$$

$$u_{2,0} = u(0.5, 0) = 0.303265,$$

$$u_{3,0} = u(0.75, 0) = 0.354275.$$

Also, $u(x, 0.3) = 1 - x$:

$$u_{1,3} = u(0.25, 0.3) = 1 - 0.25 = 0.75,$$
$$u_{2,3} = u(0.5, 0.3) = 1 - 0.5 = 0.5,$$
$$u_{3,3} = u(0.75, 0.3) = 1 - 0.75 = 0.25.$$

Applying Equation (11.6) we get the finite-difference formula, as follows:

$$k^2 u_{i+1,j} - 2(k^2 + h^2)u_{i,j} + k^2 u_{i-1,j} + h^2 u_{i,j+1} + h^2 u_{i,j-1} = 10h^2 k^2 x_i y_j,$$
$$0.01 u_{i+1,j} - 2(0.01 + 0.0625)u_{i,j} + 0.01 u_{i-1,j} + 0.0625 u_{i,j+1}$$
$$+ 0.0625 u_{i,j-1} = 0.00625 x_i y_j.$$

The finite-difference equation generates the following 6×6 system of linear equations:

$i = 1, j = 1$: $0.01 u_{2,1} - 0.145 u_{1,1} + 0.01 u_{0,1} + 0.0625 u_{1,2} + 0.0625 u_{1,0}$
$$= 0.00625(0.25)(0.1),$$
$$-0.01 u_{1,1} + 0.01 u_{2,1} + 0.0625 u_{1,2} = -0.012112.$$

$i = 2, j = 1$: $0.01 u_{3,1} - 0.145 u_{2,1} + 0.01 u_{1,1} + 0.0625 u_{2,2} + 0.0625 u_{2,0}$
$$= 0.00625(0.5)(0.1),$$
$$0.01 u_{3,1} - 0.145 u_{2,1} + 0.01 u_{1,1} + 0.0625 u_{2,2} = -0.018642.$$

$i = 3, j = 1$: $0.01 u_{4,1} - 0.145 u_{3,1} + 0.01 u_{2,1} + 0.0625 u_{3,2} + 0.0625 u_{3,0}$
$$= 0.00625(0.75)(0.1),$$
$$-0.145 u_{3,1} + 0.01 u_{2,1} + 0.0625 u_{3,2} = -0.030722.$$

$i = 1, j = 2$: $0.01 u_{2,2} - 0.145 u_{1,2} + 0.01 u_{0,2} + 0.0625 u_{1,3} + 0.0625 u_{1,1}$
$$= 0.00625(0.25)(0.2),$$
$$0.01 u_{2,2} - 0.145 u_{1,2} + 0.0625 u_{1,1} = -0.046960.$$

$i = 2, j = 2$: $0.01 u_{3,2} - 0.145 u_{2,2} + 0.01 u_{1,2} + 0.0625 u_{2,3} + 0.0625 u_{2,1}$
$$= 0.00625(0.5)(0.2),$$
$$0.01 u_{3,2} - 0.145 u_{2,2} + 0.01 u_{1,2} + 0.0625 u_{2,1} = -0.030625.$$

$i = 3, j = 2$: $0.01 u_{4,2} - 0.145 u_{3,2} + 0.01 u_{2,2} + 0.0625 u_{3,3} + 0.0625 u_{3,1}$
$$= 0.00625(0.75)(0.2),$$
$$-0.145 u_{3,2} + 0.01 u_{2,2} + 0.0625 u_{3,1} = -0.022875,$$

$$
\begin{bmatrix}
-.145 & .01 & 0 & .0625 & 0 & 0 \\
.01 & -.145 & .01 & 0 & .0625 & 0 \\
0 & .01 & -.145 & 0 & 0 & .0625 \\
.0625 & 0 & 0 & -.145 & .01 & 0 \\
0 & .0625 & 0 & .01 & -.145 & .01 \\
0 & 0 & .0625 & 0 & .01 & -.145
\end{bmatrix}
\begin{bmatrix}
u_{1,1} \\
u_{2,1} \\
u_{3,1} \\
u_{1,2} \\
u_{2,2} \\
u_{3,2}
\end{bmatrix}
=
\begin{bmatrix}
-.012112 \\
-.018642 \\
-.030722 \\
-.046960 \\
-.030625 \\
-.022875
\end{bmatrix}.
$$

This system of linear equations is solved to produce

$$u(0.25, 0.1) = u_{1,1} = 0.320110,$$

$$u(0.5, 0.1) = u_{2,1} = 0.360723,$$

$$u(0.75, 0.1) = u_{3,1} = 0.389812,$$

$$u(0.25, 0.2) = u_{1,2} = 0.491201,$$

$$u(0.5, 0.2) = u_{2,2} = 0.425095,$$

$$u(0.75, 0.2) = u_{3,2} = 0.355092.$$

11.3 LAPLACE'S EQUATION

The Laplace equation has many applications. For example, the steady-state solution of the heat equation is an event governed by the Laplace's equation. In another event, the Laplace's equation represents the potential function for the field created by electrical charges that form the electrical force of attraction or repulsion.

Technically, the Laplace's equation is the homogeneous form of the Poisson's equation, obtained by setting $w(x, y) = 0$ in the latter, as follows:

$$\nabla^2 u = \frac{\partial^2 u}{\partial x^2} + \frac{\partial^2 u}{\partial y^2} = 0. \tag{11.8}$$

In the above equation, $u(x, y)$ is in $x_0 < x < x_n$ and $y_0 < y < y_m$, where m and n are the number of vertical and horizontal intervals in the rectangular grids, respectively.

The properties of the Laplace's equation are inherited much from the Poisson's equation. These properties include the initial and boundary conditions that produce the same rectangular grids as in the Poisson's equation. Hence, the same diagrams as in Figures 11.2 and 11.3 are applied for solving the boundary-value problem. The boundary conditions for this problem are given as

$$u(x, y_0) = f_1(x) \text{ and } u(x, y_m) = f_2(x), \text{ for } x_0 < x < x_n,$$

whereas the initial conditions are

$$u(x_0, y) = g_1(y) \text{ and } u(x_n, y) = g_2(y), \text{ for } y_0 \le y \le y_m.$$

The finite-difference formula for the Laplace's equation is obtained by setting $w_{i,j} = 0$ into Equation (11.6), to produce

$$h^2 u_{i,j-1} + k^2 u_{i-1,j} - 2(k^2 + h^2)u_{i,j} + k^2 u_{i+1,j} + h^2 u_{i,j+1} = 0. \tag{11.9}$$

The above finite-difference equation for the Laplace's equation is represented by the same molecule as in Figure 11.4.

In the case of 4×4 rectangular grids, the following system of linear equations is obtained by setting $w_{i,j} = 0$:

$$
\begin{bmatrix}
-2(h^2+k^2) & k^2 & 0 & h^2 & 0 & 0 & 0 & 0 & 0 \\
k^2 & -2(h^2+k^2) & k^2 & 0 & h^2 & 0 & 0 & 0 & 0 \\
0 & k^2 & -2(h^2+k^2) & 0 & 0 & h^2 & 0 & 0 & 0 \\
h^2 & 0 & 0 & -2(h^2+k^2) & k^2 & 0 & h^2 & 0 & 0 \\
0 & h^2 & 0 & k^2 & -2(h^2+k^2) & k^2 & 0 & h^2 & 0 \\
0 & 0 & h^2 & 0 & k^2 & -2(h^2+k^2) & 0 & 0 & h^2 \\
0 & 0 & 0 & h^2 & 0 & 0 & -2(h^2+k^2) & k^2 & 0 \\
0 & 0 & 0 & 0 & h^2 & 0 & k^2 & -2(h^2+k^2) & k^2 \\
0 & 0 & 0 & 0 & 0 & h^2 & 0 & k^2 & -2(h^2+k^2)
\end{bmatrix}
\begin{bmatrix}
u_{1,1} \\ u_{2,1} \\ u_{3,1} \\ u_{1,2} \\ u_{2,2} \\ u_{3,2} \\ u_{1,3} \\ u_{2,3} \\ u_{3,3}
\end{bmatrix}
$$

$$
=
\begin{bmatrix}
-h^2 u_{1,0} - k^2 u_{0,1} \\
-h^2 u_{2,0} \\
-h^2 u_{3,0} - k^2 u_{4,1} \\
-k^2 u_{0,2} \\
0 \\
-k^2 u_{4,2} \\
-k^2 u_{0,3} - h^2 u_{1,4} \\
-h^2 u_{2,4} \\
-k^2 u_{4,3} - h^2 u_{3,4}
\end{bmatrix}.
$$

Example 11.2. Solve the Laplace's equation given by $u_{xx} + u_{yy} = 0$ for $0 < x < 1$ and $0 < y < 0.3$, with $h = \Delta x = 0.25$ and $k = \Delta y = 0.1$. The boundary values in this problem are $u(0, y) = y \sin y$, and $u(1, y) = e^{-y}$, for $0 < y < 0.3$, and the initial values are $u(x, 0) = xe^{-x}$ and $u(x, 0.3) = 1 - x$, for $0 \le x \le 1$.

Solution. The same rectangular domain as in Example 11.1 is produced and shown in Figure 11.5, as the number of intervals for x and y are $m = (0.3 - 0)/0.1 = 3$ and $n = (1 - 0)/0.25 = 4$. We get the same boundary and initial values as shown in the figure. The finite-difference equation for this problem is obtained from Equation (11.9), as follows:

$$0.01u_{i+1,j} - 2(0.01 + 0.0625)u_{i,j} + 0.01u_{i-1,j} + 0.0625u_{i,j+1} + 0.0625u_{i,j-1} = 0.$$

The finite-difference equation generates the following 6×6 system of linear equations, obtained by mapping the finite-difference molecule into the grids in Figure 11.5:

$i = 1, \ j = 1: \ 0.01u_{2,1} - 2(0.01 + 0.0625)u_{1,1} + 0.01u_{0,1} + 0.0625u_{1,2}$
$\qquad\qquad + 0.0625u_{1,0} = 0,$
$\qquad\qquad - 0.145u_{1,1} + 0.01u_{2,1} + 0.0625u_{1,2} = -0.012269.$

$i = 2, \ j = 1: \ 0.01u_{3,1} - 2(0.01 + 0.0625)u_{2,1} + 0.01u_{1,1} + 0.0625u_{2,2}$
$\qquad\qquad + 0.0625u_{2,0} = 0,$
$\qquad\qquad 0.01u_{3,1} - 0.145u_{2,1} + 0.01u_{1,1} + 0.0625u_{2,2} = -0.018954.$

$i = 3, \ j = 1: \ 0.01u_{4,1} - 2(0.01 + 0.0625)u_{3,1} + 0.01u_{2,1} + 0.0625u_{3,2}$
$\qquad\qquad + 0.0625u_{3,0} = 0,$
$\qquad\qquad - 0.1u_{3,1} + 0.145u_{2,1} + 0.0625u_{3,2} = -0.031191.$

$i = 1, \ j = 2: \ 0.01u_{2,2} - 2(0.01 + 0.0625)u_{1,2} + 0.01u_{0,2} + 0.0625u_{1,3}$
$\qquad\qquad + 0.0625u_{1,1} = 0,$
$\qquad\qquad 0.01u_{2,2} - 0.145u_{1,2} + 0.0625u_{1,1} = -0.047272.$

$i = 2, \ j = 2: \ 0.01u_{3,2} - 2(0.01 + 0.0625)u_{2,2} + 0.01u_{1,2} + 0.0625u_{2,3}$
$\qquad\qquad + 0.0625u_{2,1} = 0,$
$\qquad\qquad 0.01u_{3,2} - 0.145u_{2,2} + 0.01u_{1,2} + 0.0625u_{2,1} = -0.031250.$

$i = 3, \ j = 2: \ 0.01u_{4,2} - 2(0.01 + 0.0625)u_{3,2} + 0.01u_{2,2} + 0.0625u_{3,3}$
$\qquad\qquad + 0.0625u_{3,1} = 0,$
$\qquad\qquad - 0.145u_{3,2} + 0.01u_{2,2} + 0.0625u_{3,1} = -0.023812,$

$$
\begin{bmatrix}
-.145 & .01 & 0 & .0625 & 0 & 0 \\
.01 & -.145 & .01 & 0 & .0625 & 0 \\
0 & .01 & -.145 & 0 & 0 & .0625 \\
.0625 & 0 & 0 & -.145 & .01 & 0 \\
0 & .0625 & 0 & .01 & -.145 & .01 \\
0 & 0 & .0625 & 0 & .01 & -.145
\end{bmatrix}
\begin{bmatrix}
u_{1,1} \\
u_{2,1} \\
u_{3,1} \\
u_{1,2} \\
u_{2,2} \\
u_{3,2}
\end{bmatrix}
=
\begin{bmatrix}
-.012269 \\
-.018954 \\
-.031191 \\
-.047272 \\
-.031250 \\
-.023812
\end{bmatrix}.
$$

The above system of linear equations is solved using the Gaussian elimination method or another appropriate method to produce the following results:

$$u(0.25, 0.1) = u_{1,1} = 0.323412,$$

$$u(0.5, 0.1) = u_{2,1} = 0.367195,$$

$$u(0.75, 0.1) = u_{3,1} = 0.398015$$

$$u(0.25, 0.2) = u_{1,2} = 0.495305,$$

$$u(0.5, 0.2) = u_{2,2} = 0.433187,$$

$$u(0.75, 0.2) = u_{3,2} = 0.365675.$$

11.4 HEAT EQUATION

Heat equation is a parabolic equation that finds a lot of applications in engineering. Heat flow is best modeled as partial differential equations as it involves two or more variables. In one dimension, the heat problem can be modeled as heat flowing along a given length of a rod, whose variables are the distance from one end of the rod and time. In two dimensions, heat flow is modeled using variables whose domain comprises a bounded area. In three dimensions, heat is modeled based on a domain formed from volume.

A two-dimensional heat equation is given explicitly as

$$u_t(x, t) = c^2 u_{xx}(x, t). \tag{11.10}$$

In the above equation, $u(x, t)$ is the heat value that depends on distance x and time t, and c is a constant. The domain for this problem is $\{\Omega : x_0 < x < x_n, t > 0\}$, where n is the number of subintervals in the x-axis. The initial condition is defined as

$$u(x, 0) = f(x),$$

for $x_0 < x < x_n$, whereas the boundary conditions are

$$u(x_0, t) = g_1(t) \text{ and } u(x_n, t) = g_2(t),$$

for $t > 0$. Figure 11.6 shows a 7×11 rectangular domain for the heat problem with $m = 7$ and $n = 11$. It is obvious that, unlike the case of the Poisson's equation, values are given in three sides in the rectangle only with the top side not included.

A common approach for solving the heat problem is the finite-difference method that can be implemented either explicitly or implicitly. The explicit approach solves the problem directly from substitutions from its finite-difference equation, whereas the implicit approach requires longer steps in its solution. We will only discuss an implicit method called the Crank–Nicolson's method for solving this problem.

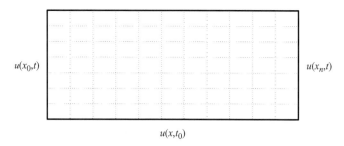

$u(x_0,t)$ $u(x_n,t)$

$u(x,t_0)$

FIGURE 11.6. Rectangular domain for the heat problem.

Crank–Nicolson's Method

The Crank–Nicolson's method is the most common implicit method for solving the heat problem. As in the Poisson's equation case, the Crank–Nicolson's method is based on two main steps. The first step consists of deriving the finite-difference equation using the central-difference rules, and the values from the given initial and boundary conditions. The second step generates the system of linear equations from the derived finite-difference formula, and this system is solved to produce the final results, $u_{i,j}$, for $i = 1, 2, \ldots, n - 1$ and $j = 1, 2, \ldots, m$.

In deriving the finite-difference formula for the heat equation problem, several considerations are made. Figure 11.7 shows the point $u_{i,j}$ and its neighbors in levels j and $j + 1$. Consider the point $u_{i,j+1/2}$, which is the middle point between $u_{i,j}$ and $u_{i,j+1}$. The heat value at this point is $u(x_i, t_j + k/2)$. From the central-difference rule, the derivative u_t at this point is

$$u_t = \frac{u_{i,j+1} - u_{i,j}}{2(k/2)} = \frac{u_{i,j+1} - u_{i,j}}{k}.$$

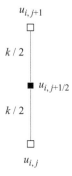

FIGURE 11.7. Position of $u_{i,j+1/2} = u(x_i, t_{j+1/2})$.

The second derivative u_{xx} at $(x_i, t_j + k/2)$ is evaluated based on its upper and lower points. An approximation is made on this derivative by taking the value as the average between the derivatives u_{xx}, at (x_i, t_j) and at (x_i, t_{j+1}), as follows:

$$u_{xx} \approx \frac{u_{xx}(x_i, t_{j+1}) + u_{xx}(x_i, t_j)}{2}$$

$$= \frac{1}{2}\left[\frac{u_{i+1,j+1} - 2u_{i,j+1} + u_{i-1,j+1}}{h^2} + \frac{u_{i+1,j} - 2u_{i,j} + u_{i-1,j}}{h^2}\right].$$

From the approximations of u_t and u_{xx}, we obtain the discrete representation of Equation (11.10), given by

$$\frac{u_{i,j+1} - u_{i,j}}{k} - \frac{\alpha^2}{2}\left(\frac{u_{i+1,j+1} - 2u_{i,j+1} + u_{i-1,j+1}}{h^2} + \frac{u_{i+1,j} - 2u_{i,j} + u_{i-1,j}}{h^2}\right) = 0.$$

This equation is simplified further to produce

$$\frac{2h^2(u_{i,j+1} - u_{i,j}) - \alpha^2 k(u_{i+1,j+1} - 2u_{i,j+1} + u_{i-1,j+1}) - \alpha^2 k(u_{i+1,j} - 2u_{i,j} + u_{i-1,j})}{2h^2 k} = 0,$$

$$2h^2(u_{i,j+1} - u_{i,j}) - \alpha^2 k(u_{i+1,j+1} - 2u_{i,j+1} + u_{i-1,j+1})$$

$$- \alpha^2 k(u_{i+1,j} - 2u_{i,j} + u_{i-1,j}) = 0.$$

Finally, we obtain the finite-difference formula for this problem,

$$- \alpha^2 k u_{i-1,j} + 2(-h^2 + \alpha^2 k)u_{i,j} - \alpha^2 k u_{i+1,j} - \alpha^2 k u_{i-1,j+1}$$

$$+ 2(h^2 + \alpha^2 k)u_{i,j+1} - \alpha^2 k u_{i+1,j+1} = 0. \qquad (11.11)$$

It is obvious from Equation (11.11) that the finite-difference formula at $u_{i,j}$ involves five other neighboring points. This equation is shown in its molecular form in Figure 11.8.

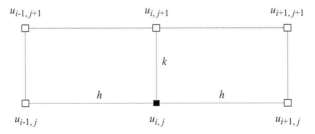

FIGURE 11.8. Molecular representation of Equation (11.11).

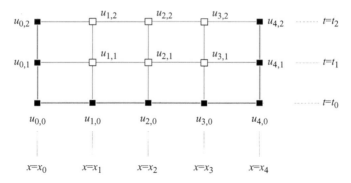

FIGURE 11.9. Rectangular grids with $m = 2$ and $n = 4$ for the heat problem.

Our objective in solving the boundary-value problem is to find the values of $u_{i,j}$ for $i = 1, 2, \ldots, n-1$ and $j = 1, 2, \ldots, m$, for a total of $m(n-1)$ unknowns. This requires a system of linear equations $Au = b$ of size $m(n-1) \times m(n-1)$, where $u = [u_{i,j}]$ for $i = 1, 2, \ldots, n-1$ and $j = 1, 2, \ldots, m$ are the unknowns. The finite-difference equation from Equation (11.11) paves the way for the formation of a system of linear equations, whose solution solves the heat problem.

The linear equations are obtained by placing the molecule of Figure 11.8 into the rectangular grids in Figure 11.9 starting with $u_{i,j}$ at $i = 1$ and $j = 0$, and this produces the first equation. The molecule is then moved one unit to the right to get the second equation, and the process continues until it reaches $i = n-1$ and $j = 0$. The molecule is then moved upward, and the process is repeated until the final position at $i = n-1$ and $j = m$ is reached. At each move, a linear equation is generated. Altogether, the moves produce $m(n-1)$ linear equations.

We discuss a case of four horizontal and two vertical intervals, as shown in Figure 11.9. The domain consists of $x_0 \le x \le x_n$ and $t_0 \le t \le t_m$, where $m = 2$ and $n = 4$. The initial condition is expressed as $u(x, t_0) = f(x)$, whereas the boundary conditions are $u(x_0, t) = g_1(t)$ and $u(x_n, t) = g_2(t)$. The given initial and boundary values in the rectangular domain are shown as black squares, whereas white squares represent the unknowns.

The side values in the rectangle are evaluated from the given initial and boundary conditions, as follows:

Bottom boundary: $u_{0,0} = u(x_0, t_0) = f(x_0)$, $u_{2,0} = u(x_2, t_0) = f(x_2)$,

$u_{3,0} = u(x_3, t_0) = f(x_3)$, $u_{4,0} = u(x_4, t_0) = f(x_4)$,

$u_{1,0} = u(x_1, t_0) = f(x_1)$.

Left boundary: $u_{0,1} = u(x_0, t_1) = g_1(t_1)$ and $u_{0,2} = u(x_0, t_2) = g_1(t_2)$.

Right boundary: $u_{n,1} = u(x_n, t_1) = g_2(t_1)$ and $u_{n,2} = u(x_n, t_2) = g_2(t_2)$.

Next, the finite-difference formula of Equation (11.11) is applied using the calculated side values. This step is achieved by placing the molecule beginning at the bottom row from left to right at $i = 1$, $j = 0$, and then upward. The steps are shown as follows:

$i = 1$, $j = 0$:

$$-\alpha^2 k u_{0,0} + (-2h^2 + 2\alpha^2 k)u_{1,0} - \alpha^2 k u_{2,0} - \alpha^2 k u_{0,1}$$

$$+ (2h^2 + 2\alpha^2 k)u_{1,1} - \alpha^2 k u_{2,1} = 0,$$

$$(2h^2 + 2\alpha^2 k)u_{1,1} - \alpha^2 k u_{2,1}$$

$$= \alpha^2 k u_{0,0} - (-2h^2 + 2\alpha^2 k)u_{1,0} + \alpha^2 k u_{2,0} + \alpha^2 k u_{0,1}.$$

$i = 2$, $j = 0$:

$$-\alpha^2 k u_{1,0} + (-2h^2 + 2\alpha^2 k)u_{2,0} - \alpha^2 k u_{3,0} - \alpha^2 k u_{1,1}$$

$$+ (2h^2 + 2\alpha^2 k)u_{2,1} - \alpha^2 k u_{3,1} = 0,$$

$$-\alpha^2 k u_{1,1} + (2h^2 + 2\alpha^2 k)u_{2,1} - \alpha^2 k u_{3,1}$$

$$= \alpha^2 k u_{1,0} - (-2h^2 + 2\alpha^2 k)u_{2,0} + \alpha^2 k u_{3,0}.$$

$i = 3$, $j = 0$:

$$-\alpha^2 k u_{2,0} + (-2h^2 + 2\alpha^2 k)u_{3,0} - \alpha^2 k u_{4,0} - \alpha^2 k u_{2,1}$$

$$+ (2h^2 + 2\alpha^2 k)u_{3,1} - \alpha^2 k u_{4,1} = 0,$$

$$-\alpha^2 k u_{2,1} + (2h^2 + 2\alpha^2 k)u_{3,1}$$

$$= \alpha^2 k u_{2,0} - (-2h^2 + 2\alpha^2 k)u_{3,0} + \alpha^2 k u_{4,0} + \alpha^2 k u_{4,1}.$$

$i = 1$, $j = 1$:

$$-\alpha^2 k u_{0,1} + (-2h^2 + 2\alpha^2 k)u_{1,1} - \alpha^2 k u_{2,1} - \alpha^2 k u_{0,2}$$

$$+ (2h^2 + 2\alpha^2 k)u_{1,2} - \alpha^2 k u_{2,2} = 0,$$

$$(-2h^2 + 2\alpha^2 k)u_{1,1} - \alpha^2 k u_{2,1} + (2h^2 + 2\alpha^2 k)u_{1,2} - \alpha^2 k u_{2,2}$$

$$= \alpha^2 k u_{0,1} + \alpha^2 k u_{0,2}.$$

$i = 2$, $j = 1$:

$$-\alpha^2 k u_{1,1} + (-2h^2 + 2\alpha^2 k)u_{2,1} - \alpha^2 k u_{3,1} - \alpha^2 k u_{1,2}$$

$$+ (2h^2 + 2\alpha^2 k)u_{2,2} - \alpha^2 k u_{3,2} = 0,$$

$$-\alpha^2 k u_{1,1} + (-2h^2 + 2\alpha^2 k) u_{2,1} - \alpha^2 k u_{3,1} - \alpha^2 k u_{1,2}$$

$$+ (2h^2 + 2\alpha^2 k) u_{2,2} - \alpha^2 k u_{3,2} = 0.$$

$i = 3, \ j = 1$:

$$-\alpha^2 k u_{2,1} + (-2h^2 + 2\alpha^2 k) u_{3,1} - \alpha^2 k u_{4,1} - \alpha^2 k u_{2,2}$$

$$+ (2h^2 + 2\alpha^2 k) u_{3,2} - \alpha^2 k u_{4,2} = 0,$$

$$-\alpha^2 k u_{2,1} + (-2h^2 + 2\alpha^2 k) u_{3,1} - \alpha^2 k u_{2,2} + (2h^2 + 2\alpha^2 k) u_{3,2}$$

$$= \alpha^2 k u_{4,1} + \alpha^2 k u_{4,2}.$$

We get the following system of linear equations:

$$
\begin{bmatrix}
2h^2 + 2\alpha^2 k & -\alpha^2 k & 0 & 0 & 0 & 0 \\
-\alpha^2 k & 2h^2 + 2\alpha^2 k & -\alpha^2 k & 0 & 0 & 0 \\
0 & -\alpha^2 k & 2h^2 + 2\alpha^2 k & 0 & 0 & 0 \\
-2h^2 + 2\alpha^2 k & -\alpha^2 k & 0 & 2h^2 + 2\alpha^2 k & -\alpha^2 k & 0 \\
-\alpha^2 k & -2h^2 + 2\alpha^2 k & -\alpha^2 k & -\alpha^2 k & 2h^2 + 2\alpha^2 k & -\alpha^2 k \\
0 & -\alpha^2 k & -2h^2 + 2\alpha^2 k & 0 & -\alpha^2 k & 2h^2 + 2\alpha^2 k
\end{bmatrix}
\begin{bmatrix}
u_{1,1} \\ u_{2,1} \\ u_{3,1} \\ u_{1,2} \\ u_{2,2} \\ u_{3,2}
\end{bmatrix}
$$

$$
=
\begin{bmatrix}
\alpha^2 k u_{0,0} - (-2h^2 + 2\alpha^2 k) u_{1,0} + \alpha^2 k u_{2,0} + \alpha^2 k u_{0,1} \\
\alpha^2 k u_{1,0} - (-2h^2 + 2\alpha^2 k) u_{2,0} + \alpha^2 k u_{3,0} \\
\alpha^2 k u_{2,0} - (-2h^2 + 2\alpha^2 k) u_{3,0} + \alpha^2 k u_{4,0} + \alpha^2 k u_{4,1} \\
\alpha^2 k u_{0,1} + \alpha^2 k u_{0,2} \\
0 \\
\alpha^2 k u_{4,1} + \alpha^2 k u_{4,2}
\end{bmatrix}.
$$

The above illustration can be generalized into the case of a $m \times n$ rectangular domain. Three algorithms are presented, one for computing the contents of $A = [a_{i,j}]$ in the system of linear equations given by $Au = b$. The second algorithm computes the contents of $b = [b_i]$ in the same system, whereas the third is the Crank–Nicolson solution.

Algorithm 11.2a. Computing the contents of $A = [a_{i,j}]$.
Given h, k, m, n and α;
For $j = 0$ to $m - 1$
    For $i = 1$ to $n - 1$
        Set $a_{i,j} = 0$;
For $j = 0$ to $m - 1$

For $i = 1$ to $n - 1$
    Let $r = i + j(n - 1)$;
    Set $a_{r,r} = 2(h^2 + \alpha^2 k)$;
    If $r\%(n - 1) <> 0$
        Set $a_{r,r+1} = a_{r+1,r} = -\alpha^2 k$;
    Endif
    If $r \leq 2(n - 1) - 1$
        If $r\%(n - 1) <> 1$
            Set $a_{n+r-2,r} = -\alpha^2 k$;
        Endif
        If $r\%(n - 1) <> 0$
            Set $a_{n+r,r} = -\alpha^2 k$;
        Endif
    Endif
    If $r \leq 2(n - 1)$
        Set $a_{n+r-1,r} = 2(-h^2 + \alpha^2 k)$;
    Endif
  Endfor
Endfor

The contents of $b = [b_i]$ in the system $Au = b$ are evaluated using Algorithm 11.2b.

Algorithm 11.2b. Computing the contents of $b = [b_i]$.
Given h, k, m, n and α;
Given $u_{i,0}, u_{i,m}, u_{0,j}$ and $u_{n,j}$ for $i = 0, 1, \ldots, n$ and $j = 0, 1, \ldots, m$;
For $j = 0$ to $m - 1$
  Let $b_i = 0$;
For $j = 0$ to $m - 1$
  For $i = 1$ to $n - 1$
    Let $r = i + j(n - 1)$;
    If $j = 0$
      If $i = 1$
        Set $b_r = \alpha^2 k u_{i-1,0} - 2u_{i,0}(-h^2 + \alpha^2 k) + \alpha^2 k u_{i+1,0} + \alpha^2 k u_{0,1}$;
      Endif
      If $i = n - 1$
        Set $b_r = \alpha^2 k u_{i-1,0} - 2u_{i,0}(-h^2 + \alpha^2 k) + \alpha^2 k u_{i+1,0} + \alpha^2 k u_{n,1}$;
      Endif
    If $j > 0$
      If $i = 1$
        Set $b_r = \alpha^2 k u_{0,j} + u_{0,j+1}$;
      Endif
      If $i = n - 1$

            Set $b_r = \alpha^2 k u_{n,j} + u_{n,j+1}$
        Endif
      Endif
    Endif
  Endfor
Endfor

Algorithm 11.3. Crank–Nicolson's method for the heat equation problem.
  Given $u_t = \alpha^2 u_{xx}$, $h = \Delta x$, $k = \Delta t$ in $x_0 \le x \le x_n$ and $t > 0$;
  Given the boundary values, $u(x_0, t) = g_1(t)$ and $u(x_n, t) = g_2(t)$;
  Given the initial values, $u(x, t_0) = f(x)$;
  Evaluate $m = (y_m - y_0)/k$ and $n = (x_n - x_0)/h$;
  Evaluate $u_{0,j}$ and $u_{n,j}$ for $j = 1, 2, \ldots, m$;
  Evaluate $u_{i,0}$ and $u_{i,m}$ for $i = 0, 1, \ldots, n$;
  Find the finite-difference formula using Equation (11.11);
  Find matrix A in $Au = b$ using Algorithm 11.2a;
  Find vector b in $Au = b$ using Algorithm 11.2b;
  Solve for u in $Au = b$;

Algorithm 11.3 is an implicit finite-difference approach for solving the heat problem. This algorithm is further illustrated through Example 11.3.

Example 11.3. Solve the heat equation, $u_t = u_{xx}$, for $0 < x < 1$ and $t > 0$, on two
levels, with the boundary values given by $u(0, t) = t \sin t$ and $u(1, t) = 1 - e^{-t}$ for
$t \ge 0$, and the initial conditions $u(x, 0) = x(1 - x)$ for $0 \le x \le 1$. The widths of the
intervals are $h = \Delta x = 0.25$ and $k = \Delta t = 0.1$.

Solution. The graphical solution is shown in Figure 11.9. There are $n = \frac{x_n - x_0}{h} =$
$\frac{1.0 - 0.0}{0.25} = 4$ subintervals in the x-axis and $m = 2$ subintervals in the vertical axis.
Applying the central-difference rules for replacing the partial derivatives,

$$u_t = \alpha^2 u_{xx},$$

$$\frac{u_{i,j+1} - u_{i,j}}{k} - \frac{\alpha^2}{2}\left(\frac{u_{i+1,j+1} - 2u_{i,j+1} + u_{i-1,j+1}}{h^2} + \frac{u_{i+1,j} - 2u_{i,j} + u_{i-1,j}}{h^2}\right) = 0,$$

$$-\alpha^2 k u_{i-1,j} + 2(-h^2 + \alpha^2 k)u_{i,j} - \alpha^2 k u_{i+1,j} - \alpha^2 k u_{i-1,j+1} + 2(h^2 + \alpha^2 k)u_{i,j+1}$$
$$- \alpha^2 k u_{i+1,j+1} = 0.$$

With $\alpha = 1$, $h = 0.25$ and $k = 0.1$, we get the finite-difference equation,

$$-0.1u_{i-1,j} + 0.075u_{i,j} - 0.1u_{i+1,j} - 0.1u_{i-1,j+1} + 0.325u_{i,j+1} - 0.1u_{i+1,j+1} = 0.$$

The finite-difference formula produces a 6×6 system of linear equations by fitting the molecule from Figure 11.8 into the rectangular grids in Figure 11.9. The process starts at $i = 1$ and $j = 0$, as follows:

$i = 1, \ j = 0: \ -0.1u_{0,0} + 0.075u_{1,0} - 0.1u_{2,0} - 0.1u_{0,1} + 0.325u_{1,1} - 0.1u_{2,1} = 0,$
$$+0.325u_{1,1} - 0.1u_{2,1} = 0.011936.$$

$i = 2, \ j = 0: \ -0.1u_{1,0} + 0.075u_{2,0} - 0.1u_{3,0} - 0.1u_{1,1} + 0.325u_{2,1} - 0.1u_{3,1} = 0,$
$$-0.1u_{1,1} + 0.325u_{2,1} - 0.1u_{3,1} = 0.018750.$$

$i = 3, \ j = 0: \ -0.1u_{2,0} + 0.075u_{3,0} - 0.1u_{4,0} - 0.1u_{2,1} + 0.325u_{3,1} - 0.1u_{4,1} = 0,$
$$-0.1u_{2,1} + 0.325u_{3,1} = 0.020454.$$

$i = 1, \ j = 1: \ -0.1u_{0,1} + 0.075u_{1,1} - 0.1u_{2,1} - 0.1u_{0,2} + 0.325u_{1,2} - 0.1u_{2,2} = 0,$
$$0.075u_{1,1} - 0.1u_{2,1} + 0.325u_{1,2} - 0.1u_{2,2} = 0.004972.$$

$i = 2, \ j = 1: \ -0.1u_{1,1} + 0.075u_{2,1} - 0.1u_{3,1} - 0.1u_{1,2} + 0.325u_{2,2} - 0.1u_{3,2} = 0.$

$i = 3, \ j = 1: \ -0.1u_{2,1} + 0.075u_{3,1} - 0.1u_{4,1} - 0.1u_{2,2} + 0.325u_{3,2} - 0.1u_{4,2} = 0,$
$$-0.1u_{2,1} + 0.075u_{3,1} - 0.1u_{2,2} + 0.325u_{3,2} = 0.027643.$$

The above linear equations are organized into a matrix form as

$$
\begin{bmatrix}
.325 & -.01 & 0 & 0 & 0 & 0 \\
-.01 & .325 & -.01 & 0 & 0 & 0 \\
0 & -.01 & .325 & 0 & 0 & 0 \\
.075 & -.01 & 0 & .325 & -.01 & 0 \\
-.01 & .075 & -.01 & -.01 & .325 & -.01 \\
0 & -.01 & .075 & 0 & -.01 & .325
\end{bmatrix}
\begin{bmatrix}
u_{1,1} \\ u_{2,1} \\ u_{3,1} \\ u_{1,2} \\ u_{2,2} \\ u_{3,2}
\end{bmatrix}
=
\begin{bmatrix}
.011936 \\ .018750 \\ .020454 \\ .004972 \\ 0 \\ .027643
\end{bmatrix}.
$$

Finally, the system is solved using the Gaussian elimination method to produce

$$u(0.25, 0.1) = u_{1,1} = 0.070352,$$

$$u(0.5, 0.1) = u_{2,1} = 0.109215,$$

$$u(0.75, 0.1) = u_{3,1} = 0.096352,$$

$$u(0.25, 0.2) = u_{1,2} = 0.058095,$$

$$u(0.5, 0.2) = u_{2,2} = 0.081452,$$

$$u(0.75, 0.2) = u_{3,2} = 0.120952.$$

11.5 WAVE EQUATION

The wave equation is a type of hyperbolic equation that has many applications involving waves, such as in modeling the vibration of a string. The movement of soliton
in water is also modeled as a wave equation. Another common application is the air
flow dynamics as a result from the aircraft movement. Wave movement is a periodic
process that can best be modeled using partial differential equations.

The general form of a non-homogeneous wave equation $u(x, t)$ has of the distance
x in the horizontal axis and time-dependent variable t as its vertical axis, given by

$$\frac{\partial^2 u}{\partial t^2} - \alpha^2 \frac{\partial^2 u}{\partial x^2} = w(x, t), \tag{11.12}$$

where $w(x, t)$ is another function of x and t and α is a real constant. A common form
of the wave equation is the linear Klein–Gordon's equation, given by

$$\frac{\partial^2 u}{\partial t^2} - a^2 \frac{\partial^2 u}{\partial x^2} = bu,$$

where a and b are constants. In this section, we will discuss the linear and homogeneous form of the wave equation in Equation (11.12), stated as

$$\frac{\partial^2 u}{\partial t^2} - \alpha^2 \frac{\partial^2 u}{\partial x^2} = 0, \tag{11.13}$$

for $u(x, t)$ in $x_0 < x < x_n$ and $t > t_0$.

Figure 11.10 shows a rectangular domain for the wave problem consisting of 7
rows and 11 columns. There are two initial values in this problem, given at $t = t_0$ as

$$u(x, t_0) = f_1(x) \text{ and } u_t(x, t_0) = f_2(x).$$

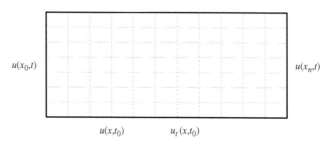

FIGURE 11.10. The domain for the wave equation problem.

The first-order partial differential equation $u_t(x, t_0) = f_2(x)$ is an additional initial condition that contributes in solving the problem. This differential form for the initial is the rate of change of the wave function with respect to time at this value.

The boundary conditions for this problem are the left and right boundaries that are functions of t, given by

$$u(x_0, t) = g_1(t) \text{ and } u(x_n, t) = g_2(t).$$

Unlike the methods in the elliptic and parabolic cases, the finite-difference solution to the wave problem does not involve the reduction of the boundary-value problem to a system of linear equations. This is because the finite-difference equation derived from the given initial and boundary conditions is inherently in an explicit form that is sufficient to solve the problem directly.

We derive the finite-difference equation for the wave equation problem. From the central-difference differentiation rules:

$$u_{tt} - \alpha^2 u_{xx} = 0,$$

$$\frac{u_{i,j+1} - 2u_{i,j} + u_{i,j-1}}{k^2} - \alpha^2 \frac{u_{i+1,j} - 2u_{i,j} + u_{i-1,j}}{h^2} = 0,$$

$$h^2(u_{i,j+1} - 2u_{i,j} + u_{i,j-1}) - k^2\alpha^2(u_{i+1,j} - 2u_{i,j} + u_{i-1,j}) = 0.$$

The above equation simplifies to

$$h^2 u_{i,j-1} - k^2\alpha^2 u_{i-1,j} + (-2h^2 + 2k^2\alpha^2)u_{i,j} - k^2\alpha^2 u_{i+1,j} + h^2 u_{i,j+1} = 0.$$

We obtain the explicit form of the finite-difference formula, given by

$$u_{i,j+1} = \frac{1}{h^2}\left[-h^2 u_{i,j-1} + k^2\alpha^2 u_{i-1,j} - (-2h^2 + 2k^2\alpha^2)u_{i,j} + k^2\alpha^2 u_{i+1,j}\right].$$
$$(11.14)$$

Figure 11.11 shows the molecular representation of the finite-difference formula. The white square is $u_{i,j+1}$, which is the left-hand term of Equation (11.14). This term is explicitly separated from other terms in the equation. This arrangement suggests $u_{i,j+1}$ can be evaluated directly once the values of $u_{i-1,j}, u_{i+1,j}, u_{i,j}$, and $u_{i,j-1}$ are known.

The second initial condition is given by $u_t(x, t_m) = f_2(x)$, which is the rate of change of the wave with respect to time. This first partial derivative can be discretized using the central-difference rule, as follows:

$$\frac{u_{i,j+1} - u_{i,j-1}}{2k} = f_2(x_i).$$

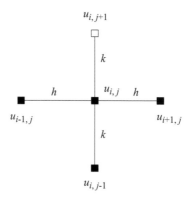

FIGURE 11.11. The molecule of the finite-difference formula for wave equation.

We consider a special case of the above equation at level $j = 0$, which produces

$$\frac{u_{i,1} - u_{i,-1}}{2k} = f_2(x_i).$$

We obtain $u_{i,-1}$, which is a virtual value as its point is not in the given rectangular domain. This value lies one level below the bottom side of the rectangle that does not really exist. Therefore, it is wise to make this value as a subject in the equation as

$$u_{i,-1} = u_{i,1} - 2k \, f_2(x_i). \tag{11.15}$$

Equation (11.15) is a useful conversion that affects all the points in the first level of the rectangular grids. We now apply the finite-difference formula in Equation (11.14) to the rectangular grids. Starting at level $j = 0$, we get

$$u_{i,1} = \frac{1}{h^2} \left[-h^2 u_{i,-1} + k^2 \alpha^2 u_{i-1,0} - (-2h^2 + 2k^2 \alpha^2) u_{i,0} + k^2 \alpha^2 u_{i+1,0} \right].$$

Substituting $u_{i,-1}$ in the above equation using Equation (11.14),

$$u_{i,1} = \frac{1}{h^2} \left[-h^2 (u_{i,1} - 2k \, f_2(x_i)) + k^2 \alpha^2 u_{i-1,0} - (-2h^2 + 2k^2 \alpha^2) u_{i,0} + k^2 \alpha^2 u_{i+1,0} \right],$$

$$u_{i,1} + \frac{h^2}{h^2} u_{i,1} = \frac{1}{h^2} \left[2h^2 k \, f_2(x_i) + k^2 \alpha^2 u_{i-1,0} - (-2h^2 + 2k^2 \alpha^2) u_{i,0} + k^2 \alpha^2 u_{i+1,0} \right].$$

We obtain the explicit representation of $u_{i,1}$ at level $j = 0$ for $i = 1, 2, \ldots, n - 1$,

$$u_{i,1} = \frac{1}{2h^2} \left[2h^2 k \, f_2(x_i) + k^2 \alpha^2 u_{i-1,0} - (-2h^2 + 2k^2 \alpha^2) u_{i,0} + k^2 \alpha^2 u_{i+1,0} \right]. \tag{11.16}$$

The values for $i = 1, 2, \ldots, n - 1$ at other levels follow immediately. They are computed upward from the second level in the order by letting $j = 1, 2, \ldots, m - 1$ into Equation (11.12), as follows:

$$j = 1: \qquad u_{i,2} = \frac{1}{h^2} \left[-h^2 u_{i,0} + k^2 \alpha^2 u_{i-1,1} - (-2h^2 + 2k^2 \alpha^2) u_{i,1} + k^2 \alpha^2 u_{i+1,1} \right],$$

$$j = 2: \qquad u_{i,3} = \frac{1}{h^2} \left[-h^2 u_{i,1} + k^2 \alpha^2 u_{i-1,2} - (-2h^2 + 2k^2 \alpha^2) u_{i,2} + k^2 \alpha^2 u_{i+1,2} \right],$$

$$\cdots$$

$$j = m - 1: u_{i,m} = \frac{1}{h^2} \left[-h^2 u_{i,m-2} + k^2 \alpha^2 u_{i-1,m-1} - (-2h^2 + 2k^2 \alpha^2) u_{i,m-1} + k^2 \alpha^2 u_{i+1,m-1} \right].$$

Algorithm 11.4 outlines the computational steps in solving the wave problem. The algorithm is further illustrated using Example 11.4.

Algorithm 11.4. Finite-difference method for the wave equation problem.
  Given $u_{tt} = \alpha^2 u_{xx}$, $h = \Delta x$, $k = \Delta t$ in $x_0 \leq x \leq x_n$ and $t > 0$;
  Given the boundary values, $u(x_0, t) = g_1(t)$ and $u(x_n, t) = g_2(t)$;
  Given the initial values, $u(x, t_0) = f_1(x)$ and $u_t(x, t_0) = f_2(x)$;
  Evaluate $m = (y_m - y_0)/k$ and $n = (x_n - x_0)/h$;
  Evaluate $u_{0,j}$ and $u_{n,j}$ for $j = 1, 2, \ldots, m$;
  Evaluate $u_{i,0}$ and $u_{i,m}$ for $i = 0, 1, \ldots, n$;
  Find $u_{i,-1}$ from Equation (11.15), for $i = 1, 2, \ldots, n - 1$;
  Find the finite-difference formula;
  Find $u_{i,j+1}$ using Equation (11.14), for $i = 1, 2, \ldots, n - 1$ and $j = 1, 2, \ldots, m$;

Example 11.4. Solve the wave equation, $u_{tt} - 4u_{xx} = 0$, for $0 < x < 1$ and $t > 0$ on two levels, with the boundary values $u(0, t) = t \sin t$ and $u(1, t) = 1 - e^{-t}$ for $t \geq 0$, and initial conditions, $u(x, 0) = x(1 - x)$ and $u_t(x, 0) = 3x^2$, for $0 \leq x \leq 1$. Assume $h = \Delta x = 0.25$ and $k = \Delta t = 0.1$.

Solution. The number of x intervals is $n = \frac{1.0-0.0}{0.25} = 4$. There are $m = 2$ vertical intervals in the problem. The rectangular domain is shown in Figure 11.12. The figure also shows the position of the virtual values $u_{i,-1}$ for $i = 0, 1, 2, 3, 4$. Applying the central-difference rules,

$$u_{tt} - \alpha^2 u_{xx} = 0,$$

$$\frac{u_{i,j+1} - 2u_{i,j} + u_{i,j-1}}{k^2} - \alpha^2 \frac{u_{i+1,j} - 2u_{i,j} + u_{i-1,j}}{h^2} = 0,$$

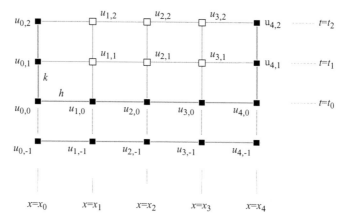

FIGURE 11.12. Rectangular domain showing the virtual values.

$$\frac{u_{i,j+1} - 2u_{i,j} + u_{i,j-1}}{0.1^2} - 4\frac{u_{i+1,j} - 2u_{i,j} + u_{i-1,j}}{0.25^2} = 0,$$

$$100(u_{i,j+1} - 2u_{i,j} + u_{i,j-1}) - 64(u_{i+1,j} - 2u_{i,j} + u_{i-1,j}) = 0.$$

We get the finite-difference formula,

$$100u_{i,j+1} = 72u_{i,j} - 100u_{i,j-1} + 64u_{i+1,j} + 64u_{i-1,j}.$$

From the differentiated initial condition, we get

$$u_t(x, 0) \approx \frac{u_{i,1} - u_{i,-1}}{2k} = 3x_i^2,$$

$$u_{i,-1} = u_{i,1} - 6kx_i^2 = u_{i,1} - 0.6x_i^2.$$

Applying the finite-difference formula to level $j = 0$,

$i = 1, \ j = 0: 100u_{1,1} = 72u_{1,0} - 100u_{1,-1} + 64u_{2,0} + 64u_{0,0},$
$\qquad\qquad\quad 100u_{1,1} = 72u_{1,0} - 100\left(u_{1,1} - 0.6x_1^2\right) + 64u_{2,0} + 64u_{0,0},$
$\qquad\qquad\quad 200u_{1,1} = 72(0.188) - 100(-0.6)(0.25^2) + 64(0.250) + 64(0),$
$\qquad\qquad\quad u_{1,1} = 0.1660.$

$i = 2, \ j = 0: 100u_{2,1} = 72u_{2,0} - 100u_{2,-1} + 64u_{3,0} + 64u_{1,0},$
$\qquad\qquad\quad 100u_{2,1} = 72u_{2,0} - 100\left(u_{2,1} - 0.6x_2^2\right) + 64u_{3,0} + 64u_{1,0},$
$\qquad\qquad\quad 200u_{2,1} = 72(0.25) - 100(-0.6)0.5^2 + 64(0.188) + 64(0.188),$
$\qquad\qquad\quad u_{2,1} = 0.2850.$

$i = 3, \ j = 0 : 100u_{3,1} = 72u_{3,0} - 100u_{3,-1} + 64u_{4,0} + 64u_{2,0},$

$\quad\quad 100u_{3,1} = 72u_{3,0} - 100(u_{3,1} - 0.6x_3^2) + 64u_{4,0} + 64u_{2,0},$

$\quad\quad 200u_{3,1} = 72(0.188) - 100(-0.6)0.75^2 + 64(0) + 64(0.25),$

$\quad\quad u_{3,1} = 0.3163.$

The values in the second level at $j = 1$ are computed directly from Equation (11.15),

$i = 1, \ j = 1 : 100u_{1,2} = 72u_{1,1} - 100u_{1,0} + 64u_{2,1} + 64u_{0,1},$

$\quad\quad 100u_{1,2} = 72(0.166) - 100(0.188) + 64(0.285) + 64(0.010),$

$\quad\quad u_{1,2} = 0.1210.$

$i = 2, \ j = 1 : 100u_{2,2} = 72u_{2,1} - 100u_{2,0} + 64u_{3,1} + 64u_{1,1},$

$\quad\quad 100u_{2,2} = 72(0.285) - 100(0.250) + 64(0.316) + 64(0.166),$

$\quad\quad u_{2,2} = 0.2640.$

$i = 3, \ j = 1 : 100u_{3,2} = 72u_{3,1} - 100u_{3,0} + 64u_{4,1} + 64u_{2,1},$

$\quad\quad 100u_{3,2} = 72(0.316) - 100(0.188) + 64(0.095) + 64(0.285),$

$\quad\quad u_{3,2} = 0.2835.$

11.6 VISUAL SOLUTION: CODE11

Code11. User Manual.

1. Select a method from the menu.
2. Enter the input according to the selected method.
3. Click the button to view the results.
Development files: `Code11.cpp`, `Code11.h`, and `MyParser.obj`.

A visual interface for problems in partial differential equations is necessary as the problem is referred extensively in many science and engineering applications. A typical problem in PDE is relatively difficult to solve as it involves a lot of input as well as a lot of steps in its solution. A friendly interface in the form of direct input from the user and other interactive activities will definitely help in understanding the topic. The solution too needs to be presented in such a way that it displays all the textual and graphical elements in the problem.

We discuss the visual interfaces for problems involving the Poisson's, heat, and wave equations. The project is called `Code11`. The basic elements for the visual solution in `Code11` consist of an input region, a region for displaying matrix A and vector b in $Au = b$, and another region for displaying the solutions in the rectangular grids.

Figure 11.13 shows the output from a problem involving the Poisson's equation. The output consists of four regions for displaying the menu, an input region, the

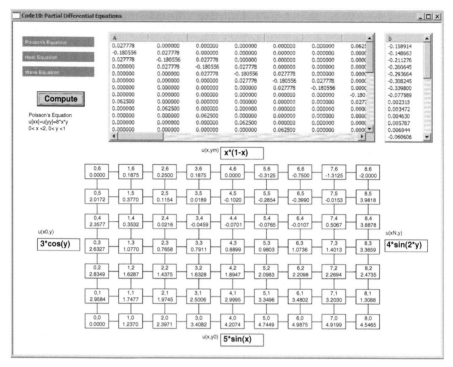

FIGURE 11.13. Visual solution for the Poisson equation problem.

rectangular grids region for displaying the results, and a list view tables region for displaying the system of linear equations generated from the problem.

The first region consists of the menu items that become activated when they are left-clicked. The second region hosts the edit boxes for the first set of input. This same region is also shared by the fourth region, which hosts the list view tables for displaying the system of linear equations generated from the problem. The third region is the rectangular grids for displaying the results from the calculations.

The menu provides a selection of items from the Poisson's, heat, and wave equations. When a selection is made, edit boxes as shown in Figure 11.14 appear. These edit boxes collect information about the initial values of the variables in the problem. The figure shows the case when the selection is the Poisson's equation. Almost similar items also appear in cases of heat and wave equations.

The edit boxes in Figure 11.14 provide input on the number of subintervals in the domain and on their start and ending values. It is important to furnish all this information as the rectangular grids are drawn according to these input values. Input for the number of vertical m and horizontal subintervals n in the edit boxes in Figure 11.14 will determine the actual sizes of the rectangular grids for displaying the results in the main window.

Once the first set of input has been completed and the push button is left-clicked, the rectangular grids appear. Input for the initial and boundary conditions are needed

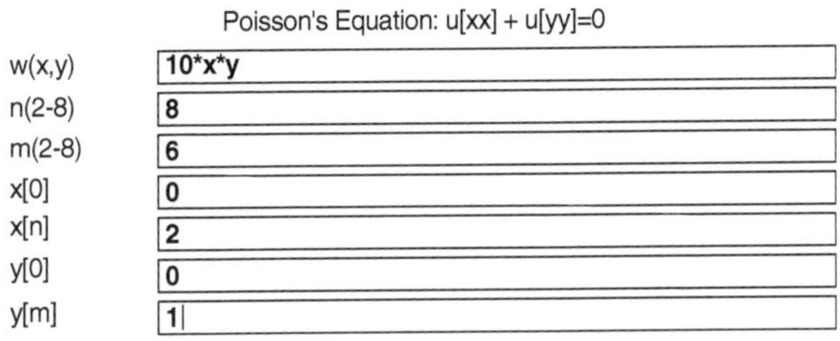

Poisson's Equation: u[xx] + u[yy]=0

w(x,y)	10*x*y
n(2-8)	8
m(2-8)	6
x[0]	0
x[n]	2
y[0]	0
y[m]	1|

FIGURE 11.14. Input for the rectangular grids in `Code11`.

in order to proceed further. The input strings representing the equations for these initial and boundary conditions can be entered in the edit boxes that are provided along the sides of the grids. When the inputs are completed, a click at the push button produces the results, which are displayed inside the rectangular grids. The values of the *A* matrix and *b* vector are also displayed in the list view tables.

Code11 consists of three files, Code11.cpp, Code11.h, and MyParser.obj. Only a single class called CCode11 is used in this application, and this class is derived from CFrameWnd. Figure 11.15 shows a schematic drawing of the main development steps in Code11. A variable called fStatus monitors the progress of the execution with fStatus=0 as its initial value. The value changes to fStatus=1 when the first set of input has been completed and fStatus=2 when the second set is completed. The final stage is at fStatus=3, which indicates the selected method is successful, and the results are displayed both in the rectangular grids and in the list view tables.

Several functions have been declared in CCode11, and they are summarized in Table 11.2. The three main functions from this list are Poisson(), Heat(), and Wave(), which represent the solutions for the Poisson's equation, heat equation, and wave equation problems, respectively. Another function called SolveSLE() solves the system of linear equations using the Gaussian elimination method when it is called from Poisson() and Heat(). Input for each problem is collected from the edit boxes using OnButton() and SecondInput(), whereas the output from each selected method is displayed in OnPaint() and ShowTable().

The main variables used in this application are organized into three structures, PT for representing the points in the real coordinates, INPUT for representing the objects in the input variables, and MENU for representing the objects in the menu. PT consists of the domain variables, x_i, y_i, and t_i, represented as pt[i].x, pt[i].y and pt[i].t, respectively, with pt as its array. The variables form the rectangular coordinates in the grids, (x_i, y_i) in the case of the Poisson's equation, and (x_i, t_i) in the heat and wave equations. The structure for supporting all these variables is declared as follows:

```
typedef struct
{
        double x,y,t;
} PT;
PT *pt;
```

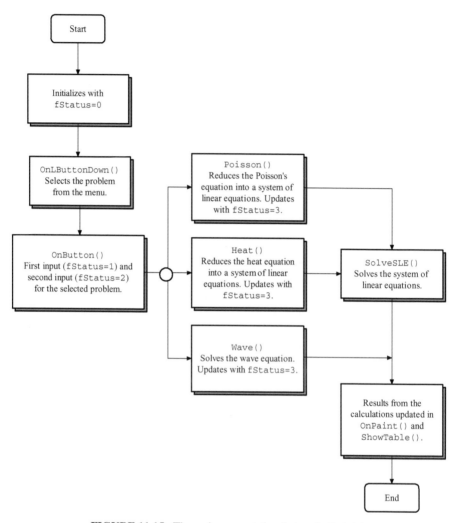

FIGURE 11.15. The main computational steps in `Code11`.

Input from the user is organized as variables in a structure called INPUT. The objects are similar to those defined earlier in the earlier chapters, as follows:

```
typedef struct
{
    CString label,item;
    CPoint hm;
    CEdit ed;
    CRect rc,display;
} INPUT;
INPUT input[maxInput+1];
```

TABLE 11.2. Functions in `Code11`

Function	Description
`Code11()`	Constructor.
`~Code11()`	Destructor.
`Poisson()`	Computes the Poisson's equation problem.
`Heat()`	Computes the heat equation problem.
`Wave()`	Computes the wave equation problem.
`SolveSLE()`	Solves the system of linear equations.
`ShowTable()`	Creates the tables for displaying the system of linear equations.
`SecondInput()`	Input on the initial and boundary values.
`InputLabels()`	Labels for the first set of inputs.
`Clear()`	Clears the named region of the window.
`OnButton()`	Message handler for the *Compute* push button. Reads the first and second inputs, and calls the corresponding function for solving the problem.
`OnLButtonDown()`	Message handler for selecting an item in the menu.
`OnPaint()`	Initial and updated display in the main window.

The menu objects are organized into a structure called MENU. These objects are also similar to those defined earlier.

```
typedef struct
{
      CString item;
      CPoint hm;
      CRect rc;
} MENU;
MENU menu[nItems+1];
```

Table 11.3 lists other main objects in Code11. They include btn for the push button and table1 and table2 for the list view tables. The objects called hGrid, rcGrid and wrc shape up the rectangular grids for displaying the output.

TABLE 11.3. Other objects in `Code11`

Object	Class	Description
`table1, table2`	CListCtrl	Objects representing matrix *A* and vector *b*, respectively, in the generated system of linear equations.
`hGrid`	CPoint	Home coordinates of the top left-hand corner rectangle of the grids.
`rcGrid`	CRect	Rectangular region of the grids.
`wrc`	CSize	Size of each rectangle in the rectangular grids.
`btn`	CButton	The push button called *Compute*.

TABLE 11.4. Main variables in `Code11`

Variable	Type	Description
`u[i][j]`	`double`	The unknown $u_{i,j}$ in the problem.
`w[i][j]`	`double`	The nonhomogeneous expression $w_{i,j}$ of the Poisson equation.
`alpha`	`double`	The given constant α in the heat and wave equations.
`h`	`double`	The width $h = \Delta x$ in the rectangular grids.
`k`	`double`	The width $k = \Delta y$ or $k = \Delta t$ in the rectangular grids.
`m`	`int`	The number of vertical subintervals in the rectangular grids.
`n`	`int`	The number of horizontal subintervals in the rectangular grids.
`a[i][j]`	`double`	Element $a_{i,j}$ in the matrix $A = [a_{i,j}]$ of the system of linear equations.
`b[i]`	`double`	Element b_i in the vector $\boldsymbol{b} = [b_i]$ of the system of linear equations.
`z[i]`	`double`	The solution z_i to the system of linear equations.
`fMenu`	`int`	The selected item in the menu with fMenu=1 for Poisson, fMenu=2 for heat, and fMenu=3 for wave.
`fStatus`	`int`	Status level of the execution indicated by a flag value, with 0 for nothing, 1 for menu selected, 2 for first input, and 3 for the second input, which produces the desired results.
`nInputItems`	`int`	Number of input items for each selected item in the menu.
`idc`	`int`	Ids for the control resources in the form of edit and static boxes.
`hSpace,` `vSpace,` `r1vspace`	`double`	Horizontal and vertical spacing for the boxes in the rectangular grids.

Table 11.4 lists the main variables in `Code11`. The variables that make up the problems include `u`, `w`, `alpha`, `h`, `k`, `m`, and `n`. The variables involved in solving the system of linear equations are the arrays `a`, `b`, and `z`. `fMenu` is a flag for indicating the selected item in the menu. `fStatus` is a variable that indicates the execution stage. The number of input items is indicated by `nInputItems`, whereas `idc` represents the control id for the edit boxes.

Three events are mapped in the application, and they are handled by their respective functions, as follows:

```
BEGIN_MESSAGE_MAP(CCode11,CFrameWnd)
      ON_WM_PAINT()
      ON_WM_LBUTTONDOWN()
      ON_BN_CLICKED(IDC_BUTTON,OnButton)
END_MESSAGE_MAP()
```

The constructor function creates the main window and initializes several global objects and variables. They include the menu items, input items, and the variables for formatting the output display. The code fragments for this function are given as

```
CCode11::CCode11()
{
      Create(NULL,"Code11: Partial Differential Equations",
           WS_OVERLAPPEDWINDOW,CRect(0,0,900,700),NULL);
      Arial80.CreatePointFont(80,"Arial");
      menu[0].hm=CPoint(10,300); hGrid=CPoint(150,285);
      menu[0].rc=CRect(menu[0].hm.x,menu[0].hm.y,
           menu[0].hm.x+139,menu[0].hm.y+500);
      menu[1].item="Poisson equation";
      menu[2].item="Heat Equation";
      menu[3].item="Wave Equation";
      for (int i=1;i<=nMenuItems;i++)
      {
           menu[i].hm=CPoint(20,30+(i-1)*30);
           menu[i].rc=CRect(menu[i].hm.x,menu[i].hm.y,
                menu[i].hm.x+150,menu[i].hm.y+20);
      }
      input[0].hm=CPoint(200,20);
      input[0].rc=CRect(input[0].hm.x,input[0].hm.y,
           input[0].hm.x+560,input[0].hm.y+220);
      for (i=1;i<=maxInput;i++)
           input[i].hm=CPoint(input[0].hm.x+10,input[0].hm.y
                +30+(i-1)*25);
      input[0].display=CRect(20,170,170,230);
      pt=new PT [M*N+1];
      fMenu=0; idc=301;
      fStatus=0; fMenu=0;
      wrc=CSize(50,30);
      hSpace=70; vSpace=50; r1vspace=40;
      rcGrid=CRect(hGrid.x-150,hGrid.y-30,hGrid.x+750,
           hGrid.y+500);
      a=new double *[M*N+1];
      b=new double [M*N+1];
      z=new double [M*N+1];
      u=new double *[N+1];
      w=new double *[N+1];
      for (int i=0;i<=N;i++)
      {
           u[i]=new double [M+1];
           w[i]=new double [M+1];
      }
```

```
        for (i=0;i<=M*N;i++)
             a[i]=new double [M*N+1];
}
```

In the above code, the input rectangular region and its home coordinates are identified as input[0].rc and input[0].hm, respectively. Another variable called input[0].display defines a small rectangular region below the *Compute* button for displaying the brief information about the currently selected problem.

OnPaint() handles and constantly updates the display in the main window. The display is updated based on the current value of fStatus whenever InvalidateRect() is called from any function within the program. With an initial value of fStatus=0, the initial display consists of items for selection in the menu shown as shaded rectangles. With fStatus=2, the display consists of the rectangular grids whose values are not calculated yet. With fStatus=3, the results from the calculations according to the selected method are shown in the rectangular grids.

```
void CCode11::OnPaint()
{
        CPaintDC dc(this);
        CString str;
        CRect rc;
        int i,j,r;
        dc.SelectObject(Arial80);
        dc.SetBkColor(RGB(150,150,150));
        dc.SetTextColor(RGB(255,255,255));
        for (i=1;i<=nMenuItems;i++)
        {
                dc.FillSolidRect(&menu[i].rc,RGB(150,150,150));
                dc.TextOut(menu[i].hm.x+5,menu[i].hm.y+5,menu[i].
                        item);
        }
        dc.SetBkColor(RGB(255,255,255));
        dc.SetTextColor(RGB(100,100,100));
        dc.Rectangle(input[0].rc);
        if (fMenu==1)
                str="u[xx] + u[yy]=0";
        if (fMenu==2)
                str="u[t] - alpha 2*u[xx]=0";
        if (fMenu==3)
                str.Format("u[tt] - alpha^2*u[xx]=0");
        dc.TextOut(input[0].hm.x+150,input[0].hm.y+5,menu[fMenu].
                item+": "+str);
        if (fMenu>=0)
                for (i=1;i<=nInputItems;i++)
```

```
                    dc.TextOut(input[i].hm.x+10,input[i].hm.y,
                        input[i].label);
if (fStatus>=2)
{
     for (j=m;j>=0;j--)
     {
          dc.MoveTo(hGrid.x+wrc.cx/2,hGrid.
               y+wrc.cy/2+j*vSpace);
          dc.LineTo(hGrid.x+wrc.cx/2+n*hSpace,
               hGrid.y+wrc.cy/2+j*vSpace);
}
for (i=0;i<=n;i++)
{
     dc.MoveTo(hGrid.x+i*hSpace+wrc.cx/2,
          hGrid.y+wrc.cy/2);
     dc.LineTo(hGrid.x+i*hSpace+wrc.cx/2,
          hGrid.y+wrc.cy/2+m*vSpace);
}
dc.TextOut(hGrid.x-100,hGrid.y+m/2*vSpace-20,
     ((fMenu==1)?"u(x0,y)":"u(x0,t)"));
dc.TextOut(hGrid.x+(n+1)*hSpace-10,hGrid.y+m/2*vSpace-20,
     ((fMenu==1)?"u(xN,y)":"u(xN,t)"));
dc.TextOut(hGrid.x+n/2*hSpace-40,250,
     ((fMenu==1)?"u(x,ym)":""));
if (fMenu==2 || fMenu==3)
{
     dc.TextOut(hGrid.x,hGrid.y+(m+1)*vSpace-10,
          "u(x,t0)");
     if (fMenu==3)
          dc.TextOut(hGrid.x+200,hGrid.
               y+(m+1)*vSpace-10,"uD(x,t0)");
}
if (fMenu==1)
     dc.TextOut(hGrid.x+n/2*hSpace-40,hGrid.y
          +(m+1)*vSpace-10,"u(x,y0)");
for (j=m;j>=0;j--)
     for (i=0;i<=n;i++)
     {
          rc=CRect(hGrid.x+i*hSpace,hGrid.y+j*vSpace,
          hGrid.x+i*hSpace+wrc.cx,hGrid.
               y+j*vSpace+wrc.cy);
          dc.Rectangle(rc);
          str.Format("%d,%d",i,m-j);
          dc.TextOut(hGrid.x+wrc.cx/3+i*hSpace,
```

```
                              hGrid.y+2+j*vSpace,str);
            }
      }
      if (fStatus==3)
      {
            for (j=m;j>=0;j--)
                  for (i=0;i<=n;i++)
                  {
                        str.Format("%.4lf",u[i][j]);
                        dc.TextOut(hGrid.x+i*hSpace+5,
                              hGrid.y+15+(m-j)*vSpace,str);
                  }
            if (fMenu==1)
            {
                  dc.TextOut(30,180,"Poisson equation");
                  dc.TextOut(30,195,"u[xx]+u[yy]
                        ="+input[1].item);
                  dc.TextOut(30,210,input[4].
                        item+"< x <"+input[5].item
                        +", "+input[6].item+"< y <"+input[7].
                              item);
            }
            if (fMenu==2)
            {
                  str.Format("%lf",pow(alpha,2));
                  dc.TextOut(30,180,"Heat Equation");
                  dc.TextOut(30,195,"u[t]-"+str+" u[xx]=0");
                  dc.TextOut(30,210,input[4].item+"< x
                        <"+input[5].item
                        +", "+input[6].item+"< t <"
                        +input[7].item);
            }
            if (fMenu==3)
            {
                  str.Format("%lf",pow(alpha,2));
                  dc.TextOut(30,180,"Wave Equation");
                  dc.TextOut(30,195,"u[tt]-"+str+" u[xx]=0");
                  dc.TextOut(30,210,input[4].item+"< x
                        <"+input[5].item+", "+input[6].
                              item+"< t <"+input[7].item);
            }
      }
}
```

A left-click on the mouse is an event that is mapped as ON_WM_LBUTTONDOWN and is handled by OnLButtonDown(). Two flag variables, fMenu and fStatus, have been assigned to monitor the status of execution starting from this event. Both variables are initially set to 0, and their values will change during the course of the execution. fMenu changes its value once one of the items in the menu is left-clicked, with fMenu=1 when the selection is Poisson, fMenu=2 with heat, and fMenu=3 with wave.

The menu is selected through a conditional test using menu[k].rc.PtInRect(pt). A selection on the menu displays the titles for the input items through InputLabels(), and creates a push button and the corresponding edit boxes for collecting the input. The code fragments for OnLButtonDown() consist of

```
void CCode11::OnLButtonDown(UINT nfStatuss,CPoint pt)
{
      int i,k;
      for (k=1;k<=nMenuItems;k++)
             if (menu[k].rc.PtInRect(pt))
             {
                   fMenu=k; fStatus=1;
                   for (i=1;i<=maxInput;i++)
                         input[i].ed.DestroyWindow();
                   table1.DestroyWindow();
                   table2.DestroyWindow();
                   Clear(input[0].display);
                   Clear(rcGrid);
                   InvalidateRect(input[0].rc);
                   btn.DestroyWindow();
                   btn.Create("Compute",WS_CHILD |  WS_VISIBLE
                         | BS_DEFPUSHBUTTON,
                         CRect(50,140,150,170),this,IDC_BUTTON);
                   for (int i=1;i<=maxInput-1;i++)
                         input[i].ed.DestroyWindow();
                   switch(k)
                   {
                         case 1:
                         fMenu=1; InputLabels(); break;
                         case 2:
                         fMenu=2; InputLabels(); break;
                         case 3:
                         fMenu=3; InputLabels(); break;
                   }
                   for (i=1;i<=nInputItems;i++)
                         input[i].ed.Create(WS_CHILD |  WS_VISIBLE
                               | WS_BORDER,
                               CRect(input[i].hm.x+100,input[i].
```

```
                              hm.y,input[i].hm.x+520,
                       input[i].hm.y+20),this,idc++);
              }
}
```

Both sets of input are made inside OnButton(). This function is activated when the push button *Compute* is left-clicked. The first set is made possible through the assigned value in the status control flag, with fStatus=1. The second input happens through SecondInput() when this function is called from the selected method. Upon the completion of the inputs, the flag status is updated to fStatus=2 by calling the corresponding function for the selected method. With this progress, the input boxes for the initial and boundary conditions are created. The code for these operations is shown below:

```
void CCode11::OnButton()
{
      CRect rc;
      int i,j;
      if (fStatus==3)
            fStatus=2;
      if (fMenu==1)
      {
            if (fStatus==2)
                  Poisson();
            if (fStatus==1)
            {
                  for (i=1;i<=nInputItems;i++)
                        input[i].ed.GetWindowText(input[i].item);
                  n=atoi(input[2].item); n=((n<=N)?n:N);
                  m=atoi(input[3].item); m=((m<=M)?m:M);
                  pt[0].x=atof(input[4].item);
                  pt[n].x=atof(input[5].item);
                  pt[0].y=atof(input[6].item);
                  pt[m].y=atof(input[7].item);
                  h=(double)1/n*(pt[n].x-pt[0].x);
                  k=(double)1/m*(pt[m].y-pt[0].y);
                  input[nInputItems+1].ed.DestroyWindow();
                  input[nInputItems+2].ed.DestroyWindow();
                  input[nInputItems+1].ed.Create(WS_CHILD |
                        WS_VISIBLE
                        | WS_BORDER, CRect(CPoint(hGrid.x-100,
                        hGrid.y+m/2*vSpace),
                        CSize(90,25)),this,idc++);
                  input[nInputItems+2].ed.Create(WS_CHILD |
```

```
                    WS_VISIBLE
                    | WS_BORDER, CRect(CPoint(hGrid.x+(n+1)
                    *hSpace-10,hGrid.y+m/2*vSpace),
                        CSize(90,25)),this,idc++);
            input[nInputItems+3].ed.DestroyWindow();
            input[nInputItems+4].ed.DestroyWindow();
            input[nInputItems+3].ed.Create(WS_CHILD |
                    WS_VISIBLE
                    | WS_BORDER, CRect(CPoint(hGrid.x+n/
                    2*hSpace,hGrid.y
                    +(m+1)*vSpace-10),CSize(90,25)),this,
                    idc++);
            input[nInputItems+4].ed.Create(WS_CHILD |
                    WS_VISIBLE
                    | WS_BORDER, CRect(CPoint(hGrid.x+n/
                    2*hSpace,250),
                    CSize(90,25)),this,idc++);
            fStatus=2;
        }
}
if (fMenu==2)
{
        if (fStatus==2)
            Heat();
        if (fStatus==1)
        {
            for (i=1;i<=nInputItems;i++)
                input[i].ed.GetWindowText(input[i].item);
            n=atoi(input[2].item); n=((n<=N)?n:N);
            m=atoi(input[3].item); m=((m<=M)?m:M);
            pt[0].x=atof(input[4].item); pt[n].
                    x=atof(input[5].item);
            pt[0].t=atof(input[6].item); pt[m].
                    t=atof(input[7].item);
            h=(double)1/n*(pt[n].x-pt[0].x);
            k=(double)1/m*(pt[m].t-pt[0].t);
            input[nInputItems+1].ed.DestroyWindow();
            input[nInputItems+2].ed.DestroyWindow();
            input[nInputItems+1].ed.Create(WS_CHILD |
                    WS_VISIBLE
                    | WS_BORDER, CRect(CPoint(hGrid.x-100,
                    hGrid.y+m/2*vSpace),
                    CSize(90,25)),this,idc++);
            input[nInputItems+2].ed.Create(WS_CHILD |
                    WS_VISIBLE
```

```
                    | WS_BORDER, CRect(CPoint(hGrid.
                    x+(n+1)*hSpace
                    -10,hGrid.y+m/2*vSpace),CSize(90,25)),
                    this,idc++);
            input[nInputItems+3].ed.DestroyWindow();
            input[nInputItems+3].ed.Create(WS_CHILD |
                    WS_VISIBLE | WS_BORDER,
                    CRect(CPoint(hGrid.x+n/
                    2*hSpace,
                    hGrid.y+(m+1)*vSpace-10),CSize(90,25)),
                    this,idc++);
            fStatus=2;
        }
    }
}
if (fMenu==3)
{
        if (fStatus==2)
                Wave();
        if (fStatus==1)
        {
                for (i=1;i<=nInputItems;i++)
                        input[i].ed.GetWindowText(input[i].item);
                n=atoi(input[2].item); n=((n<=N)?n:N);
                m=atoi(input[3].item); m=((m<=M)?m:M);
                pt[0].x=atof(input[4].item); pt[n].
                        x=atof(input[5].item);
                pt[0].t=atof(input[6].item); pt[m].
                        t=atof(input[7].item);
                h=(double)1/n*(pt[n].x-pt[0].x);
                k=(double)1/m*(pt[m].t-pt[0].t);
                input[nInputItems+1].ed.DestroyWindow();
                input[nInputItems+2].ed.DestroyWindow();
                input[nInputItems+1].ed.Create(WS_CHILD |
                        WS_VISIBLE
                        | WS_BORDER, CRect(CPoint(hGrid.x-100,
                        hGrid.y+m/2*vSpace),
                        CSize(90,25)),this,idc++);
                input[nInputItems+2].ed.Create(WS_CHILD |
                        WS_VISIBLE
                        | WS_BORDER, CRect(CPoint(hGrid.
                        x+(n+1)*hSpace
                        -10,hGrid.y+m/2*vSpace),CSize(90,25)),
                        this,idc++);
                input[nInputItems+3].ed.DestroyWindow();
```

```
                input[nInputItems+4].ed.DestroyWindow();
                input[nInputItems+3].ed.Create(WS_CHILD |
                    WS_VISIBLE
                    | WS_BORDER, CRect(CPoint(hGrid.x+40,
                    hGrid.y
                    +(m+1)*vSpace-10),CSize(90,25)),
                    this,idc++);
                input[nInputItems+4].ed.Create(WS_CHILD |
                    WS_VISIBLE
                    | WS_BORDER, CRect(CPoint(hGrid.x+240,
                    hGrid.y
                    +(m+1)*vSpace-10),CSize(90,25)),
                    this,idc++);
                fStatus=2;
            }
        }
        rc=CRect(CPoint(hGrid.x+(n+1)*hSpace-10,hGrid.y
            +(m+1)*vSpace-10),CSize(100,25));
        InvalidateRect(rcGrid);
}
```

The next step in the execution is a call to the respective function of the selected problem, with fMenu=1 to Poisson(), fMenu=2 to Heat(), and fMenu=3 to Wave(). This is achieved when the push button is left-clicked at an instance when fStatus=2. We discuss the the code for each problem.

Solution to Poisson's Equation

The solution to the Poisson's equation problem is given in a function called Poisson(). The code fragments for this function are written as

```
void CCode11::Poisson()
{
        int i,j,r,sA;
        int psi[4];
        double psv[4];
        SecondInput();
        pt[1].y=pt[0].y+k;
        for (j=1;j<=m-1;j++)
        {
                psi[1]=24; psv[1]=pt[j].y;
                u[0][j]=parse(input[nInputItems+1].item,1,psv,psi);
                u[n][j]=parse(input[nInputItems+2].item,1,psv,psi);
                pt[j+1].y=pt[j].y+k;
        }
```

```
for (i=0;i<=n;i++)
{
      psi[1]=23; psv[1]=pt[i].x;
      u[i][0]=parse(input[nInputItems+3].item,1,psv,psi);
      u[i][m]=parse(input[nInputItems+4].item,1,psv,psi);
      if (i<n)
            pt[i+1].x=pt[i].x+h;
}
for (j=0;j<=m;j++)
      for (i=0;i<=n;i++)
      {
            psi[1]=23; psv[1]=pt[i].x;
            psi[2]=24; psv[2]=pt[j].y;
            w[i][j]=parse(input[1].item,2,psv,psi);
      }

// form the SLE
sA=(m-1)*(n-1);
for (i=1;i<=sA;i++)
{
      b[i]=0;
      for (j=1;j<=sA;j++)
            a[i][j]=0;
}
for (r=1;r<=sA;r++)
{
      a[r][r]=-2*(h*h+k*k);
      if (r<=2*(n-1))
      {
            a[r][n-1+r]=h*h;
            a[n-1+r][r]=h*h;
      }
      a[1][2]=k*k; a[2][1]=k*k;
      if (r%(n-1)!=0 && r<(m-1)*(n-1))
      {
            a[r][r+1]=k*k;
            a[r+1][r]=k*k;
      }
}
for (j=1;j<=m-1;j++)
      for (i=1;i<=n-1;i++)
      {
            // compute b[r]
            r=i+(j-1)*(n-1);
```

```
            b[r]=h*h*k*k*w[i][j];
            if (j==1)
            {
                    b[r]  -= h*h*u[i][0];
                    if (i==1)
                            b[r]  -= k*k*u[0][1];
                    if (i==n-1)
                            b[r]  -= k*k*u[n][1];
            }
            if (j>1 && j<m-1)
            {
                    if (i==1)
                            b[r]  -= k*k*u[0][j];
                    if (i==n-1)
                            b[r]  -= k*k*u[n][j];
            }
            if (j==m-1)
            {
                    b[r]  -= h*h*u[i][m];
                    if (i==1)
                            b[r]  -= k*k*u[0][m-1];
                    if (i==n-1)
                            b[r]  -= k*k*u[n][m-1];
            }
        }
    fStatus=3;
    Clear(input[0].rc);
    ShowTable();
    SolveSLE();
    InvalidateRect(rcGrid);
    InvalidateRect(input[0].display);
}
```

Poisson() collects its initial input data from OnButton() and its input from the initial and boundary conditions from SecondInput(). The strings from the initial and boundary conditions are converted to double values and are evaluated through parse(). The parsed values are assigned to their respective variables in the code segment given by

```
for (j=1;j<=m-1;j++)
{
        psi[1]=24; psv[1]=pt[j].y;
        u[0][j]=parse(input[nInputItems+1].item,1,psv,psi);
        u[n][j]=parse(input[nInputItems+2].item,1,psv,psi);
        pt[j+1].y=pt[j].y+k;
}
```

```
for (i=0;i<=n;i++)
{
      psi[1]=23; psv[1]=pt[i].x;
      u[i][0]=parse(input[nInputItems+3].item,1,psv,psi);
      u[i][m]=parse(input[nInputItems+4].item,1,psv,psi);
      if (i<n)
            pt[i+1].x=pt[i].x+h;
}
for (j=0;j<=m;j++)
      for (i=0;i<=n;i++)
      {
            psi[1]=23; psv[1]=pt[i].x;
            psi[2]=24; psv[2]=pt[j].y;
            w[i][j]=parse(input[1].item,2,psv,psi);
      }
```

The biggest challenge in Poisson() is in forming the system of linear equations. The system has a size of sA=(m-1)*(n-1). In forming the system, the arrays a and b are first initialized by assigning them with zeros, through

```
sA=(m-1)*(n-1);
for (i=1;i<=sA;i++)
{
      b[i]=0;
      for (j=1;j<=sA;j++)
            a[i][j]=0;
}
```

Next, the coefficient matrix A is formed using Equation (11.7a). This code segment takes care of the nonzero values of A only, as those not filled with these values have already been assigned with zeros earlier.

```
// form the SLE
for (r=1;r<=sA;r++)
{
      a[r][r]=-2*(h*h+k*k);
      if (r<=2*(n-1))
      {
            a[r][n-1+r]=h*h;
            a[n-1+r][r]=h*h;
      }
      a[1][2]=k*k; a[2][1]=k*k;
      if (r%(n-1)!=0 && r<(m-1)*(n-1))
      {
```

```
            a[r] [r+1]=k*k;
            a[r+1] [r]=k*k;

    }
}
```

The entries for the **b** vector follow similar steps as in *A*, using Equation (11.7b). The coding is longer here because of the irregular nature of the initial and boundary values.

```
for (j=1;j<=m-1;j++)
    for (i=1;i<=n-1;i++)
    {
        // compute b[r]
        r=i+(j-1)*(n-1);
        b[r]=h*h*k*k*w[i][j];
        if (j==1)
        {
            b[r] -= h*h*u[i][0];
            if (i==1)
                b[r] -= k*k*u[0][1];
            if (i==n-1)
                b[r] -= k*k*u[n][1];
        }
        if (j>1 && j<m-1)
        {
            if (i==1)
                b[r] -= k*k*u[0][j];
            if (i==n-1)
                b[r] -= k*k*u[n][j];
        }
        if (j==m-1)
        {
            b[r] -= h*h*u[i][m];
            if (i==1)
                b[r] -= k*k*u[0][m-1];
            if (i==n-1)
                b[r] -= k*k*u[n][m-1];
        }
    }
```

With **A** and **b** formed, the rest of the code in Poisson() involves solving the system of linear equations through SolveSLE(), and displaying the results in the main window.

```
Clear(input[0].rc);
ShowTable();
```

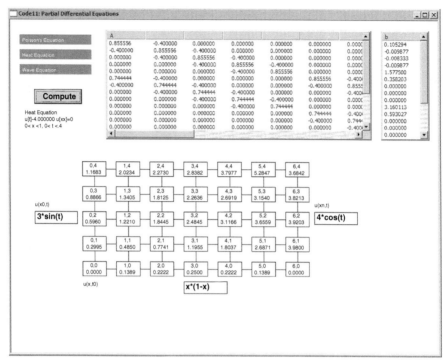

FIGURE 11.16. Visual solution for the heat problem.

```
SolveSLE();
InvalidateRect(rcGrid);
InvalidateRect(input[0].display);
```

Solution to the Heat Equation

The solution to the heat equation is handled by the function Heat() according to Algorithm 11.4. Figure 11.16 shows an output from Heat().

The contents of Heat() consist of the finite-difference method that generates a system of linear equations, and solves this system by calling SolveSLE(). The function is given, as follows:

```
void CCode11::Heat()
{
        int i,j,r,sA;
        int psi[4];
        double psv[4];
        SecondInput();
        alpha=atof(input[1].item);
```

```
pt[1].t=pt[0].t+k;
for (j=1;j<=m;j++)
{
      psi[1]=19; psv[1]=pt[j].t;
      u[0][j]=parse(input[nInputItems+1].item,1,psv,psi);
      u[n][j]=parse(input[nInputItems+2].item,1,psv,psi);
      if (j<m)
            pt[j+1].t=pt[j].t+k;
}
for (i=0;i<=n;i++)
{
      psi[1]=23; psv[1]=pt[i].x;
      u[i][0]=parse(input[nInputItems+3].item,1,psv,psi);
      if (i<n)
            pt[i+1].x=pt[i].x+h;
}
sA=m*(n-1);
for (i=1;i<=sA;i++)
{
      b[i]=0;
      for (j=1;j<=sA;j++)
            a[i][j]=0;
}
for (j=0;j<=m-1;j++)
      for (i=1;i<=n-1;i++)
      {
            r=i+j*(n-1);
            // compute A
            a[r][r]=2*(h*h+alpha*alpha*k);
            if (r%(n-1)!=0 && r<m*(n-1))
            {
                  a[r][r+1]=-alpha*alpha*k;
                  a[r+1][r]=a[r][r+1];
            }
            if (r<=2*(n-1)+1)
                  if (r%(n-1)!=1)
                        a[n+r-2][r]=-alpha*alpha*k;
            if (r<=2*(n-1))
                  a[n-1+r][r]=2*(-h*h+alpha*alpha*k);
            if (r<=2*(n-1)-1)
                  if (r%(n-1)!=0)
                        a[n+r][r]=-alpha*alpha*k;
            // compute b[r]
            if (j==0)
```

```
                        {
                                b[r]=pow(alpha,2)*k*u[i-1][0]-2*u[i][0]*
                                        (-pow(h,2)+k*pow(alpha,2))
                                        +pow(alpha,2)*k*u[i+1][0];
                                if (i==1)
                                        b[r] += pow(alpha,2)*k*u[0][1];
                                if (i==n-1)
                                        b[r] += pow(alpha,2)*k*u[n][1];
                        }
                        if (j>0)
                        {
                                if (i==1)
                                        b[r]=pow(alpha,2)*k*(u[0][j]+u[0]
                                                [j+1]);
                                if (i==n-1)
                                        b[r]=pow(alpha,2)*k*(u[n][j]+u[n]
                                                [j+1]);
                        }
                }
        fStatus=3;
        Clear(input[0].rc);
        ShowTable();
        SolveSLE();
        InvalidateRect(rcGrid);
        InvalidateRect(input[0].display);
}
```

The initial and boundary values are read in Heat() in the same manner as in Poisson(). The input strings are converted to the initial and boundary values through parse(). The code segment for this job is given by

```
for (j=1;j<=m;j++)
{
        psi[1]=19; psv[1]=pt[j].t;
        u[0][j]=parse(input[nInputItems+1].item,1,psv,psi);
        u[n][j]=parse(input[nInputItems+2].item,1,psv,psi);
        if (j<m)
                pt[j+1].t=pt[j].t+k;
}
for (i=0;i<=n;i++)
{
        psi[1]=23; psv[1]=pt[i].x;
        u[i][0]=parse(input[nInputItems+3].item,1,psv,psi);
        if (i<n)
                pt[i+1].x=pt[i].x+h;
}
```

Again, the biggest challenge in Heat() is in forming matrix A and vector b. These two tasks are performed using Algorithms 11.2a and 11.2b, respectively. The code segment is

```
sA=m*(n-1);
for (i=1;i<=sA;i++)
{
      b[i]=0;
      for (j=1;j<=sA;j++)
            a[i][j]=0;
}
for (j=0;j<=m-1;j++)
      for (i=1;i<=n-1;i++)
      {
            r=i+j*(n-1);
            // compute A
            a[r][r]=2*(h*h+alpha*alpha*k);
            if (r%(n-1)!=0 && r<m*(n-1))
            {
                  a[r][r+1]=-alpha*alpha*k;
                  a[r+1][r]=a[r][r+1];
            }
            if (r<=2*(n-1)+1)
                  if (r%(n-1)!=1)
                        a[n+r-2][r]=-alpha*alpha*k;
            if (r<=2*(n-1))
                  a[n-1+r][r]=2*(-h*h+alpha*alpha*k);
            if (r<=2*(n-1)-1)
                  if (r%(n-1)!=0)
                        a[n+r][r]=-alpha*alpha*k;
            // compute b[r]
            if (j==0)
            {
                  b[r]=pow(alpha,2)*k*u[i-1][0]-2*u[i][0]*
                        (-pow(h,2)+k*pow(alpha,2))
                        +pow(alpha,2)*k*u[i+1][0];
                  if (i==1)
                        b[r] += pow(alpha,2)*k*u[0][1];
                  if (i==n-1)
                        b[r] += pow(alpha,2)*k*u[n][1];
            }
            if (j>0)
            {
                  if (i==1)
```

```
                    b[r]=pow(alpha,2)*k*(u[0][j]+u[0][j+1]);
            if (i==n-1)
                    b[r]=pow(alpha,2)*k*(u[n][j]+u[n][j+1]);
        }
    }
```

The last part in Heat() is solving the system of linear equations and is displaying the results in the rectangular grids as well as in the list view tables.

Solution to the Wave Equation

The wave equation problem is slightly different from the Poisson's and heat equation problems as it is not a system of linear equations. The fundamental tool is still the finite-difference formula, but the equation generated is in explicit form. Figure 11.17 shows an output produced from Wave().

The solution to the wave equation problem is provided in Wave(). The function is written as follows:

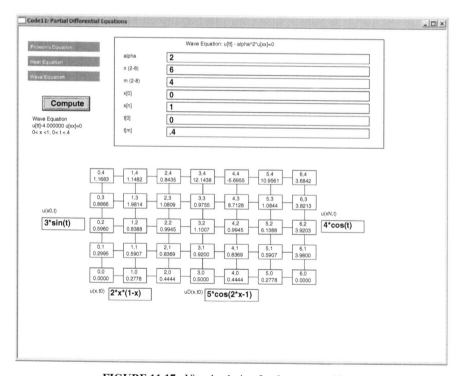

FIGURE 11.17. Visual solution for the wave problem.

```
void CCode11::Wave()
{
      int i,j,r,sA;
      int psi[2];
      double psv[2];
      double uD[N+1];
      SecondInput();
      alpha=atof(input[1].item);
      pt[1].t=pt[0].t+k;
      psi[1]=19;
      for (j=1;j<=m;j++)
      {
            psv[1]=pt[j].t;
            u[0][j]=parse(input[nInputItems+1].item,1,psv,psi);
            u[n][j]=parse(input[nInputItems+2].item,1,psv,psi);
            if (j<m)
                  pt[j+1].t=pt[j].t+k;
      }
      psi[1]=23;
      for (i=0;i<=n;i++)
      {
            psv[1]=pt[i].x;
            u[i][0]=parse(input[nInputItems+3].item,1,psv,psi);
            uD[i]=parse(input[nInputItems+4].item,1,psv,psi);
            if (i<n)
                  pt[i+1].x=pt[i].x+h;
      }
      for (i=1;i<=n-1;i++)
            u[i][1]=1/(2*pow(h,2))
                  *(2*pow(h,2)*k*uD[i]+pow(k,2)*pow(alpha,2)
                  *u[i-1][0]
                  -(-2*pow(h,2)+2*pow(k,2)*pow(alpha,2))*u[i][0]
                  +pow(k,2)*pow(alpha,2)*u[i+1][0]);
      for (j=1;j<=m-1;j++)
            for (i=1;i<=n-1;i++)
                  u[i][j+1]=1/pow(h,2)*(-pow(h,2)*u[i][j-1]
                  +pow(k,2)*pow(alpha,2)*u[i-1][j]
                  -(-2*pow(h,2)+2*pow(k,2)*pow(alpha,2))*u[i][j]
                  +pow(k,2)*pow(alpha,2)*u[i+1][j]);
      fStatus=3;
      InvalidateRect(rcGrid);
      InvalidateRect(input[0].display);
}
```

The input data in `Wave()` is read in `OnButton()` and `SecondInput()`. `Wave()` processed the input strings and converted them into initial and boundary values according to the code fragments given by

```
for (j=1;j<=m;j++)
{
        psv[1]=pt[j].t;
        u[0][j]=parse(input[nInputItems+1].item,1,psv,psi);
        u[n][j]=parse(input[nInputItems+2].item,1,psv,psi);
        if (j<m)
                pt[j+1].t=pt[j].t+k;
}
psi[1]=23;
for (i=0;i<=n;i++)
{
        psv[1]=pt[i].x;
        u[i][0]=parse(input[nInputItems+3].item,1,psv,psi);
        uD[i]=parse(input[nInputItems+4].item,1,psv,psi);
        if (i<n)
                pt[i+1].x=pt[i].x+h;
}
```

The second initial value is read and processed through `parse()` as the array uD, which represents $u_t(x_i, t_0)$ for $i = 0, 1, \ldots, n$. As mentioned, this array is needed in substituting the values of the virtual values $u_{i,-1}$ when its finite-difference method is applied.

The solution to the wave equation is provided by the explicit finite-difference formula of Equation (11.14). This formula is written in `Wave()` according to

```
for (i=1;i<=n-1;i++)
        u[i][1]=1/(2*pow(h,2))
                *(2*pow(h,2)*k*uD[i]+pow(k,2)*pow(alpha,2)*u[i-1][0]
                -(-2*pow(h,2)+2*pow(k,2)*pow(alpha,2))*u[i][0]
                +pow(k,2)*pow(alpha,2)*u[i+1][0]);
for (j=1;j<=m-1;j++)
        for (i=1;i<=n-1;i++)
                u[i][j+1]=1/pow(h,2)*(-pow(h,2)*u[i][j-1]
                        +pow(k,2)*pow(alpha,2)*u[i-1][j]
                        -(-2*pow(h,2)+2*pow(k,2)*pow(alpha,2))*u[i][j]
                        +pow(k,2)*pow(alpha,2)*u[i+1][j]);
```

11.7 SUMMARY

This chapter discusses the numerical solutions to the partial differential equations involving initial- and boundary-value problems. Four topics are covered, namely, Laplace's, Poisson's, heat, and wave equations. The problems are solved using the finite-difference method over rectangular grids that have uniform horizontal and vertical widths. We discuss the numerical solution for each topic and develop the visual interface for solving each problem.

The computing platform for the partial differential equations problems has been very challenging. This is because of the nature of the problem, which involves three variables on three dimensions. The domain in each problem has been assumed to be rectangular with the initial and boundary values given along the horizontal and vertical boundaries. Therefore, the quantities to be evaluated, such as the magnetic field, heat, and wave values can be visualized as the surfaces over the rectangular domain.

The idea from the visual solutions provided in this chapter can be extended to create projects from several applications in science and engineering. The finite-difference method is a fundamental tool in solving the boundary-value problems involving partial differential equations. The method produces good approximations to the problems, which are very close to the exact solutions. However, the requirement that the grids in the rectangular domain must have equal width confines the method to some applications only. The method is not applicable in cases where the domain is not in the form of rectangle, or in grids with non-equal width.

Nevertheless, the concepts in the finite-difference method provide the fundamentals to several advanced methods, such as the finite-element and boundary-element methods. The reader should study the methods and visual solutions discussed in this chapter and should apply them in projects involving modeling and simulation in partial differential equations.

NUMERICAL EXERCISES

1. Solve the following boundary-value problems involving the Poisson's and Laplace's equations:

   a. $u_{xx} + u_{yy} = 3x^2 y$, for $0 < x < 1$ and $0 < y < 0.3$, where $h = \Delta x = 0.25$ and $k = \Delta y = 0.1$. The initial conditions are $u(x, 0) = 10$ and $u(x, 0.3) = 20$, whereas the boundary conditions are $u(0, y) = 5$ and $u(1, y) = 15$.

   b. $u_{xx} + u_{yy} = 10 \sin xy$, for $0 < x < 1$ and $0 < y < 0.3$, where $h = \Delta x = 0.25$ and $k = \Delta y = 0.1$. The initial conditions are $u(x, 0) = 5 \sin x$ and $u(x, 0.3) = 4 \cos x$, whereas the boundary conditions are $u(0, y) = 4 \sin 2y$ and $u(1, y) = 3 \cos 3y$.

   c. $u_{xx} + u_{yy} = 0$, for $0 < x < 1$ and $0 < y < 0.1$, where $h = \Delta x = 0.25$ and $k = \Delta y = 0.025$. The initial conditions are $u(x, 0) = 2 \sin 2x$ and $u(x, 0.1) = 3 \cos x$, whereas the boundary conditions are $u(0, y) = 4 \sin y$ and $u(1, y) = 2 \cos 2y$.

Check the results by running `Code11`.

2. Solve the following boundary-value problems involving the heat equation:

   a. $u_t - 4u_{xx} = 0$, for $0 < x < 1$ and $t > 0$, at $t = 0.1$, where $h = \Delta x = 0.25$ and $k = \Delta t = 0.1$. Given the initial value of $u(x, 0) = 10$, and the boundary values of $u(0, t) = 30$ and $u(1, t) = 50$.

   b. $u_t - 9u_{xx} = 0$, for $0 < x < 1$ and $t > 0$, at $t = 0.1$, where $h = \Delta x = 0.25$ and $k = \Delta t = 0.1$. Given the initial value of $u(x, 0) = 3 \sin x$, and the boundary values of $u(0, t) = 3 \cos t$ and $u(1, t) = 4e^{-t}$.

   c. $u_t - 2u_{xx} = 0$, for $0 < x < 1$ and $t > 0$, at $t = 0.1$, where $h = \Delta x = 0.25$ and $k = \Delta t = 0.1$. Given the initial value of $u(x, 0) = 3 \cos 2x$, and the boundary values of $u(0, t) = 3 \cos 2t$ and $u(1, t) = 4e^{-t}$.

Check the results by running `Code11`.

3. Solve the following boundary-value problems involving the wave equation:

   a. $u_{tt} - 4u_{xx} = 0$, for $0 < x < 1$ and $t > 0$, at $t = 0.1$, where $h = \Delta x = 0.25$ and $k = \Delta t = 0.1$. Given the initial values of $u(x, 0) = 10$ and $u_t(x, 0) = 15$. The boundary values are $u(0, t) = 30$ and $u(1, t) = 50$.

   b. $u_{tt} - 9u_{xx} = 0$, for $0 < x < 1$ and $t > 0$, at $t = 0.1$, where $h = \Delta x = 0.25$ and $k = \Delta t = 0.1$. Given the initial values of $u(x, 0) = 3 \sin x$ and $u_t(x, 0) = 5 \cos x$. The boundary values are $u(0, t) = 3 \cos t$ and $u(1, t) = 4e^{-t}$.

Check the results by running `Code11`.

4. Find the finite-difference formula for the non-homogeneous wave equation, given by Equation (11.12) using the central-difference rules.

5. By referring to Table 11.1, apply the central-difference rules to derive the finite-difference formula for each of the following equations:

   a. Diffusion equation

   b. Helmholtz's equation

   c. Klein–Gordon's equation

   d. Modified Louiville's equation

   e. Vibrating membrane

PROGRAMMING CHALLENGES

1. The Laplace's equation is the homogeneous form of the Poisson's equation, obtained by setting $w_{i,j} = 0$ in Equation (11.5). Describe the difference between the Laplace's equation and the Poisson's equation. Hence, develop this new module as an item in the menu in `Code11`.

2. Provide flexibility to `Code11` by adding several new features, such as file open and retrieve options. These features are important as the problem generates new

matrices and vectors, as well as solutions in rectangular grids. The generated data can be very massive as it is dependent on the input.

3. Modify `Code11` to include the non-homogeneous wave equation of Equation (11.12). Study the initial and boundary conditions, and derive the finite-difference formula for this problem.

4. The Helmholtz's equation given in Table 11.1 by $\frac{\partial^2 u}{\partial x^2} + \frac{\partial^2 u}{\partial y^2} + f(x, y)u = g(x, y)$ is an elliptic equation that does not differ much from the Poisson's equation. Study the initial and boundary conditions for this problem, and develop it as a new module in `Code11`.

5. The diffusion equation given in Table 11.1 by $\sigma \frac{\partial^2 u}{\partial x^2} - \frac{\partial u}{\partial t} = 0$ has a lot of applications, such as in the movement of fluids from one medium to another. Study the initial and boundary conditions for this problem, and develop it as a new module in `Code11`.